Food chemical safety
Volume 2: Additives

Related titles from Woodhead's food science, technology and nutrition list:

Food chemical safety Volume 1: Contaminants (ISBN: 1 85573 462 1)

This volume provides comprehensive information about contaminants in the food industry. The book opens with an explanation of risk analysis and analytical methods used for detecting contaminants in food products. This is followed by full details of relevant EU and USA regulations. The second part of the book provides information about specific contaminants.

Instrumentation and sensors in the food industry Second edition (ISBN: 1 85573 560 1)

The first edition of this book quickly established itself as a standard work in its field, providing an authoritative and practical guide to the range of instrumentation and sensors available to the food industry professional. This new edition has been comprehensively revised to include new developments and techniques.

Making the most of HACCP (ISBN: 1 85573 504 0)

Based on the experience of those who have successfully implemented HACCP systems, this book will meet the needs of food processing businesses at all stages of HACCP system development. The collection is edited by two internationally recognised HACCP experts and includes chapters reflecting the experience of both major companies such as Cargill, Heinz and Sainsbury and the particular challenges facing SMEs. The scope of the book is truly international with experiences of HACCP implementation in countries such as Thailand, India, China and Poland.

Details of these books and a complete list of Woodhead's food science, technology and nutrition titles can be obtained by:

- visiting our web site at www.woodhead-publishing.com
- contacting Customer services (email: sales@woodhead-publishing.com; fax: +44 (0) 1223 893694; tel.: +44 (0) 1223 891358 ext. 30; address: Woodhead Publishing Limited, Abington Hall, Abington, Cambridge CB1 6AH, England)

If you would like to receive information on forthcoming titles in this area, please send your address details to: Francis Dodds (address, tel. and fax as above; e-mail: francisd@woodhead-publishing.com). Please confirm which subject areas you are interested in.

Food chemical safety

Volume 2: Additives

Edited by
David H. Watson

CRC Press
Boca Raton Boston New York Washington, DC

WOODHEAD PUBLISHING LIMITED
Cambridge England

Published by Woodhead Publishing Limited
Abington Hall, Abington
Cambridge CB1 6AH
England
www.woodhead-publishing.com

Published in North America by CRC Press LLC
2000 Corporate Blvd, NW
Boca Raton FL 33431
USA

First published 2002, Woodhead Publishing Limited and CRC Press LLC
© 2002, Woodhead Publishing Limited
The authors have asserted their moral rights.

This book contains information obtained from authentic and highly regarded sources. Reprinted material is quoted with permission, and sources are indicated. Reasonable efforts have been made to publish reliable data and information, but the authors and the publishers cannot assume responsibility for the validity of all materials. Neither the authors nor the publishers, nor anyone else associated with this publication, shall be liable for any loss, damage or liability directly or indirectly caused or alleged to be caused by this book.

Neither this book nor any part may be reproduced or transmitted in any form or by any means, electronic or mechanical, including photocopying, microfilming and recording, or by any information storage or retrieval system, without permission in writing from the publishers.

The consent of Woodhead Publishing Limited and CRC Press LLC does not extend to copying for general distribution, for promotion, for creating new works, or for resale. Specific permission must be obtained in writing from Woodhead Publishing Limited or CRC Press LLC for such copying.

Trademark notice: Product or corporate names may be trademarks or registered trademarks, and are used only for identification and explanation, without intent to infringe.

British Library Cataloguing in Publication Data
A catalogue record for this book is available from the British Library.

Library of Congress Cataloging-in-Publication Data
A catalog record for this book is available from the Library of Congress.

Woodhead Publishing Limited ISBN 1 85573 563 6
CRC Press ISBN 0-8493-1211-6
CRC Press order number: WP1211

Cover design by The ColourStudio
Project managed by Macfarlane Production Services, Markyate, Hertfordshire (macfarl@aol.com)
Typeset by MHL Typesetting Limited, Coventry, Warwickshire
Printed by TJ International, Padstow, Cornwall, England

Contents

List of contributors		xi
Part I General issues		1
1 Introduction		3
D. Watson, *Food Standards Agency, London*		
1.1	Background	3
1.2	Controls on additives	5
1.3	Future work on additives	7
1.4	Acknowledgement and dedication	11
1.5	References	11
2 The regulation of additives in the EU		12
D. W. Flowerdew, *Consultant (formerly Leatherhead Food RA)*		
2.1	Introduction	12
2.2	The key directives	13
2.3	Future developments	27
2.4	Sources of further information and advice	29
2.5	References	30
	Appendix: List of E numbers of permitted additives	32
3 The regulation of additives in the USA		42
P. Curtis, *North Carolina State University, Raleigh*		
3.1	Introduction	42
3.2	Food additive laws and amendments	43
3.3	Federal agencies responsible for enforcement	45

3.4	Generally recognized as safe (GRAS) substances	46
3.5	Prior sanctioned substances	47
3.6	Color additives	47
3.7	Pesticide residues	48
3.8	Setting tolerance levels	49
3.9	The approval process	50
3.10	References	57

Part II Analysing additives 59

4 Risk analysis of food additives 61
D. R. Tennant, Consultant, UK

4.1	Introduction	61
4.2	Hazard identification in the food chain	62
4.3	Dose-response characterisation	63
4.4	Exposure analysis	64
4.5	Risk evaluation	75
4.6	Methods for risk management	75
4.7	Risk communication	76
4.8	Future trends	76
4.9	Conclusion	77
4.10	Sources of further information and advice	77
4.11	References	78

5 Analytical methods: quality control and selection 79
R. Wood, Food Standards Agency, Norwich

5.1	Introduction	79
5.2	Legislative requirements	80
5.3	FSA surveillance requirements	83
5.4	Laboratory accreditation and quality control	83
5.5	Proficiency testing	89
5.6	Analytical methods	95
5.7	Standardised methods of analysis for additives	99
5.8	The future direction for methods of analysis	102
5.9	References	103
	Appendix: Information for potential contractors on the analytical quality assurance requirements for food chemical surveillance exercises	104

6 New methods in detecting food additives 111
C. J. Blake, Nestlé Research Centre, Lausanne

6.1	Introduction	111
6.2	Reference or research methods	112
6.3	Rapid or alternative methods	127
6.4	Future trends	131

6.5	Sources of further information and advice		132
6.6	References		134

7 Adverse reactions to food additives ... 145
M. A. Kantor, University of Maryland

7.1	Introduction		145
7.2	Consumer attitudes about food additives		146
7.3	Reporting adverse reactions		147
7.4	Controversial food additives		150
7.5	Summary and conclusions		162
7.6	Future trends and directions		165
7.7	Sources of further information and advice		165
7.8	References		166

Part III Specific additives ... 171

8 Colorants ... 173
F. Jack Francis, University of Massachusetts, Amherst

8.1	Introduction		173
8.2	Food, drug and cosmetic colorants		174
8.3	Carotenoid extracts		178
8.4	Lycopene		181
8.5	Lutein		182
8.6	Annatto and saffron		183
8.7	Paprika		184
8.8	Synthetic carotenoids		186
8.9	Anthocyanins		187
8.10	Betalains		190
8.11	Chlorophylls		191
8.12	Tumeric		192
8.13	Cochineal and carmine		193
8.14	Monascus		195
8.15	Iridoids		196
8.16	Phycobilins		198
8.17	Caramel		199
8.18	Brown polyphenols		200
8.19	Titanium dioxide		201
8.20	Carbon black		202
8.21	Miscellaneous colorants		202
8.22	Outlook		203
8.23	References		203

viii Contents

9 Safety assessment of flavor ingredients 207
K. R. Schrankel and P. L. Bolen, International Flavors & Fragrances, Union Beach and R. Petersen, International Flavors & Fragrances, Dayton
9.1   Introduction: definition and use of flavoring substances 207
9.2   The range and sources of flavoring ingredients 208
9.3   Basic principles of safety evaluation 209
9.4   Regulatory groups 216
9.5   References 224

10 Sweeteners   228
G. von Rymon Lipinski, Nutrovina Nutrition Specialities and Food Ingredients GmbH, Frankfurt
10.1   Introduction 228
10.2   Definitions 229
10.3   Functionality and uses 229
10.4   The available sweeteners 232
10.5   Sweetener safety testing 233
10.6   Case study: Acesulfame K 234
10.7   Other sweeteners 236
10.8   Regulatory status 242
10.9   Analytical methods 244
10.10 Outlook 244
10.11 Summary 245
10.12 Bibliography 245
10.13 References 246

11 Food additives, other than colours and sweeteners 249
Brian Whitehouse, Consultant, UK
11.1   Introduction: classifying the range of additives 249
11.2   The regulatory background 254
11.3   Acceptable daily intake (ADI) 255
11.4   JECFA safety evaluation 255
11.5   Summary 257
       Appendix 1: Specifications for food additives 258
       Appendix 2: Acceptable daily intake (ADI) values and
       references by additive category 266
       Appendix 3: Acceptable daily intake (ADI) values and
       references by additive name 274
11.6   Bibliography 282

12 Case study: antioxidant preservatives 283
K. Míková, Institute of Chemical Technology, Prague
12.1   Introduction 283
12.2   Toxicological aspects 284

12.3	The *Codex Alimentarius*	286
12.4	The regulation of antioxidants in the EU	288
12.5	The regulation of antioxidants in the USA	291
12.6	The regulation of antioxidants in Australia	295
12.7	The regulation of antioxidants in Japan	296
12.8	Future trends	296
12.9	Sources of further information and advice	298
12.10	References	298

Index 300

Contributors

Chapter 1

Dr David Watson
Food Standards Agency
Room 516C
Aviation House
125 Kingsway
London WC2B 6NH

Tel: +44 (0)20 7276 8537
Fax: +44 (0)20 7276 8514
E-mail: david.watson@foodstandards.
  gsi.gov.uk

Chapter 2

Dr D. W. Flowerdew
8 Broomhurst Court
Ridgeway Road
Dorking RH4 3EY

Tel/Fax: +44 (0)1306 886626
E-mail: d.flowerdew@foodlaw.
  demon.co.uk

Chapter 3

Professor Patricia Curtis
Department of Food Science
North Carolina State University
Box 7624
Raleigh
NC 27695-7624
USA

Tel: 001 919 515 2956
Fax: 001 919 515 7124
E-mail: Pat_Curtis@ncsu.edu

Chapter 4

Dr David Tennant
Food Chemical Risk Analysis
14 St Mary's Square
Brighton
East Sussex BN2 1FZ

Tel: +44 (0)1273 241753
Fax: +44 (0)1273 276358
E-mail: david-t@dircon.co.uk

Chapter 5

Dr Roger Wood
Food Standards Agency
c/o Institute of Food Research
Norwich Research Park
Colney
Norwich NR4 7UA

Tel: +44 (0)1603 255298
Fax: +44 (0)1603 507723
E-mail: roger.wood@foodstandards.
   gsi.gov.uk

Chapter 6

Christopher Blake
Manager Micronutrients and
   Additives Group
Quality and Safety Assurance Dept
Nestlé Research Centre
Lausanne 26
CH-1000, Switzerland

Tel: (021) 785 8348
E-mail: christopher-john.blake@rdls.
   nestle.com

Chapter 7

Professor Mark A. Kantor
Associate Professor & Extension
   Specialist
University of Maryland
Department of Nutrition and Food
   Science
3306 Marie Mount Hall
College Park, Maryland
20742-7521
USA

Tel: 001 301 405-1018
Fax: 001 301 314-9327
E-mail: mk4@umail.umd.edu

Chapter 8

Professor F. Jack Francis
Department of Food Science
University of Massachusetts
Amherst
MA 01003
USA

Fax: 001 413 545 1262
E-mail: ffrancis@foodsci.umass.edu

Chapter 9

Dr Kenneth R. Schrankel
Vice President (IFF-US)
Director, Flavor & Fragrance Safety
   Assurance
1515 State Highway 36
Union Beach
NJ 07735-3597
USA

Tel: 001 732 355 2305
Fax: 001 732 335 2599
E-mail: ken.schrankel@iff.com

Chapter 10

Professor Gert von Rymon Lipinski
Nutrinova Nutrition Specialities and
   Food Ingredients GmbH
Scientific Services and Regulatory
   Affairs
D-65926 Frankfurt am Main
Germany

E-mail: Rymon@nutrinova.com

Chapter 11

Brian Whitehouse
6 Church Bank
Richmond Road
Bowdon
Cheshire WA14 3NW

Tel: +44 (0)161 928 6681
Fax: +44 (0)8700 568 568
E-mail:
    brian@churchbank.demon.co.uk

Chapter 12

Professor Kamila Míková
Department of Food Chemistry and
    Analysis
Faculty of Food and Biochemical
    Technology
Prague Institute of Chemical
    Technology
Technická Street 5
CZ-166 28
Prague 6
Czech Republic

Tel: +4202 7277 0370
Fax: +4202 6731 1483
E-mail: kamila-mikova@boneco.cz

Part I

General issues

1
Introduction

D. Watson, Food Standards Agency, London

1.1 Background

The control of food additives has been much discussed in newspapers and other parts of the media and yet, rather surprisingly, there have been few scientific books about additives. There have been detailed reports of the work of expert committees such as the Joint FAO/WHO Expert Committee on Food Additives (e.g. WHO, 2000) and informative commentaries on individual additives and legislative controls, notably those in the European Union (EU) (e.g. Leatherhead Food Research Association, 2000). There have also been articles on work to harmonise worldwide control of additives (Keefe et al., 2000).

This book reviews one extensive group of substances, flavourings (Chapter 9), that is being brought into international controls on additives. It also considers three major groups of widely-controlled additives in detail: artificial sweeteners (Chapter 10), substances used as colourings (Chapter 8), and antioxidants (Chapter 12). A more general review of the other additives and how they are controlled is presented in Chapter 11.

The contrasts in public perception about different groups of additives are very interesting. Undoubtedly these differences influence controls applied to different additives in food. For example, concern about whether artificial colourings might mislead purchasers of processed food is continuing to influence policy on their use in some countries. Another example is the particular difficulty experienced by the producers of some artificial sweeteners in convincing others that new artificial sweeteners are safe, despite extensive toxicological work. Lastly the use of antioxidants does not seem to have benefited from the emerging evidence that natural antioxidants may play a major role in health. The core of any good control system on additives must include an assessment of whether the additive is needed, a quantitative assessment of its safety and the foods and consumers which should not be exposed to the given additive. An example of the latter factor is the exclusion of most additives from baby foods in the EU.

There has been much debate about several aspects of additives. Their safety has been questioned, often as a result of much-publicised reports of adverse reactions. Indeed the need for additives has been questioned. The chemical nature of additives has also led to debate. Some members of the public in this and probably several other countries firmly believe that additives are 'a bad thing'. This book touches on these topics, but it does not and cannot give definitive answers. Perhaps the shortage of books about this scientific work has contributed to public distrust. But it is not the purpose of this work in some way to re-educate the public. That would be both arrogant and trying to turn back history! The purpose of this volume is to demonstrate that there is considerable scientific work to enable the effective protection of consumers' safety.

The main groups of additives used in food share the following characteristics:

- They are intentionally added to food. This raises the question about whether they are needed. In the UK a 'case of need' has to be considered before an additive is approved. Nevertheless there is a growing movement to sell foods containing few if any additives.
- They are added for specific purposes (e.g. to preserve or sweeten food). Growing pressure to use 'natural' substances to meet such needs has led to surprisingly little discussion about the relative safety of naturally-occurring substitutes as opposed to synthetic additives. For example, although synthetic additives should be more easy to produce in purified form, naturally-occurring chemicals may be more acceptable to some members of the public.
- They must be safe. Illness is likely to result if consumers ingest too much of them, like contaminants (see Watson, 2001). Therefore numerical limits are set for many additives in EU law. Consumers can also become ill if some additives are not used. For example preservatives such as nitrite have to be biologically-active to work. The question is: are they biologically-active against us? A lot of scientific work had to be done before many preservatives were allowed for use in the European Community (now the EU). Preservatives were one of the first groups of additives to have their levels controlled in food so that consumers did not consume unsafe amounts of them.
- There is a balance to be drawn between risks. Looking at nitrite again, this ion has been used since the Middle Ages to preserve meat. It is the preservative in saltpetre which helps preserve cured meats and stop those who eat them getting botulism. There has been much research to find an alternative to nitrite as it can react with some amines to form carcinogenic nitrosamines. Risk analysis, reviewed in Chapter 4, is an important tool in controlling the use of additives.

1.2 Controls on additives

As the number of different controls on additives grows it is becoming very clear that more work is needed on harmonising these controls. The UK Food Standards Agency has been pressing for this to be done in the EU. Indeed the Food Standards Agency is pressing for consumer protection to be taken forward more effectively by submitting the legislation on additives under a Simpler Legislation in the Internal Market initiative.

The list of EU legislation on additives was as follows, in January 2001:

- Food Additives Framework Directive 89/107/EEC
- Directive 94/34/EC – first amendment to Directive 89/107/EEC
- Commission Working Document WGA/003/00 rev.1 – second amendment to Directive 89/107/EEC

- Extraction Solvents Framework Directive 88/344/EEC
- Directive 92/115 – first amendment to Directive 88/344/EEC
- Directive 94/52 – second amendment to Directive 88/344/EEC
- Directive 97/60 – third amendment to Directive 88/344/EEC

- Flavourings Framework Directive 88/388/EEC
- Commission Working Document SANCO/1184/2000 rev.1 – amendment to Directive 88/388/EEC

- Colours Directive 94/36/EC

- Sweeteners Directive 94/35/EC
- Directive 96/83/EC – first amendment to Directive 94/35/EC
- Directive 198/2000 – second amendment to Directive 94/35/EC
- Commission Working Document WGA/010/00 – third amendment to Directive 94/35/EC

- Miscellaneous Food Additives Directive 95/2/EC (e.g. preservatives, antioxidants, emulsifiers, stabilisers, thickeners, flavour enhancers, acidity regulators, etc.)
- Directive 96/85/EC – first amendment to Directive 95/2/EC
- Directive 98/72/EC – second amendment to Directive 95/2/EC
- Council Working Document 11911/00 (Common Position) – third amendment to Directive 95/2/EC
- Commission Working Document WGA/005/rev.1 – fourth amendment to Directive 95/2/EC

- Council Decision 88/389/EEC on inventory of the source materials and substances used in the preparation of flavourings
- Regulation 2232/96 laying down a Community procedure for flavourings
- Commission Directive 91/71/EEC completing Council Directive 88/388/EEC
- Commission Decision 1999/217/EC adopting a register of flavourings
- Commission Decision 2000/489/EC amending Commission Decision 1999/217/EC

6 Food chemical safety

- Related on-going work on evaluation of >2000 flavourings for positive list
- Commission proposals on labelling of food additives and flavourings from GM sources (amendment to Commission Regulation 50/2000 awaited)
- Commission Working Document SANCO/1796/2000 amending purity criteria Directives for colours, miscellaneous food additives and sweeteners (i.e. evaluation of additives and flavourings from GM sources)
- Commission Working Document WGF/016/00 on food additives used in flavourings
- Commission Directive 95/45/EC laying down specific purity criteria for colours
- Commission Directive 1999/75/EC – first amendment to Directive 95/45/EC
- Commission Working Document WGA/009/00 rev.1 – second amendment to Directive 95/45/EC
- Commission Directive 96/77/EC laying down specific purity criteria for miscellaneous food additives
- Commission Directive 98/86/EC – first amendment to Directive 96/77/EC
- Commission Directive 2000/63/EC – second amendment to Directive 96/77/EC
- Commission Working Document SANCO/2622/2000 – third amendment to Directive 96/77/EC
- Commission Working Document WGA/011/00 – fourth amendment to Directive 96/77/EC
- Commission Directive 95/31/EC laying down specific purity criteria for sweeteners
- Commission Directive 98/66/EC – first amendment to Directive 95/31/EC
- Commission Working Document WGA/008/00 rev.1 – second amendment to Directive 95/31/EC

The Food Standards Agency is also contributing to harmonising controls on additives across the world. This is gradually developing at the Codex Committee on Food Additives and Contaminants (CCFAC). They are actively developing a Codex Standard for Food Additives. Work on this standard is drawing together the many controls that are applied in different parts of the world. This is a very important step in developing a common approach to both consumer protection and trade. Several main trading blocks have already made major strides in developing their own standards – the EU, the USA, the Southern Common Market (MERCOSUR) in South America, and Australia and New Zealand together under the aegis of their joint Food Authority (ANZFA). The general work of these trading blocks in developing food standards has been described recently in Rees and Watson (2000). Work in the EU and the USA is perhaps furthest advanced, and so is reviewed in this book in some detail (Chapters 2 and 3).

1.3 Future work on additives

The next major stage in controlling food additives will be to develop numerical standards that are consistent with what different countries and trading blocks do across the world. This will take some time, not least because many countries have no alliance with any of the trading blocks and yet share with them a strong interest in the control of additives. Many of these 'non-aligned' countries export raw materials used to make some additives, and of course they must also protect their consumers, some of whom suffer malnutrition. Delay in agreeing standards would send out the wrong signals about the utility of what is a potentially very valuable Codex Standard.

There is also still considerable scientific work to do on additives. Chapters 5 and 6 review progress in developing practical methodology. A lot of effort has been put successfully into the toxicology of additives, although much remains to be done in understanding adverse reactions to additives (Chapter 7). There is also a lot of work in progress to improve control systems. For example, methods are not widely available to test for many approved additives in food. Indeed for such tests to be applied they must be quick and easy to use. Research on screening techniques such as immunoassay is in progress (e.g. see Food Standards Agency, 2000). Whether or not testing is speeded up by using an initial screen to test food samples, control authorities will also need definitive confirmatory tests.

There is also effort to develop practical ways of measuring intake of additives. The direct measurement of intake continues to be problematical. Duplicate diet studies require a lot of resources, and there are not enough analytical methods available to test for all the additives of interest. Total diet studies, also known as market basket studies, provide a very general picture which can be a useful start to more detailed work on intake, but they too suffer from the same shortage of analytical methods. Biomarker studies are currently problematical. It is difficult to identify metabolites that are both unique to particular additives and can be readily measured in urine. The estimation of additive intake by calculation is still the preferred method, although it requires a large amount of information on both additive levels in food and much data on food consumption. The latter is difficult to obtain without using a lot of resources – many of us have very varied diets!

The purpose of the Food Standards Agency's research programme on food additives is to ensure that the use of food additives does not prejudice food safety. Much of the current work aims to develop and validate appropriate methodology to measure levels of additives in foods (Table 1.1). This information is required to enable the UK to fulfil its EU obligation to monitor intakes and usage of additives. The programme also provides technical information to underpin UK work in EU negotiations on specific topics, for example criteria of purity. Current projects range from feasibility studies to acquire a better understanding of factors affecting additive intakes to the development of appropriate test protocols. The Food Standards Agency's

8   Food chemical safety

Table 1.1   The UK Food Standards Agency's research programme on food additives (as of January 2001)

Project title	Objective
Development of a surveillance method for Class III caramels.	To develop a method capable of extracting Class III caramel from foods, and to apply it to the quantification of Class III caramel in a small range of foods by capillary electrophoresis.
Development and validation of procedures to detect and quantify the use of dimethyl dicarbonate in non-alcoholic drinks.	To confirm the rapid degradation of DMDC (dimethyl dicarbonate) (E242) in non-alcoholic drinks and to develop and validate methods of analysis for the detection and quantification of the use of the sterilant.
Biomarker assessment of benzoate consumption.	To develop a method to determine urinary biomarkers for benzoates. The method involves the use of gas chromatography-combustion-isotope ratio mass spectrometry. It is proposed that the procedure developed will allow differentiation of natural benzoates from synthetic ones.
Urinary biomarkers for selected food colours.	To select and determine the amount of food colours commonly found in test foods and to develop ELISA methods using 24 hr urinary samples to examine feasibility of urinary biomarkers.
Food colourant analysis: characterisation of caramel colours and discrimination of malts from malt extracts.	To provide data on the composition and heterogeneity of caramels Class I and IV and to provide test results on concentrated colour and flavour properties.
Method development and analysis of foods for annatto.	To develop a validated method for the determination of the food colour annatto in permitted foods with reference to the Colours in Food Additives Regulations 1995, as amended.
Novel approaches to the identification and quantification of caramels in food.	To assess the use of HPLC and fluorophore assisted carbohydrate gel electrophoresis as a means of profiling caramel colour types III and IV.
Literature review of food additive interactions.	The work will review the literature on the reactions of food additive and food constituents and with other additives in foods. The literature data will be used as a training tool and it is intended that this research will help inform and guide the setting of priorities on which additives may merit experimental attention in the future.
Identification of fatty acid adducts of BHA and BHT.	To prepare fatty acid or triglyceride BHA and BHT adduct standards to enable mass spectra to be obtained for these compounds. These

Introduction 9

Project title	Objective
	will be used to assist in the positive identification of suspected adducts in vegetable oil.
Identification of the causative agents responsible for mineral hydrocarbon toxicity.	To conduct feeding studies on female F344 rats and to locate accumulated hydrocarbon material in tissue section and also to determine the characteristics of HC absorbed through the gut and deposited in body tissues.
Method development to determine levels of caramels in soft drinks.	To develop a validated method for the quantitative analysis of Class IV caramels in soft drinks.
Validation of ELISAs for the detection and quantification of gums in foodstuffs.	To produce validated methods for the identification and quantification of the following gums in foodstuffs; *Acacia senegal* gum, *A. seyal* gum, *A. combretum* gum, *carrageenan* and alginates.
Coumarin and safrole in foods.	Development and validation of a solid phase microextraction method for the determination of coumarin and safrole in foods.
Identification of the raw materials and methods of manufacture of EU-permitted food additives.	To identify and collect data for incorporation into the Food Additives Information System.
Development of a method to determine the levels of additive caramels in foodstuffs.	Optimise and validate methodology for extraction and detection of Class III and IV caramels in foods.
Statistical evaluation of the critical factors influencing intake calculations.	Assess suitability of methods used for dietary intake estimations of different types of survey data by using Uncertainty Budget to quantify error incurred for each step.
Urinary biomarkers for caffeine.	To investigate whether one or more metabolites of caffeine can be used as a urinary biomarker to assess caffeine intake.
HPLC method for simultaneous determination in intense sweeteners in foodstuffs.	To provide a HPLC method for the simultaneous determination of permitted sweeteners in a range of foodstuffs.
Development of the Food Additives Information System (FAIS).	Update FAIS information on colours, sweeteners and miscellaneous additives, with a view to expanding general, toxicological and legislative data.
Development of methods for the analysis of antioxidants in a range of foods.	Select and optimise techniques suitable for the extraction and determination of antioxidants in a range of foods. Subject chosen techniques to limited ring trial.
Intakes of sulphur containing additives and their stability in food during storage.	Develop and validate methodology suitable to determine fate of sulphiting agents in foodstuffs under varying conditions.

research programme is carefully focused. It is also carefully managed to ensure value for money in both the science it produces and its relevance to the work of the Food Standards Agency. Projects are awarded to contractors following open competition. The Working Party on Food Additives has the specific role of providing independent advice on what research is needed and how it is carried out. To ensure that the projects relate to current practice the membership of the Working Party includes experts from industry as well as consumer and regulatory interests. The activities of the Working Party are structured around two meetings a year, in the Spring and Autumn. The Spring meeting reviews possible future research and progress in legislation on food additives. The Autumn meeting reviews current research projects.

Work to research and control additives in food continues to come up against differences in perception about the risk they present to consumers. Many have opinions: the media, the general public, those who believe they or their children may have reacted adversely to additives, the medical profession, scientists, environmental groups. ... The views expressed to me by lay groups, over the last twenty years, seem to place additives near to or at the top of the list of substances in food that cause concern. In summary the ranking, from highest to lowest risk, appears to be as follows:

1. Additives and residues of pesticides
2. Environmental chemicals (e.g. dioxins)
3. Natural toxins including fungal toxins; chemical migration from packaging; veterinary drug residues.

The major influence behind this seems to have been the extent of coverage in the media, where problems with food rather than the successes tend to be reported. Scientific perception of risk is, not unexpectedly, rather different. Indeed scientists may not agree amongst themselves about where additives should be placed in any ranked list of chemicals in food. But most would probably agree that they should be near the bottom. My own ranking would be as follows, from greatest to least risk:

1. Toxins from higher plants
2. Fungal toxins
3. Marine toxins
4. Environmental chemicals in food
5. Chemical migration from packaging into food
6. Residues of pesticides
7. Residues of veterinary drugs
8. Additives.

How these different perceptions can be made more similar is a very major unresolved question. Unless there can be some progress on this paradoxical situation it will be very difficult to communicate work on many of the areas covered in this book to the general public.

1.4 Acknowledgement and dedication

The hard work of all the authors has made this book possible. Their contributions, like the one above represent their own views and are not necessarily those of the organisations that employ them or with which they are affiliated. My personal thanks go to Mark Willis and Glynis Griffiths, both key members of the Food Standards Agency, for their respective work on the compilations shown in Table 1.1 and the list of Directives given above.

My work on this book would not have been possible without the support of my wife Linda.

1.5 References

FOOD STANDARDS AGENCY (2000) *Working Party on Food Additives: annual review of current research projects,* available from lucy.foster @foodstandards.gsi.gov.uk

KEEFE, D., KUZNESOF, P., CARBERRY, S. and RULIS, A. (2000) *The Codex General Standard for Food Additives – a work in progress,* pp. 171–194 in Rees, N. and Watson, D. (eds), *International Standards for Food Safety*, Aspen Publishers Inc., Gaithersburg, Maryland.

LEATHERHEAD FOOD RESEARCH ASSOCIATION (2000) *Essential Guide to Food Additives* (Saltmarsh, M. (ed.)). Leatherhead Food R.A. Publishing, Randalls Road, Leatherhead.

REES, N. and WATSON, D. (2000) *International Standards for Food Safety*, Aspen Publishers Inc., Gaithersburg, Maryland.

WATSON, D. (ed.) (2001) *Food Chemical Safety, Volume 1: Contaminants*, Woodhead Publishing, Cambridge.

WHO (2000) *Evaluation of Certain Food Additives.* 51st report of the Joint FAO/WHO Expert Committee on Food Additives, WHO Technical Report Series 891, WHO, Geneva.

2

The regulation of additives in the EU

D. W. Flowerdew, Consultant (formerly Leatherhead Food RA)

2.1 Introduction

A directive that sets safety and need criteria for the use of additives in food and makes provisions for further specific rules to be drawn up provides the basic general controls concerning food additives in EU countries. Three further detailed directives covering authorisation and conditions of use for sweeteners, colours and food additives other than colours and sweeteners (commonly called miscellaneous additives), form a 'comprehensive' directive that controls the specific use in food of all direct food additives in the EU (except flavourings). These directives result from years of discussions to bring about agreement between the EU member states, whose laws on food additives were previously disparate. Both the additives permitted and the degree of control on their use varied greatly throughout the EU countries, from no positive lists for certain additives, through precise lists with conditions of use to authorisation only with specific permission from the authorities. As expected, countries with strict controls have been unwilling to relax these, so that the resultant directives have veered towards more stringent control of the use of additives in food products. All these directives aim primarily to take account of safety and human health protection, stating that economic and technical needs are within the limits required for the protection of health. The Scientific Committee for Food must be consulted before health-related provisions are introduced. All listed food additives have been evaluated for their safety and the recommendations of the Scientific Committee for Food are taken into account during considerations for the inclusion and conditions of use in food of every food additive.

This chapter also summarises EU laws applicable to flavourings and extraction solvents (which are not within the scope of Directive 89/107/EEC), and labelling requirements for food additives, including specific requirements relating to the use of packaging gases, sweeteners and genetically modified (GM) additives and flavourings in food. Some implementation problems are discussed and relevant recent developments following the Commission's review of food safety legislation, including intended work on food additives are outlined.

2.2 The key directives

2.2.1 Directive 89/107/EEC

Directive 89/107/EEC,[1] sometimes known as a 'framework' directive, contains a number of general safety and authorising measures concerning food additives. There is a definition of 'food additive' as follows:

> any substance not normally consumed as a food in itself and not normally used as a characteristic ingredient of food whether or not it has nutritive value, the intentional addition of which to food for a technological purpose in the manufacture, processing, preparation, treatment, packaging, transport or storage of such food results, or may be reasonably expected to result, in it or its by-products becoming directly or indirectly a component of such foods.

The definition does not include processing aids, but includes an explanation of a processing aid as being a substance (not consumed as a food ingredient by itself) that is intentionally used for the processing of raw materials, foods or ingredients to fulfil a certain technological purpose during treatment or processing. It may result in unintentional but technically unavoidable residues of the substance or its derivatives in the final product, but these residues must not have any technological effect on the final product, and they must not present any health risk. There is no EU positive list of processing aids. Plant health products, flavourings covered by Directive 88/388/EEC and substances used as nutrients (e.g. vitamins, trace elements, minerals) in foodstuffs are not regarded as additives.

The directive makes provision for rules regarding most categories of additive to be made in a further 'comprehensive' directive. These additive categories, listed in Annex I of the Directive, are shown in Table 2.1.

The Council and the Commission must adopt positive lists of additives, authorised to the exclusion of all others, lists of foods in which the additives may be used with conditions of use, and rules on carrier solvents. The Standing Committee on Foodstuffs must approve adoption of purity criteria for listed additives and, if necessary, methods of analysis to verify these purity criteria, sampling procedures and methods of analysis for food additives in food. The directive includes provisions that a member state may temporarily suspend or restrict application of an authorisation if it has detailed grounds for considering that the use of the additive in food, although permitted, endangers health. The

14   Food chemical safety

Table 2.1   Categories of additives listed in annex I of Directive 89/107/EEC

Colour	Modified starch
Preservative	Sweetener
Antioxidant	Raising agent
Emulsifier	Anti-foaming agent
Emulsifying salt	Glazing agent
Thickener	Flour treatment agent
Gelling agent	Firming agent
Stabiliser	Humectant
Flavour enhancer	Sequestrant
Acid	Enzyme
Acidity regulator	Bulking agent
Anti-caking agent	Propellant gas and packaging gas

Commission and the Standing Committee on Foodstuffs must examine this action and may make amendments to the directives to ensure protection of health. The directive also includes rules regarding temporary authorisation of an additive by a member state for reasons of scientific and technical development.

Annex II of the Directive sets out general criteria concerning technological need for, safety and use of food additives. Food additives may only be approved if a technological need is demonstrated and the purpose cannot be achieved by other technological means, they do not present a hazard to health at the proposed levels of use, judged on the available scientific evidence, and they do not mislead the consumer. Additives may only be used for certain purposes including improvement of keeping quality and stability of food and to aid in the manufacture, processing and treatment of food, when they must not mask the use of faulty raw materials or undesirable practices. Additives must undergo appropriate toxicological testing and evaluation in order to assess possible harmful effects of the additive or its derivatives. The evaluation must take into account any cumulative, synergistic or potentiating effect of its use and of human intolerance to foreign substances. All food additives must be kept under continual observation and must be re-evaluated when necessary. They must comply with approved purity criteria. The approval for food additives must specify the foodstuffs to which additives may be added and the conditions of their use, be limited to the lowest level of use needed to achieve the intended effect and take account of any acceptable daily intake, or equivalent assessment, established for the food additive and the likely intake of it from all sources. Intake of additives by special groups of consumers must also be considered.

Directive 94/34/EC,[1] an amendment to Directive 89/107/EEC, allows member states to prohibit in their country certain classes of additives in the production and sale of foodstuffs that they consider as traditional. However, the production and sale of all foodstuffs that are considered as non-traditional must be allowed. The list of 'traditional' foods was published in Commission Decision No. 292/97.[2]

2.2.2 The 'comprehensive' directive

The three directives forming the 'comprehensive' directive concern sweeteners,[3] colours[4] and all other classes of additives[5] (or miscellaneous additives) permitted in food and drink products. They are usually referred to by their specific names or as the 'specific' directives. These technical directives are of key interest to the food manufacturer since they list the authorised additives, the foods in which these additives may be used and, for many listed additives, maximum levels of use. In each directive the stated restrictions and conditions of use take account of known scientific and toxicological information on these additives. For many additives which are considered safe under envisaged conditions of use it is not always considered necessary to prescribe a maximum quantity. In these cases the term *quantum satis* is used. This term means that the additive may be used in food in accordance with good manufacturing practice, at a level not higher than that needed to achieve the intended technological purpose, and provided that the consumer is not misled.

2.2.3 Sweeteners

Directive 94/35/EC[3] on sweeteners for use in foodstuffs applies to food additives that are used to impart a sweet taste to foodstuffs or used as table-top sweeteners. Most authorisations are for foods with no added sugar or foods that are energy-reduced. The term 'with no added sugar' means without any added mono- or disaccharides or any other foodstuff used for its sweetening properties, and the term 'energy-reduced' means with an energy value reduced by at least 30% compared with the original foodstuff or a similar product. Foods that have sweetening properties (such as honey) are not regarded as sweeteners. Sweeteners are not permitted in food including dietary food (unless otherwise specified) for infants and young children covered by Directive 89/398/EEC. Table 2.2 shows the sweeteners permitted for use in foodstuffs or for sale to the consumer.

Table 2.2 List of permitted sweeteners

E 420	Sorbitol: (i) sorbitol, (ii) sorbitol syrup
E 421	Mannitol
E 953	Isomalt
E 965	Maltitol: (i) maltitol, (ii) maltitol syrup
E 966	Lactitol
E 967	Xylitol
E 950	Acesulfame K
E 951	Aspartame
E 952	Cyclamic acid, sodium and calcium cyclamates (use levels expressed as free acid)
E 954	Saccharin, sodium, potassium and calcium saccharin (use levels expressed as free imide)
E 957	Thaumatin
E 959	Neohesperidine DC

E 420, E 421, E 953, E 965, E 966 and E 967 are polyhydric alcohols and they are allowed to *quantum satis* in: certain foods that are energy-reduced or with no added sugar, including certain dessert products, breakfast cereals, edible ices, jams and marmalades, certain confectionery products, fine bakery wares and sandwich spreads; in chewing gum with no added sugar; and in sauces, mustard, products for particular nutritional uses and solid supplements/dietary integrators.

E 951, E 952, E 954, E 957 and E 959 are intensely sweet substances and conditions of use are imposed, with maximum limits prescribed in certain foods for each sweetener. A few examples of authorised uses of these sweeteners are stated in Table 2.3.

2.2.4 Colours

The directive on colours for use in foodstuffs, 94/36/EC,[4] replaced a previous positive list of colours, but it goes much further to harmonise completely the laws of the EU member countries in respect of food colours, since, for many colours, the foods in which certain colours are permitted and maximum levels of use are prescribed. Colours are defined as: 'substances which add or restore colour in a food, and include natural constituents of foodstuffs and natural sources which are normally not consumed as foodstuffs as such and not normally used as characteristic ingredients of food'.

The definition includes substances obtained from foods and other natural sources by physical and/or chemical extraction that results in selective extraction of the pigments, but it does not include foodstuffs, flavourings that have a secondary colouring effect, such as paprika, turmeric and saffron, or colours used on external inedible parts of foods, such as cheese coatings and sausage casings. The colours that may be used for health marking and other required marking on meat products and for decorative colouring of eggshells are prescribed. The main technical details are contained in five annexes to the directive.

Annex I contains the list of permitted colours for food. Only colours included on this list may be sold directly to consumers, except that E 123, E 127, E 128, E 154, E 160b, E 161g, E 173 and E 180 may not be sold directly to consumers. The complete list of authorised colours is stated in Table 2.4.

Annex II lists foodstuffs that may not contain added colours unless these are expressly permitted by other annexes or they are present because of legitimate carry-over in an ingredient. The list includes unprocessed foods and processed foods that would not be expected to contain colours, also some processed foods listed in subsequent annexes which may contain only a few colours. The list includes bottled waters, milk, cream, oils and fats, eggs and egg products, flour, bread, pasta, sugar, processed fruit and vegetables, extra jam, coffee and tea and preparations of these, salt, honey, certain spirits, and wine covered by Regulation (EEC) No. 822/87.

Annex III lists foods in which only certain colours are permitted, detailing the specific permitted colours with maximum levels for each food. Examples include

Table 2.3 Examples of uses of authorised sweeteners

Food product	Maximum limit of use					
	E 950	E 951	E 952	E 954	E 957	E 959
Water-based flavoured drinks, energy reduced or with no added sugar	350 mg/l	600 mg/l	400 mg/l	80 mg/l	–	30 mg/l
Confectionery with no added sugar	500 mg/kg	1,000 mg/kg	500 mg/kg	500 mg/kg	50 mg/kg	100 mg/kg
Energy-reduced beer	25 mg/l	25 mg/l	–	–	–	10 mg/kg
Edible ices, energy reduced or with no added sugar	800 mg/kg	800 mg/kg	250 mg/kg	100 mg/kg	50 mg/kg	50 mg/kg
Food supplements/dietary integrators based on vitamins and/or mineral elements, syrup-type or chewable	2,000 mg/kg	5,500 mg/kg	1,250 mg/kg	1,200 mg/kg	400 mg/kg	–
Breakfast cereals with a fibre content more than 15%, containing at least 20% bran, energy reduced or with no added sugar	1,200 mg/kg	1,000 mg/kg	–	100 mg/kg	–	50 mg/kg
Energy-reduced soups	110 mg/l	110 mg/l	–	110 mg/l	–	50 mg/l
Breath-freshening micro-sweets, with no added sugar	2,500 mg/kg	6,000 mg/kg	2,500 mg/kg	3,000 mg/kg	–	400 mg/kg

18 Food chemical safety

Table 2.4 List of authorised colours (annex 1 to Directive 94/36/EC)[1]

E number	Colour	Colour Index number
E 100	Curcumin	75300
E 101	(i) Riboflavin	
	(ii) Riboflavin-5'phosphate	
E 102	Tartrazine	19140
E 104	Quinoline Yellow	47005
E 110	Sunset Yellow FCF	15985
	Orange Yellow S	
E 120	Cochineal, Carminic acid, Carmines	75470
E 122	Azorubine, Carmoisine	14720
E 123	Amaranth	16185
E 124	Ponceau 4R, Cochineal Red A	16255
E 127	Erythrosine	45430
E 128	Red 2G	18050
E 129	Allura Red AC	16035
E 131	Patent Blue V	42051
E 132	Indigotine, Indigo carmine	73015
E 133	Brilliant Blue FCF	42090
E 140	Chlorophylls and	75810
	Chlorophyllins:	75815
	(i) Chlorophylls	
	(ii) Chlorophyllins	
E 141	Copper complexes of chlorophylls and chlorophyllins	75815
	(i) Copper complexes of chlorophylls	
	(ii) Copper complexes of chlorophyllins	
E 142	Green S	44090
E 150a	Plain caramel[2]	
E 150b	Caustic sulphite caramel	
E 150c	Ammonia caramel	
E 150d	Sulphite ammonia caramel	
E 151	Brilliant Black BN, Black PN	28440
E 153	Vegetable carbon	
E 154	Brown FK	
E 155	Brown HT	20285
E 160a	Carotenes:	
	(i) Mixed carotenes	75130
	(ii) Beta-carotene	40800
E 160b	Annatto, bixin, norbixin	75120
E 160c	Paprika extract, capsanthin, capsorubin	
E 160d	Lycopene	
E 160e	Beta-apo-8'-carotenal (C30)	40820
E 160f	Ethyl ester of beta-apo-8'-carotenic acid (C30)	40825
E 161b	Lutein	
E 161g	Canthaxanthin	
E 162	Beetroot Red, betanin	

E number	Colour	Colour Index number
E 163	Anthocyanins	Prepared by physical means from fruits and vegetables
E 170	Calcium carbonate	77220
E 171	Titanium dioxide	77891
E 172	Iron oxides and hydroxides	77491, 77492 77499
E 173	Aluminium	
E 174	Silver	
E 175	Gold	
E 180	Litholrubine BK	

Notes: [1]Aluminium lakes of listed colours are also permitted.
[2]Caramel means the brown products intended for colouring, not the sugary product obtained by heating sugars which is used for flavouring food such as confectionery, pastry and alcoholic drinks.

margarine to which E 160a and E 100 may be added *quantum satis*, and E 160b max. 10 mg/kg; vinegar and liqueur wines, which may only contain E 150a–d *quantum satis*; breakfast sausages with a minimum cereal content of 6% and burger meat with a minimum vegetable and/or cereal content of 4%, which may contain E 129 max. 25 mg/kg, E 120 max. 100 mg/kg and E 150a–d *quantum satis*; dried potato granules which may only contain E 100 *quantum satis*.

Annex IV lists colours with very restricted food use: E 123, E 127, E 128, E 154, E 161g, E 173, E 174, E 175, E 180 and E 160b. Examples of permitted uses include E 123 which may be added to aperitif wines, spirit drinks including products with less than 15% alcohol by volume, max. 30 mg/l, and fish roe, max. 30 mg/kg; E 154 which may be added to kippers, max. 20 mg/kg; E 174 and E 175 which may be used for external coating of confectionery, decoration of chocolates and in liqueurs, *quantum satis*; and E 180 which may be used for edible cheese rind, *quantum satis*.

Annex V is the most complicated annex and lists the most common uses of colours in foodstuffs. Part 1 contains a list of colours, mostly of natural derivation, that may be added generally to foods at *quantum satis*, unless this is prohibited by earlier annexes. These are: E 101, E 140, E 141, E 150a–d, E 153, E 160a, E 160c, E 162, E 163, E 170, E 171 and E 172. Part 2 contains a list of colours that may be used, singly or in combination, to the maximum level in foods specified in an accompanying table. However, amounts of each of the colours E 110, E 122, E 124 and E 155 may not exceed 50 mg/kg or mg/l in non-alcoholic flavoured drinks, edible ices, desserts, fine bakery wares and confectionery. The part 2 list of colours is: E 100, E 102, E 104, E 110, E 120, E 122, E 124, E 129, E 131, E 132, E 133, E 142, E 151, E 155, E 160d, E 160e, E 160f and E 161b. The foods listed in the table to this annex include non-alcoholic flavoured drinks max. 100 mg/l; edible ices max. 150 mg/kg; desserts max. 150

mg/kg; smoked fish max. 100 mg/kg; dry, savoury potato, cereal or starch-based snack products (extruded or expanded) max. 200 mg/kg, (others) max. 100 mg/kg; liquid food supplements/dietary integrators max. 100 mg/l; solid food supplements/dietary integrators max. 300 mg/kg; spirituous beverages, aromatised wines and similar products (unless mentioned in annexes II or III), fruit wines, cider, perry max. 200 mg/l.

2.2.5 Food additives other than colours and sweeteners

Directive 95/2/EC, as amended by Directives 96/85/EC, 98/72/EC and 2001/5/EC,[5] contains the rules for use in food of other classes of food additives, except flavourings. This directive is the most complicated of the specific directives, and makes provisions for the use of a large number of food additives, with varying degrees of control. There is a long list of additives that may be used to *quantum satis*, lists of antioxidants and preservatives and several other additives that are very strictly controlled, a list of foods in which only certain additives may be used and lists of foods for infants and young children that may contain certain additives. The provisions of the directive apply generally to corresponding foods intended for particular nutritional uses covered by Directive 89/398/EEC except such foods for infants and young children where specific authorisations are made. The directive does not apply to enzymes, except invertase and lysozyme.

The categories of additives covered by Directive 95/2/EC are: preservatives, antioxidants, carriers including carrier solvents, acids, acidity regulators, anti-caking agents, anti-foaming agents, bulking agents, emulsifiers, emulsifying salts, firming agents, flavour enhancers, foaming agents, gelling agents, glazing agents, humectants, chemically modified starches, packaging gases, propellants, raising agents, sequestrants, stabilisers and thickeners. Each of these categories is defined. Substances that are not regarded as additives for the purposes of this directive are also listed: substances used for treatment of drinking water, liquid pectin, chewing gum bases, dextrins, starches and physically modified starches (however, chemically modified starches are regarded as food additives), ammonium chloride, blood plasma, protein hydrolysates, gelatin, milk protein, gluten, most amino acids that have no additive function, casein and caseinates and inulin. Six annexes are appended to the directive, and these contain the major technical provisions.

Annex I contains a long list of additives permitted generally in food products to *quantum satis*. These additives include acetic acid and its potassium, sodium and calcium salts (E 260, E 261, E 262), ascorbic acid and its sodium and calcium salts (E 300, E 301, E 302), citric acid and its sodium, potassium and calcium salts (E 330, E 331, E 332, E 333), alginic acid and alginates (E 400–E 404), gums, pectins, fatty acids and salts and esters of fatty acids, sodium, potassium, ammonium and magnesium carbonates, glucono-delta-lactone and sodium, potassium and calcium gluconates, the packaging gases argon, helium, nitrogen, nitrous oxide and oxygen, polydextrose and chemically modified starches. These additives may not be used in unprocessed foods, honey, non-emulsified animal or vegetable oils, butter, milk, cream, fermented milk products, natural mineral water, spring water, coffee and coffee extracts, unflavoured leaf tea, sugars, pasta and natural unflavoured buttermilk, or in the foods listed in annex II or in foods for infants and young children, unless specific provisions are made in other annexes for such use. However, carbon dioxide and other packaging gases may be used in these foods.

Annex II lists foodstuffs for which only certain of the annex I additives may be used. Such foodstuffs include cocoa products and chocolate products, fruit juices and nectars, jam, jellies and marmalades and partially dehydrated and dehydrated milk, which are the subjects of EU vertical standards, and a number of other foods including frozen unprocessed fruit and vegetables, quick-cook rice, non-emulsified oils and fats, canned and bottled fruit and vegetables, bread made with basic ingredients only, fresh pasta and beer.

Annex III lays down the conditions of use for permitted preservatives and antioxidants, with lists of foods and maximum levels in each case. Part A lists the sorbates, benzoates and p-hydroxybenzoates, E 200–E 219; part B lists sulphur dioxide and the sulphites, E 220–E 228; part C lists other preservatives with their uses, including nisin, dimethyl dicarbonate and substances allowed for surface treatment of certain fruits, E 249 potassium nitrite, E 250 sodium nitrite, E 251 sodium nitrate and E 252 potassium nitrate, E 280–E 283 propionic acid and the propionates; part D lists the antioxidants E 320 butylated hydroxyanisole (BHA), E 321 butylated hydroxytoluene (BHT), E 310 propyl gallate, E 311 octyl gallate, E 312 dodecyl gallate, E 315 erythorbic acid and E 316 sodium erythorbate.

Annex IV contains a long list of other additives that are permitted with restrictions in named foods. This list includes E 297 fumaric acid, phosphates (E 338, E 339, E 340, E 341, E 343, E 450, E 451 and E 452), E 405 propane-1,2-diol alginate, E 416 karaya gum, sorbitol and other polyols (when not used for sweetening purposes), polysorbates (E 432–E 436), E 473 sucrose esters of fatty acids and E 474 sucroglycerides, E 481 sodium stearoyl-2-lactylate and E 482 calcium stearoyl-2-lactylate, (E 491–E 495) sorbitan esters of fatty acids, E 551–E 559 silicon dioxide and certain silicates, (E 620–E 635) glutamic acid, glutamates, guanylic acid and guanylates, inosinic acid and inosinates, ribonucleotides, and acesulfame-K, aspartame, thaumatin and neohesperidine DC when used as flavour enhancers, E 999 quillaia extract and E 1505 triethyl citrate.

Annex V lists permitted carriers and carrier solvents, some of which may only be used for restricted purposes.

Annex VI lists the additives permitted in infant foods and foods for young children. Part 1 lists the few additives allowed in infant formulae for infants in good health, part 2 those allowed in follow-on formulae for infants in good health, and part 3 the additives permitted in weaning foods for infants and young children in good health. Part 4 applies the lists in parts 1–3 to foods for infants and young children for special medical purposes.

It is to be expected that a directive that covers so many additives would be amended frequently, and several amendments have already been made to take account of technical developments in the field of food additives since the original Directive 95/2/EC was adopted. Directive 96/85/EC adds E 407a processed eucheuma seaweed to the annex I list of generally permitted additives. Directive 98/72/EC includes E 920 L-cysteine in the annex I list of generally permitted additives, but only for use as a flour treatment agent. E 469 enzymatically hydrolysed carboxy methyl cellulose, E 1103 invertase and E 1451 acetylated oxidised starch are also added to annex I. The further annexes are also amended by the addition of several new additives and extended uses of existing additives, with conditions of use laid down. Further authorisations are listed in annex VI, including a list of additives permitted with restrictions in specific foods for infants and young children for special medical purposes.

Amending Directive 2001/5/EC contains several new authorisations including the addition of E 949 hydrogen to annex I, and E 943a butane, E 943b iso-butane and E 944 propane to annex IV for use as vegetable oil pan sprays (professional use only) and as water-based emulsion sprays, *quantum satis*. E 650 zinc acetate is also added to annex IV for use in chewing gum, max. 1000 mg/kg.

2.2.6 Purity criteria for additives

As required by Directive 89/107/EEC, criteria of purity have been drawn up for all the listed food additives (with a couple of exceptions). Purity criteria for all the permitted sweeteners have been prescribed in Directive 95/31/EC,[6] as amended, and criteria for all the permitted colours are contained in Directive 95/45/EC,[7] as amended. Directives that prescribe purity criteria for all the additives authorised under Directive 95/2/EC have been drawn up in stages. Directive 96/77/EC[8] containing purity criteria for antioxidants and preservatives is amended by Directives 98/86/EC which lays down purity criteria for emulsifiers, stabilisers and thickeners and 2000/63/EC which contains purity criteria for most additives numbered E 500 and above, and for certain other additives not covered in the earlier directives. Purity criteria for most of the few remaining permitted miscellaneous additives are contained in Directive 2001/30/EC; however, purity criteria for E 1201 polyvinylpyrrolidone and E 1202 polyvinylpolypyrrolidone are still being considered by the Scientific Committee on Food. Some methods of analysis for verifying prescribed purity criteria have been developed at EU level; these are contained in Directive 81/712/EEC.[9]

2.2.7 Flavourings

Harmonisation of laws relating to flavourings used in food production was considered before Directive 89/107/EEC[1] and the subsequent specific directives concerning additives were drawn up. Directive 88/388/EEC[10] relates to flavourings and to source materials for their preparation. It is a framework directive which prescribes some general requirements for the food use of flavourings, taking account of human health requirements and, within this limit, of economic and technical needs. The directive requires that further appropriate provisions are drawn up, including further laws regarding various types of flavourings, additives needed for production, use and storage of flavourings and procedures for sampling and analysis of flavourings in or on foods.

The directive does not apply to edible substances to be consumed as such, substances with exclusively a sweet, sour or salt taste or to material of vegetable or animal origin that has flavouring properties but is not used as a flavouring source. It includes definitions of 'flavouring', 'flavouring substance', 'flavouring preparation', 'process flavouring' and 'smoke flavouring'.

Some general purity criteria are listed, and limits for certain substances in foods resulting from the use of flavourings are prescribed. Annex I prescribes a limit for 3,4 benzopyrene of 0.03 μg/kg in foods and beverages, and Annex II details limits for other substances, including beta asarone, coumarin, hydrocyanic acid, pulegone and safrole. Currently, the definition of 'process flavouring' and some of the limits for undesirable substances in foods are being reviewed.

Council Decision 88/389/EEC[11] required the Commission to draw up an inventory of source materials and substances used in the preparation of flavourings, and to update this inventory regularly. Regulation 2232/96,[12] the first specific measure drawn up under Directive 88/388/EEC, aims to harmonise national laws on flavourings and eventually to draw up an exclusive list of chemical flavouring substances for food use. The laws will take account primarily of protection of human health and, within this requirement, of economic and technical needs. For health reasons some conditions of use of flavourings may be prescribed. The Regulation sets out a staged procedure whereby an exclusive list of chemical flavouring substances for food use is to be drawn up. It applies to the following types of flavouring substances: those obtained by physical, enzymatic or microbiological processes from vegetable or animal raw materials; chemically synthesised or chemically isolated flavouring substances that are chemically identical to flavouring substances naturally present in foods or in herbs and spices normally considered as foods; chemically synthesised or chemically isolated flavouring substances that are chemically identical to flavouring substances naturally present in vegetable or animal raw materials that are not normally considered as foods; other chemically synthesised or chemically isolated flavouring substances. The annex to the regulation prescribes safety and use criteria for these flavouring substances. Flavouring substances may be authorised provided they present no risk to the health of the consumer, as assessed by the Scientific Committee for Food, and their use does not mislead the consumer. They must be subjected to appropriate toxicological evaluation; for those substances consisting of or containing GM organisms the evaluation must take account of environmental safety. All flavouring substances must be monitored constantly and re-evaluated when necessary.

The regulation contains details of the procedure for drawing up a positive list of flavouring substances. Member states must first notify the Commission of a list of such flavouring substances that might be used in or on foods marketed nationally; the notification should include relevant information such as the nature of the flavouring substance, the foods in or on which the flavouring is used and the level of compliance with the limits prescribed in Directive 88/388/EEC for certain undesirable substances in flavourings. On the basis of this information the Commission must draw up a register of notified flavouring substances, the legal use of which in one member state must be recognised by the other member states. This part of the procedure has been completed and the register of available information is contained in Commission Decision 1999/217/EC.[13] An evaluation programme must then be adopted and this must define in particular the order of priorities for evaluation of the flavouring substances taking account of their uses, the time limits and the flavouring substances that are to be the subject of scientific cooperation. Several substances have been selected for priority evaluation (including caffeine). The positive list of flavourings authorised to the exclusion of all others should be drawn up within five years of adoption of the evaluation programme.

2.2.8 Extraction solvents

The extraction solvents that are authorised for use in food production in the EU are not covered by Directive 89/107/EEC,[1] since harmonising legislation was in force before that directive was drawn up. Directive 88/344/EEC, as amended,[14] concerning extraction solvents used in the production of food and food ingredients, was made to harmonise laws in this area with a view to free movement of foodstuffs, taking account primarily of issues concerning public health, but also of economic and technical needs. An extraction solvent means a solvent that is used in an extraction procedure during the processing of raw materials, foodstuffs or of components or ingredients of these, and which is removed but may result in the unintentional but technically unavoidable presence of residues or derivatives in the final food or ingredients. Water, foods with solvent properties and ethanol may be used as extraction solvents generally in the production of foods and food ingredients.

The annex to Directive 84/344/EEC, as amended, which contains the technical details, includes three lists of extraction solvents allowed for food processing:

- Part I: extraction solvents allowed for all uses according to good manufacturing practice: propane, butane, ethyl acetate, ethanol, carbon dioxide, acetone (not for production of olive-pomace oil), nitrous oxide.
- Part II: extraction solvents for which conditions of use/maximum residue limits, are specified: hexane, methanol, propan-2-ol, methyl acetate, ethyl-methylketone, dichloromethane.
- Part III: extraction solvents that may be used for the preparation of flavourings from natural flavouring materials: diethyl ether, hexane, methyl acetate, butan-1-ol, butan-2-ol, ethylmethyl ketone, dichloromethane,

propan-1-ol, cyclohexane, 1,1,1,2-tetrafluoroethane. In Part III maximum residue limits in the final food are stated for each solvent. In Parts II and III the combined use of hexane and ethylmethyl ketone is not allowed.

The directive has been amended three times to take account of the recommendations of the Scientific Committee for Food and the need for certain solvents by the food industry. General criteria of purity are prescribed for the listed extraction solvents. Extraction solvents need not be listed in the ingredients list of food products but the directive includes labelling requirements for extraction solvents sold as such for business purposes.

2.2.9 Labelling requirements for additives
General requirements for labelling of additives

Generally food additives that are used in the manufacture of food products must be identified as prescribed in Directive 2000/13/EC,[15] (formerly 79/112/EEC, as amended). The correct category name, representing the function of the additive in the food, must be stated, followed by the specific name or E number of the additive or additives present. If an additive performs more than one function, the category that represents its principal function in that food must be named. The list of category names is provided in Table 2.5.

The E number need not be stated in the case of modified starches; however, if a modified starch could contain gluten, the category name must be accompanied by an indication of the specific vegetable origin of the starch. The category name 'emulsifying salt' may only be used for processed cheese and processed cheese products. The indications of additives must be placed in the list of ingredients of the product in the correct position in order by weight (greatest first).

There are certain exemptions from these rules. Some food ingredients need only be identified by a generic term, and additives used in such ingredients need not be named. Additives contained in ingredients of a food need not be listed, provided that they do not perform a technological function in the final foodstuff, and additives used as processing aids need not be listed. The constituents of compound ingredients that constitute less than 25% of the product need not be

Table 2.5 Permitted category names for food additives in food labelling

Acid	Flour treatment agent
Acidity regulator	Gelling agent
Anti-caking agent	Glazing agent
Anti-foaming agent	Humectant
Antioxidant	Modified starch
Bulking agent	Preservative
Colour	Propellant gas
Emulsifier	Raising agent
Emulsifying salts	Stabiliser
Firming agent	Sweetener
Flavour enhancer	Thickener

stated. Additives present in a compound ingredient must usually be stated as described above, but if such an additive does not perform a significant technological function in the final product, it need not be identified in the ingredients list. However, where a flour treatment agent has been used for bread manufacture, it must always be identified, in the form described above.

Flavourings are not included in the categories listed above. When used in food products they must be designated by the term 'flavourings', or by a more specific name or description of the flavouring. Additional rules apply if the flavouring is claimed to be 'natural'.

For consumer sales of additives sold as such, information similar to labelling particulars for foods generally is required, and for business sales of such additives specific requirements are set out in Directive 89/107/EEC. Labelling requirements for flavourings sold as such are contained in Directive 88/388/EEC, completed by Directive 91/71/EEC.[16]

Specific requirements for identification of certain additives in foods
Packaging gases
The requirement to identify additives in the list of ingredients does not apply to packaging gases. Nevertheless, consumers should be aware when these substances are used to prolong the shelf-life of the food. Directive 94/54/EC[17] states that where the durability of a foodstuff has been extended by means of packaging gases (as permitted), the label should bear the statement 'packaged in a protective atmosphere'.

Sweeteners/sugars, aspartame, polyols
Table-top sweeteners must be named '... -based table-top sweetener', including the name(s) of the sweetener(s) used.

Directive 94/54/EC has been amended by Directive 96/21/EC[17] which includes further specific requirements relating to the presence of sweeteners and warnings in respect of certain sweeteners. The statements, which must be made in addition to the declarations required in the list of ingredients, are detailed as follows.

Where a foodstuff contains a sweetener or sweeteners, as allowed, the statement 'with sweetener/s' must accompany the name of the product. Where a foodstuff contains both added sugar and sweeteners, as allowed, the statement 'with sugar/s and sweetener/s' must accompany the name of the product. Foodstuffs that contain aspartame must bear the statement 'contains a source of phenylalanine'. Foodstuffs that contain more than 10% added polyols must bear the statement 'excessive consumption may produce laxative effects'.

Labelling requirements for foods and ingredients that contain genetically modified additives and flavourings (GM additives and flavourings)

EU Regulation No. 50/2000[18] prescribes the conditions under which foods and food ingredients (for consumer sale or sale to mass caterers) that contain GM additives and flavourings must be identified. The required terms are 'produced from genetically modified ...' in parenthesis immediately after the indication in the ingredients list of the additive or flavouring in question or, if appropriate, 'genetically modified ...' immediately after the indication in the ingredients list of the relevant additive or flavouring. Alternatively, the phrases may appear in a prominent asterisked footnote to the ingredients list. The font of the footnote must be at least the same size as that used for the list of ingredients. For such foods where no list of ingredients is required the wording quoted above must appear prominently on the label.

2.2.10 Implementation of the EU additives directives

Directives must be implemented nationally before they have legal force; the directives concerning food additives have been implemented into the food law of all the existing EU member states. In general this means that trade problems in this area should be eliminated, but certain difficulties may still remain. For instance there is flexibility as to how national implementation is carried out, provided the technical requirements are met. In the UK regulations have been made on sweeteners, colours, miscellaneous additives, flavourings, extraction solvents and labelling of food additives when sold as such and when used in foodstuffs, and these adopt the EU laws, including the relevant technical requirements of Directive 89/107/EEC and references to criteria of purity for the relevant additives. However, in other member states, the implementation takes different forms, with many countries listing food groups and the additives that are permitted in the various foods. In practice the manufacturer should always check the national laws as well as the EU laws, and should be prepared for varying formats of national regulations, and possibly different interpretations of the EU requirements. Aspects such as variation in the use of transitional times before amendments are implemented in the various member states, the transition of laws in Central and East European countries and other countries seeking for trade reasons to harmonise their laws with those of the EU also mean that regulations still vary in detail. Manufacturers should seek to keep up to date with food law developments on a global basis, including the changing views and developments in the EU, the USA and Codex Alimentarius. As an example, safety considerations could result in alterations to permitted levels of an additive or even its removal from the authorised lists when producers would need to remove such an additive from products at short notice.

2.3 Future developments

In a further move to achieve a high degree of consumer protection and to provide adequate information, the Commission is addressing the issue of specific identification of substances in food that cause allergies and symptoms of intolerance and hypersensitivity in certain consumers. A proposed Directive[19] will require declaration of these substances (listed in Annex IIIa) under all circumstances. The 25% exemption from ingredients listing for compound ingredients (see 2.2.9) is reduced to 5%. Where food and drink products (including alcoholic drinks) contain any substance listed in Annex IIIa the substance must be specifically indicated; there are no exemptions. The list includes cereals, eggs, fish, milk, nuts and products of them, and also sulphite at concentrations of at least 10 mg/kg. A further draft Commission Directive[20] concerns specific identification of caffeine or of quinine when used as a flavouring in foods and drinks. Where the product has a high content of added caffeine this must be indicated and followed by a statement of the caffeine content of the foodstuff.

The European Commission's White Paper on Food Safety[21] reviews in detail the existing body of food law in the EU and sets out the intentions of the Commission regarding better co-ordination and integration of food safety legislation with a view to achieving the highest possible level of health protection. The Annex to the White Paper summarises an 84-point action plan, with priority measures that include the setting up of a European Food Authority, a proposal for a General Food Law that lays down common legislation principles with food safety as the primary objective, and a proposal for safeguard measures relating to food safety including a Rapid Alert System to deal with food emergencies arising in member states. A proposed regulation that includes these priority aspects has been published[22] and has reached Common Position agreement (not published at time of writing). The extensive proposal lays down general definitions and principles governing food law, particularly food safety; it establishes a European Food Authority with provisions regarding the mission of the Authority, its tasks, organisation, operation and financial considerations; the regulation includes measures for setting up a Rapid Alert System that provides an information network in event of food safety emergencies, plans for crisis management in the field of food safety and emergency measures to deal with food originating in the Community and food imported from a third country, that constitutes a serious risk to human health.

Two proposals containing revised laws to control the safety, authorisation and labelling of GM food, including ingredients and additives, are being developed. A proposal for a regulation on genetically modified food and feed[23] sets out revised safety and authorisation requirements for GM organisms for food use, food and food ingredients including additives and flavourings that consist of or contain GM organisms and food produced from or containing ingredients produced from GM organisms. The procedures will come under the responsibility of the European Food Authority and are intended to assure a high level of protection of human health. They include labelling requirements that will identify GM foods and ingredients and name the GM organisms. The requirements do not apply where the GM material is present at a level of 1% or lower, provided its presence is adventitious and unavoidable and the authorities consider that there is no risk to human health or to the environment. A second proposal[24] contains traceability and further labelling requirements for GM products and foods and food ingredients including additives and flavourings produced from GM organisms that are placed on the market, with the same proviso covering threshold levels.

In respect of food additives the Commission has made the following recommendations in the White Paper:[21]

- Directive 89/107/EEC should be amended to confer implementing powers on the Commission to maintain lists of authorised additives and lay down specifications in respect of enzymes
- the EU lists of sweeteners, colours and other food additives need to be updated
- purity criteria for sweeteners, colours and other food additives need to be updated and completed, and for novel additives appropriate purity criteria and a requirement for a new safety evaluation need to be laid down
- further work is needed to reflect innovation in the field of flavourings and investigate toxicological effects of natural substances in flavourings
- the flavourings Directive 88/388/EC will be amended, the register of flavourings will be updated, in the programme for evaluation of flavourings priorities and time limits will be laid down, a list of additives authorised for use in flavourings will be drawn up and conditions for the production of smoke flavourings prescribed
- the detailed methods of analysis laid down in Directive 81/712/EEC will be replaced with a set of general principles
- provisions relating to additives produced by genetic engineering will be clarified, and the labelling of food ingredients produced without genetic modification ('GM-free') will be considered.

2.4 Sources of further information and advice

Food Standards Agency
Food Additives Unit
Room 516C
Aviation House
125 Kingsway
London WC2B 6NH

Leatherhead Food Research Association
Randalls Road
Leatherhead
Surrey KT22 7RY

Food Additives and Ingredients Association
Executive Secretary
10 Whitchurch Close
Maidstone
Kent ME16 8UR

2.5 References

1. Council Directive 89/107/EEC of 21 December 1988 on the approximation of the laws of the member states concerning food additives authorised for use in foodstuffs intended for human consumption (*Official Journal of the European Communities* (L40) of 11 February 1989, pp. 27–33), as amended by Directive 94/34/EC of 30 June 1994 (*OJ* (L237) of 10 September 1994, pp. 1–2).
2. Decision No. 292/97/EC of the European Parliament and of the Council of 19 December 1996 on the maintenance of national laws prohibiting the use of certain additives in the production of certain specific foodstuffs (*Official Journal of the European Communities* (L48) of 19 February 1997, pp. 13–15).
3. European Parliament and Council Directive 94/35/EC of 30 June 1994 on sweeteners for use in foodstuffs (*Official Journal of the European Communities* (L237) of 10 September 1994, pp. 3–12), as amended by Directive 96/83/EC of 19 December 1996 (*OJ* (L48) of 19 February 1997, pp. 16–19).
4. European Parliament and Council Directive 94/36/EC of 30 June 1994 on colours for use in foodstuffs (*Official Journal of the European Communities* (L237) of 10 September 1994, pp. 13–29).
5. European Parliament and Council Directive 95/2/EC of 20 February 1995 on food additives other than colours and sweeteners (*Official Journal of the European Communities* (L61) of 18 March 1995, pp. 1–40), as amended by Directives 96/85/EC of 19 December 1996 (*OJ* (L86) of 28 March 1997, p. 4), 98/72/EC of 15 October 1998 (*OJ* (L295) of 4 November 1998, pp. 18–30) and 2001/5/EC of 12 February 2001 (*OJ* (L55) of 24 February 2001, pp. 59–61).
6. Commission Directive 95/31/EC of 5 July 1995 laying down specific criteria of purity concerning sweeteners for use in foodstuffs (*Official Journal of the European Communities* (L178) of 28 July 1995, pp. 1–19), as amended by Directives 98/66/EC of 4 September 1998 (*OJ* (L257) of 19 September 1998, pp. 35–6), 2000/51/EC of 26 July 2000 (*OJ* (L198) of 4 August 2000, pp. 41–3) and 2001/52/EC of 3 July 2001 (*OJ* (L190) of 12 July 2001, pp. 18–20).
7. Commission Directive 95/45/EC of 26 July 1995 laying down specific purity criteria concerning colours for use in foodstuffs (*Official Journal of the European Communities* (L226) of 22 September 1995, pp. 1–45), as amended by Directives 1999/75/EC of 22 July 1999 (*OJ* (L206) of 5 August 1999, pp. 19–21), and 2001/50/EC of 3 July 2001 (*OJ* (L190) of 12 July 2001, pp. 14–17), Corrigenda (*OJ* (L217) of 11 August 2001, p. 18).
8. Commission Directive 96/77/EC of 2 December 1996 laying down specific purity criteria on food additives other than colours and sweeteners (*Official Journal of the European Communities* (L339) of 30 December 1996, pp. 1–69), as amended by Directives 98/86/EC of 11 November

1998 (*OJ* (L334) of 9 December 1998, pp. 1–63), 2000/63/EC of 5 October 2000 (*OJ* (L277) of 30 October 2000, pp. 1–61) and 2001/30/EC of 2 May 2001 (*OJ* (L146) of 31 May 2001, pp. 1–23).
9. First Commission Directive 81/712/EEC of 18 July 1981 laying down Community methods of analysis for verifying that certain additives used in foodstuffs satisfy criteria of purity (*Official Journal of the European Communities* 1981 (L257), pp. 1–27).
10. Council Directive 88/388/EEC of 22 June 1988 on the approximation of the laws of the member states relating to flavourings for use in foodstuffs and to source materials for their production (*Official Journal of the European Communities* (L184) of 15 July 1988, pp. 61–6).
11. Council Decision 88/389/EEC of 22 June 1988 on the establishment, by the Commission, of an inventory of the source materials and substances used in the preparation of flavourings (*Official Journal of the European Communities* (L184) of 15 July 1988, p. 67).
12. Regulation (EC) No. 2232/96 of the European Parliament and of the Council of 28 October 1996 laying down a Community procedure for flavouring substances used or intended for use in or on foodstuffs (*Official Journal of the European Communities* (L299) of 23 November 1996, pp. 1–4).
13. Commission Decision 1999/217/EC of 23 February 1999 adopting a register of flavouring substances used in or on foodstuffs drawn up in application of Regulation (EC) No. 2232/96 of the European Parliament and of the Council of 28 October 1996 (*Official Journal of the European Communities* (L84) of 27 March 1999, pp. 1–137).
14. Council Directive 88/344/EEC of 13 June 1988 on the approximation of the laws of the member states on extraction solvents used in the production of foodstuffs and food ingredients (*Official Journal of the European Communities* (L157) of 24 June 1988, pp. 28–33), as amended by Directives 92/115/EEC of 17 December 1992 (*OJ* (L409) of 31 December 1992, pp. 31–2), 94/52/EC of 7 December 1994 (*OJ* (L331) of 21 December 1994, p. 10) and 97/60/EC of 27 October 1997 (*OJ* (L331) of 3 December 1997, pp. 7–9).
15. Directive 2000/13/EC of 20 March 2000 of the European Parliament and of the Council of 20 March 2000 on the approximation of the laws of the member states relating to the labelling, presentation and advertising of foodstuffs (*Official Journal of the European Communities* (L109) of 6 May 2000, pp. 29–42); Corrigendum to Directive 2000/13/EC (*OJ* (L124) of 25 May 2000, p. 66).
16. Commission Directive 91/71/EEC of 16 January 1991 completing Council Directive 88/388/EEC on the approximation of the laws of the member states relating to flavourings for use in foodstuffs and to source materials for their production (*Official Journal of the European Communities* (L42) of 15 February 1991, pp. 25–6).
17. Commission Directive 94/54/EC of 18 November 1994 concerning the compulsory indication on the labelling of certain foodstuffs of particulars

18. Commission Regulation (EC) No. 50/2000 of 10 January 2000 on the labelling of foodstuffs and food ingredients containing additives and flavourings that have been genetically modified or have been produced from genetically modified organisms (*Official Journal of the European Communities* (L6) of 11 January 2000, pp. 15–17).
19. Proposal for a Directive of the European Parliament and of the Council amending Directive 2000/13/EC as regards indication of the ingredients present in foodstuffs (*Official Journal of the European Communities* (C332E) of 27 November 2001, pp. 257–9).
20. Draft Commission Directive on the labelling of foodstuffs containing quinine, and of foodstuffs containing caffeine, SANCO/2902/01.
21. Commission of the European Communities White Paper on Food Safety, COM (1999) 719 final, Brussels, 12 January 2000.
22. Proposal for a Regulation of the European Parliament and of the Council laying down the general principles and requirements of food law, establishing the European Food Authority, and laying down procedures in matters of food (*Official Journal of the European Communities* (C96 E) of 27 March 2001, pp. 247–68).
23. Proposal for a Regulation of the European Parliament and of the Council on genetically modified food and feed (*Official Journal of the European Communities* (C304 E) of 30 October 2001, pp. 221–40).
24. Proposal for a Regulation of the European Parliament and of the Council concerning traceability and labelling of genetically modified organisms and traceability of food and feed products produced from genetically modified organisms and amending Directive 2001/18/EC (*Official Journal of the European Communities* (C304 E) of 30 October 2001, pp. 327–30).

Appendix: List of E numbers of permitted additives

E 100	Curcumin
E 101	(i) Riboflavin
	(ii) Riboflavin-5'-phosphate
E 102	Tartrazine
E 104	Quinoline Yellow
E 110	Sunset Yellow FCF, Orange Yellow S
E 120	Cochineal, carminic acid, carmines
E 122	Azorubine, carmoisine
E 123	Amaranth
E 124	Ponceau 4R, Cochineal Red A
E 127	Erythrosine

E 128	Red 2G
E 129	Allura Red AC
E 131	Patent Blue V
E 132	Indigotine, indigo carmine
E 133	Brilliant Blue FCF
E 140	Chlorophylls and chlorophyllins:
	(i) Chlorophylls
	(ii) Chlorophyllins
E 141	Copper complexes of chlorophylls and chlorophyllins:
	(i) Copper complexes of chlorophylls
	(ii) Copper complexes of chlorophyllins
E 142	Green S
E 150a	Plain caramel
E 150b	Caustic sulphite caramel
E 150c	Ammonia caramel
E 150d	Sulphite ammonia caramel
E 151	Brilliant Black BN, Black PN
E 153	Vegetable carbon
E 154	Brown FK
E 155	Brown HT
E 160a	Carotenes:
	(i) Mixed carotenes
	(ii) Beta-carotene
E 160b	Annatto, bixin, norbixin
E 160c	Paprika extract, capsanthin, capsorubin
E 160d	Lycopene
E 160e	Beta-apo-8'-carotenal (C30)
E 160f	Ethyl ester of beta-apo-8'-carotenic acid (C30)
E 161b	Lutein
E 161g	Canthaxanthin
E 162	Beetroot Red, betanin
E 163	Anthocyanins
E 170	Calcium carbonates:
	(i) Calcium carbonate
	(ii) Calcium hydrogen carbonate
E 171	Titanium dioxide
E 172	Iron oxides and hydroxides
E 173	Aluminium
E 174	Silver
E 175	Gold
E 180	Litholrubine BK
E 200	Sorbic acid
E 202	Potassium sorbate
E 203	Calcium sorbate
E 210	Benzoic acid

E 211 Sodium benzoate
E 212 Potassium benzoate
E 213 Calcium benzoate
E 214 Ethyl p-hydroxybenzoate
E 215 Sodium ethyl p-hydroxybenzoate
E 216 Propyl p-hydroxybenzoate
E 217 Sodium propyl p-hydroxybenzoate
E 218 Methyl p-hydroxybenzoate
E 219 Sodium methyl p-hydroxybenzoate
E 220 Sulphur dioxide
E 221 Sodium sulphite
E 222 Sodium hydrogen sulphite
E 223 Sodium metabisulphite
E 224 Potassium metabisulphite
E 226 Calcium sulphite
E 227 Calcium hydrogen sulphite
E 228 Potassium hydrogen sulphite
E 230 Biphenyl, diphenyl
E 231 Orthophenyl phenol
E 232 Sodium orthophenyl phenol
E 234 Nisin
E 235 Natamycin
E 239 Hexamethylene tetramine
E 242 Dimethyl dicarbonate
E 249 Potassium nitrite
E 250 Sodium nitrite
E 251 Sodium nitrate
E 252 Potassium nitrate
E 260 Acetic acid
E 261 Potassium acetate
E 262 Sodium acetates:
(i) Sodium acetate
(ii) Sodium hydrogen acetate (sodium diacetate)
E 263 Calcium acetate
E 270 Lactic acid
E 280 Propionic acid
E 281 Sodium propionate
E 282 Calcium propionate
E 283 Potassium propionate
E 284 Boric acid
E 285 Sodium tetraborate (borax)
E 290 Carbon dioxide
E 296 Malic acid
E 297 Fumaric acid
E 300 Ascorbic acid

E 301   Sodium ascorbate
E 302   Calcium ascorbate
E 304   Fatty acid esters of ascorbic acid:
         (i) Ascorbyl palmitate
         (ii) Ascorbyl stearate
E 306   Tocopherol-rich extract
E 307   Alpha-tocopherol
E 308   Gamma-tocopherol
E 309   Delta-tocopherol
E 310   Propyl gallate
E 311   Octyl gallate
E 312   Dodecyl gallate
E 315   Erythorbic acid
E 316   Sodium erythorbate
E 320   Butylated hydroxyanisole (BHA)
E 321   Butylated hydroxytoluene (BHT)
E 322   Lecithins
E 325   Sodium lactate
E 326   Potassium lactate
E 327   Calcium lactate
E 330   Citric acid
E 331   Sodium citrates:
         (i) Monosodium citrate
         (ii) Disodium citrate
         (iii) Trisodium citrate
E 332   Potassium citrates:
         (i) Monopotassium citrate
         (ii) Tripotassium citrate
E 333   Calcium citrates:
         (i) Monocalcium citrate
         (ii) Dicalcium citrate
         (iii) Tricalcium citrate
E 334   Tartaric acid (L(+)-)
E 335   Sodium tartrates:
         (i) Monosodium tartrate
         (ii) Disodium tartrate
E 336   Potassium tartrates:
         (i) Monopotassium tartrate
         (ii) Dipotassium tartrate
E 337   Sodium potassium tartrate
E 338   Phosphoric acid
E 339   Sodium phosphates:
         (i) Monosodium phosphate
         (ii) Disodium phosphate
         (iii) Trisodium phosphate

E 340   Potassium phosphates:
          (i) Monopotassium phosphate
          (ii) Dipotassium phosphate
          (iii) Tripotassium phosphate
E 341   Calcium phosphates:
          (i) Monocalcium phosphate
          (ii) Dicalcium phosphate
          (iii) Tricalcium phosphate
E 343   Magnesium phosphates
          (i) Monomagnesium phosphate
          (ii) Dimagnesium phosphate
E 350   Sodium malates:
          (i) Sodium malate
          (ii) Sodium hydrogen malate
E 351   Potassium malate
E 352   Calcium malates:
          (i) Calcium malate
          (ii) Calcium hydrogen malate
E 353   Metatartaric acid
E 354   Calcium tartrate
E 355   Adipic acid
E 356   Sodium adipate
E 357   Potassium adipate
E 363   Succinic acid
E 380   Triammonium citrate
E 385   Calcium disodium ethylene diamine tetra-acetate (Calcium disodium EDTA)
E 400   Alginic acid
E 401   Sodium alginate
E 402   Potassium alginate
E 403   Ammonium alginate
E 404   Calcium alginate
E 405   Propane-1,2-diol alginate
E 406   Agar
E 407   Carrageenan
E 407a   Processed eucheuma seaweed
E 410   Locust bean gum
E 412   Guar gum
E 413   Tragacanth
E 414   Acacia gum (gum arabic)
E 415   Xanthan gum
E 416   Karaya gum
E 417   Tara gum
E 418   Gellan gum
E 420   Sorbitol:

	(i) Sorbitol
	(ii) Sorbitol syrup
E 421	Mannitol
E 422	Glycerol
E 425	Konjac:
	(i) Konjac gum
	(ii) Konjac glucomannane
E 431	Polyoxyethylene (40) stearate
E 432	Polyoxyethylene sorbitan monolaurate (polysorbate 20)
E 433	Polyoxyethylene sorbitan monooleate (polysorbate 80)
E 434	Polyoxyethylene sorbitan monopalmitate (polysorbate 40)
E 435	Polyoxyethylene sorbitan monostearate (polysorbate 60)
E 436	Polyoxyethylene sorbitan tristearate (polysorbate 65)
E 440	Pectins:
	(i) Pectin
	(ii) Amidated pectin
E 442	Ammonium phosphatides
E 444	Sucrose acetate isobutyrate
E 445	Glycerol esters of wood rosins
E 450	Diphosphates:
	(i) Disodium diphosphate
	(ii) Trisodium diphosphate
	(iii) Tetrasodium diphosphate
	(v) Tetrapotassium diphosphate
	(vi) Dicalcium diphosphate
	(vii) Calcium dihydrogen diphosphate
E 451	Triphosphates:
	(i) Pentasodium triphosphate
	(ii) Pentapotassium triphosphate
E 452	Polyphosphates:
	(i) Sodium polyphosphate
	(ii) Potassium polyphosphate
	(iii) Sodium calcium polyphosphate
	(iv) Calcium polyphosphate
E 459	Beta-cyclodextrine
E 460	Cellulose:
	(i) Microcrystalline cellulose
	(ii) Powdered cellulose
E 461	Methyl cellulose
E 463	Hydroxypropyl cellulose
E 464	Hydroxypropyl methyl cellulose
E 465	Ethyl methyl cellulose
E 466	Carboxy methyl cellulose
	Sodium carboxy methyl cellulose
E 468	Cross linked sodium carboxy methyl cellulose

E 469	Enzymatically hydrolysed carboxy methyl cellulose
E 470a	Sodium, potassium and calcium salts of fatty acids
E 470b	Magnesium salts of fatty acids
E 471	Mono- and diglycerides of fatty acids
E 472a	Acetic acid esters of mono- and diglycerides of fatty acids
E 472b	Lactic acid esters of mono- and diglycerides of fatty acids
E 472c	Citric acid esters of mono- and diglycerides of fatty acids
E 472d	Tartaric acid esters of mono- and diglycerides of fatty acids
E 472e	Mono- and diacetyl tartaric acid esters of mono- and diglycerides of fatty acids
E 472f	Mixed acetic and tartaric acid esters of mono- and diglycerides of fatty acids
E 473	Sucrose esters of fatty acids
E 474	Sucroglycerides
E 475	Polyglycerol esters of fatty acids
E 476	Polyglycerol polyricinoleate
E 477	Propane-1,2-diol esters of fatty acids
E 479b	Thermally oxidised soya bean oil interacted with mono- and diglycerides of fatty acids
E 481	Sodium stearoyl-2-lactylate
E 482	Calcium stearoyl-2-lactylate
E 483	Stearyl tartrate
E 491	Sorbitan monostearate
E 492	Sorbitan tristearate
E 493	Sorbitan monolaurate
E 494	Sorbitan monooleate
E 495	Sorbitan monopalmitate
E 500	Sodium carbonates: (i) Sodium carbonate (ii) Sodium hydrogen carbonate (iii) Sodium sesquicarbonate
E 501	Potassium carbonates: (i) Potassium carbonate (ii) Potassium hydrogen carbonate
E 503	Ammonium carbonates: (i) Ammonium carbonate (ii) Ammonium hydrogen carbonate
E 504	Magnesium carbonates: (i) Magnesium carbonate (ii) Magnesium hydroxide carbonate
E 507	Hydrochloric acid
E 508	Potassium chloride
E 509	Calcium chloride
E 511	Magnesium chloride
E 512	Stannous chloride

E 513	Sulphuric acid
E 514	Sodium sulphates:
	(i) Sodium sulphate
	(ii) Sodium hydrogen sulphate
E 515	Potassium sulphates:
	(i) Potassium sulphate
	(ii) Potassium hydrogen sulphate
E 516	Calcium sulphate
E 517	Ammonium sulphate
E 521	Aluminium sodium sulphate
E 522	Aluminium potassium sulphate
E 523	Aluminium ammonium sulphate
E 524	Sodium hydroxide
E 525	Potassium hydroxide
E 526	Calcium hydroxide
E 527	Ammonium hydroxide
E 528	Magnesium hydroxide
E 529	Calcium oxide
E 530	Magnesium oxide
E 535	Sodium ferrocyanide
E 536	Potassium ferrocyanide
E 538	Calcium ferrocyanide
E 541	Sodium aluminium phosphate, acidic
E 551	Silicon dioxide
E 552	Calcium silicate
E 553a	(i) Magnesium silicate
	(ii) Magnesium trisilicate
E 553b	Talc
E 554	Sodium aluminium silicate
E 555	Potassium aluminium silicate
E 556	Calcium aluminium silicate
E 558	Bentonite
E 559	Aluminium silicate (Kaolin)
E 570	Fatty acids
E 574	Gluconic acid
E 575	Glucono-delta-lactone
E 576	Sodium gluconate
E 577	Potassium gluconate
E 578	Calcium gluconate
E 579	Ferrous gluconate
E 585	Ferrous lactate
E 620	Glutamic acid
E 621	Monosodium glutamate
E 622	Monopotassium glutamate
E 623	Calcium diglutamate

E 624	Monoammonium glutamate
E 625	Magnesium diglutamate
E 626	Guanylic acid
E 627	Disodium guanylate
E 628	Dipotassium guanylate
E 629	Calcium guanylate
E 630	Inosinic acid
E 631	Disodium inosinate
E 632	Dipotassium inosinate
E 633	Calcium inosinate
E 634	Calcium 5'-ribonucleotides
E 635	Disodium 5'-ribonucleotides
E 640	Glycine and its sodium salt
E 650	Zinc acetate
E 900	Dimethyl polysiloxane
E 901	Beeswax, white and yellow
E 902	Candelilla wax
E 903	Carnauba wax
E 904	Shellac
E 912	Montan acid esters
E 914	Oxidised polyethylene wax
E 920	L-cysteine
E 927b	Carbamide
E 938	Argon
E 939	Helium
E 941	Nitrogen
E 942	Nitrous oxide
E 943a	Butane
E 943b	Iso-butane
E 944	Propane
E 948	Oxygen
E 949	Hydrogen
E 950	Acesulfame K
E 951	Aspartame
E 952	Cyclamic acid and its Na and Ca salts
E 953	Isomalt
E 954	Saccharin and its Na, K and Ca salts
E 957	Thaumatin
E 959	Neohesperidine DC
E 965	Maltitol: (i) Maltitol (ii) Maltitol syrup
E 966	Lactitol
E 967	Xylitol
E 999	Quillaia extract

E 1103   Invertase
E 1105   Lysozyme
E 1200   Polydextrose
E 1201   Polyvinylpyrrolidone
E 1202   Polyvinylpolypyrrolidone
E 1404   Oxidised starch
E 1410   Monostarch phosphate
E 1412   Distarch phosphate
E 1413   Phosphated distarch phosphate
E 1414   Acetylated distarch phosphate
E 1420   Acetylated starch
E 1422   Acetylated distarch adipate
E 1440   Hydroxy propyl starch
E 1442   Hydroxy propyl distarch phosphate
E 1450   Starch sodium octenyl succinate
E 1451   Acetylated oxidised starch
E 1505   Triethyl citrate
E 1518   Glyceryl triacetate (triacetin)
E 1520   Propane-1,2-diol (propylene glycol)

3
The regulation of additives in the USA
P. Curtis, North Carolina State University, Raleigh

3.1 Introduction

Science and technology have converted the American food supply system from one of subsistence farming and direct exchange between producer and consumer to a complex supply chain involving specialization, centralized production and processing, and widespread distribution. These changes have increased the need to ensure the safety of our food supply. It is this need for safety that has prompted the growing number of food laws and regulations. Food additives play an important role in today's food supply. They may extend shelf life, serve as a processing aid, add color, flavor and/or texture. Before going any further, it is important to clarify legally what is meant by a 'food additive.' The Food Additives Amendment authorizes the use of two types of substances: those whose use is restricted by regulations and are considered 'food additives', and those which are not food additives and are exempt from food additive requirements. True 'food additives' are defined as

> any substance, the intended use of which may reasonably be expected to result, directly or indirectly, in its becoming a component or otherwise affecting the characteristics of any food.

It is important to note that not only are directly added (intentional) substances, such as preservatives, considered food additives, but indirectly added (unintentional) substances, such as sanitizers used on processing equipment, may also be categorized as food additives. There are four groups of substances that are exempted from the Food Additives Amendment. They are:

1. generally recognized as safe substances (GRAS) (see Section 3.4)
2. prior sanctioned or approved substances (see Section 3.5)

3. pesticide chemicals (see Section 3.7)
4. color additives (see Section 3.6).

3.2 Food additive laws and amendments

There are a number of different US laws and amendments that are relevant to 'food additives.' To truly understand what is and is not considered a food additive, you must be aware of the various laws and their amendments that are discussed in some detail below.

The present system of regulation is the product of an evolutionary process marked by five major legislative enactments:

- the Pure Food and Drug Act of 1906
- the Federal Food, Drug, and Cosmetic Act of 1938
- the Food Additives Amendment of 1958
- the Color Additives Amendments of 1960 and
- the Food Quality Protection Act of 1996.

The 1906 Act merely deemed 'adulterated' any food containing an 'added poisonous or other added deleterious ingredient which may render the food injurious to health.' This standard was a great advance at the time because before 1906 there had been no generally applicable federal standard by which unsafe foods could be removed from the market. Under the 1906 Act, however, the burden of proof rested on the government. In such actions, the government bore the burden of proving that there was at least 'a reasonable possibility' that the allegedly adulterated food would be injurious to health. This no doubt seemed appropriate and adequate at the time from a safety standpoint because the 1906 Act had been enacted largely in response to very obvious food adulteration that was easy to detect and prove.

The 1938 Act added to the law a standard for 'adulteration' caused by 'naturally occurring,' as opposed to 'added,' substances. This standard deemed such substances to render the food adulterated if, by their presence, the food would be 'ordinarily injurious' to the consumer. For these 'naturally occurring' substances, as for 'added' substances, the burden of proof remained on the government. The assignment of a less rigorous safety standard ('ordinarily injurious' versus 'injurious') to naturally occurring substances seems to reflect an early 'risk-benefit' judgement by Congress. Recognizing the inherent value of natural ingredients, Congress may have concluded that the government should be able to eliminate such ingredients only upon a strong showing of actual danger to consumers from the natural commodity. This Act contained an additional potential conflict between ensuring the safety of food and ensuring an abundant, economical food supply. Congress established in section 406 of the Act the general rule that *any amount* of an avoidable poisonous or deleterious substance would render food adulterated. For 'added' poisonous or deleterious substances that are 'unavoidable' through good manufacturing practices or 'necessary' in the production of food, however, Congress provided authority for the promulgation of tolerance limits. These 'section 406' tolerances are set at the level deemed necessary to protect the public health, taking into account, however, the 'extent to which' the substances are 'unavoidable' or 'necessary' (Taylor, 1984).

The current era of food and color additive safety assessment and regulation began with the enactment of the Food Additives Amendment of 1958 and the Color Additives Amendments of 1960. The Food Additives Amendment of 1958 (72 Stat. 1785) has been codified as Section 348 in Title 21 of the United States Code. The main purpose of the amendment was to ensure the safety of the components of food which are covered only generally by provisions of Section 346 (Tolerances for poisonous and deleterious substances in food), and not included under the Pesticide Chemical Amendment (Section 346a), or in the Color Additives Amendments of 1960 (Section 376). These enactments did two fundamentally important things. They shifted the burden of proof on the issue of the safety of most food substances and coloring materials from the government to the manufacturer or other sponsor of the substance; and they established the basic safety standard for food and color additives that remain in effect today. The reallocation of the burden of proof to the sponsor of a new food additive or color additive brought with it a system of premarket approval for such substances.

Companies wishing to market a new substance must conduct tests in order to assess its safety. The Food and Drug Administration (FDA) uses the results from these tests to either write regulations allowing the use of the substance or to ban the substance. It prohibits the use of any substance which causes cancer in man or animal or when they are used to deceive the consumer or results in an adulterated or misbranded product. FDA regulations prescribe the type and extent of premarket testing that must be conducted, depending on the legal requirements applicable to the particular product and on the technology available to fulfill those requirements. Testing may include physical and chemical studies, non-clinical laboratory studies, animal tests, and clinical trials on humans. The importance of the toxicological and other data derived from such investigations demands that they be conducted according to scientifically sound protocols and procedures. The FDA has published regulations in Title 21, Code of Federal Regulations, Part 58 (21 CFR 58) prescribing good laboratory practices for conducting non-clinical research. Inquiries should be addressed to the Food and Drug Administration, Bioresearch Monitoring Program Coordinator (HFC-230), 5600 Fishers Lane, Rockville, Md. 20857 (FDA website).

On August 3, 1996, President Clinton signed into law the Food Quality Protection Act (FQPA). The new law required major changes in pesticide regulation and afforded the Environmental Protection Agency (EPA) unprecedented opportunities to provide greater health and environmental protection, particularly for infants and children. The FQPA required the EPA to review the more than 9,700 tolerances established before August 3, 1996 (the day the new law was enacted) within ten years, to ensure that the data used to evaluate them meets current scientific standards. The tolerance reassessment involves a thorough review of potential risks to humans and the environment (EPA website).

3.3 Federal agencies responsible for enforcement

3.3.1 Food and Drug Administration

The FDA enforces the tolerances in marketed food, whether raw or processed, once tolerances are established. Authority for enforcement is through the Federal Food, Drug, and Cosmetic Act, specifically through Sections 342(a)(1) and 342(a)(2), in respect to adulterated food, Section 346 regarding tolerances for poisonous or deleterious substances in food, and applicable portions of Section 346a and 348. Inspectors for the Food and Drug Administration take samples of food that have been or will be shipped interstate and submit them to laboratories to determine if residues when present are within established tolerances. This agency cooperates with the Department of Agriculture concerning the sampling and testing of meat and poultry, and products thereof, and similarly with state officials regarding intrastate food shipments (Schultz, 1981).

3.3.2 US Department of Agriculture's Food Safety and Inspection Service

Food Safety and Inspection Service (FSIS) sets standards for food safety and inspects meat, poultry, and egg products produced domestically and imported. The Service inspects animals and birds at slaughter and processed products at various stages of the production process, and analyzes products for microbiological and chemical adulterants. FSIS also informs the public about meat, poultry, and egg product food safety issues. FSIS works with the Research, Education and Economics mission area on food safety research issues and the Animal and Plant Health Inspection Service on instances where animal diseases impact food safety. FSIS also facilitates the management of US activities pertaining to the Codex Alimentarius Commission, an international organization created by the United Nations, to promote the health and economic interests of consumers while encouraging fair international trade in food.

3.3.3 Environmental Protection Agency

This agency, among other duties, has full authority to enforce the Federal Insecticide, Fungicide, and Rodenticide Act, and, therefore, regulates the formulation and use of products intended to control pests. The EPA also administers the tolerance-setting provisions of the Federal Food, Drug, and Cosmetic Act. Thus, this agency receives and processes applications and petitions for registration and the establishment of exemption from tolerances of pesticide chemicals in or on raw agricultural commodities. Prior to 1970, the Food and Drug Administration carried out this function (Schultz, 1981).

Among its policy-related activities, the EPA provides support to the work of the Codex Alimentarius Commission, a joint program of the United Nations Food and Agriculture Organization and World Health Organization, that sets international tolerances for pesticide residues in foods. The EPA's work with Codex and in World Trade Organization committees which focus on human, animal and plant health protection standards, is designed both to promote harmonization and to ensure that international agreements are consistent with the high level of protection afforded by US standards (EPA website).

3.4 Generally recognized as safe (GRAS) substances

GRAS substances are those which are generally recognized by experts as being safe for use in food. When the Food Additives Amendment was passed in 1958, ingredients that were in use at the time and considered to be safe by the FDA and USDA were granted GRAS status. The regulations explaining GRAS stated 'It is impracticable to list all substances that are generally recognized as safe for their intended use. However, by way of illustration, the Commissioner regards such common food ingredients as salt, pepper, vinegar, baking powder, and monosodium glutamate as safe for their intended use. This part includes additional substances that, when used for the purposes indicated, in accordance with good manufacturing practice, are regarded by the Commissioner as generally recognized as safe for such uses' (21 CFR Sec. 182.1(a)). Some ingredients were granted GRAS status because they fall into a certain category of ingredients. These categories are:

1. spices and other natural seasonings and flavorings;
2. essential oils, oleoresins (solvent-free), and natural extractives (including distillates);
3. natural extractives (solvent-free) used in conjunction with spices, seasonings and flavorings;
4. certain other spices, seasonings, essential oils, oleoresins, and natural extracts;
5. synthetic flavoring substances and adjuvants;
6. substances migrating from cotton and cotton fabrics used in dry food packaging;
7. substances migrating to food from paper and paperboard products; and
8. adjuvants for pesticide chemicals.

Other ingredients were granted GRAS status individually. All of these GRAS ingredients are included in the regulations found in 21 CFR Part 182.

The FDA has continuously reevaluated GRAS substances for safety. The evaluation process looks at all scientific data available on the ingredient's safety as well as the historical use of the ingredient. Substances that have been reaffirmed as GRAS have undergone the rule-making process and ultimately been published in the Code of Federal Regulations. These reaffirmed substances are removed from 21 CFR Part 182 and published in 21 CFR Part 184 (directly added ingredients) and 186 (indirectly added ingredients). Evidence for the ingredient's safety is based either on its long-term use without any side effects or on published scientific data. People may also petition the FDA to grant new ingredients GRAS status. These companies must go through the petition process required for food additives.

Unless otherwise stated in the regulation, all GRAS substances must be used under good manufacturing practices (GMPs). According to 21 CFR Sec. 184.1, this means that:

1. the ingredient is food grade;
2. that it is prepared and handled as a food ingredient; and
3. the quantity of the ingredient added to food does not exceed the amount reasonably required to accomplish the intended physical, nutritional or other technical effect in food.

3.5 Prior sanctioned substances

The 1958 amendment contains a 'grandfather clause' which exempts substances sanctioned for use before the 1958 amendment from the safety testing requirements. Prior to the Food Additives Amendment, the FDA had granted companies permission to use certain ingredients through official letters. In regard to these sanctions, the regulations state that 'A prior sanction shall exist only for a specific use of a substance for which there was explicit approval by the FDA or USDA prior to September 6, 1958' (21 CFR Sec. 181.5). The regulations reserve the right to remove any substance that is found to be harmful under the law. The prior sanction requirements are in 21 CFR Sec. 181.1 and 181.5. The list of prior sanctioned substances is in 21 CFR Sec. 181.22–34.

3.6 Color additives

The Federal Food, Drug, and Cosmetic Act provides that foods, drugs, cosmetics, and some medical devices are adulterated if they contain color additives that have not been approved safe to the satisfaction of the Food and Drug Administration for their particular use. A color additive is a dye, pigment, or other substance, whether synthetic or derived from a vegetable, animal, mineral, or other source, which imparts a color when added or applied to a food, drug, cosmetic, or the human body (Sec. 201(t)). Regulations (21 CFR 73, 74 and 81) list approved color additives and the conditions under which they may be used safely, including the amounts that may be used when limitations are necessary. Separate lists are provided for color additives for use in or on foods, drugs, medical devices, and cosmetics. Some colors may appear on more than one list.

Testing and certification by the Food and Drug Administration of each batch of color is required before that batch can be used, unless the color additive is specifically exempted by regulation. The manufacturer who wants to use color additives in foods, drugs, devices or cosmetics should check the regulations to ascertain which colors have been listed for various uses. Before using a color a manufacturer should read the product label, which is required to contain sufficient information to assure safe use, such as 'for food use only,' directions for use where tolerances are imposed, or warnings against use, such as 'Do not use in products used in the area of the eye.'

Manufacturers of certifiable colors may address requests for certification of batches of such colors to the Food and Drug Administration, Office of Cosmetics and Colors (HFS-105), 200 C Street, SW, Washington, DC 20204. Certification is not limited to colors made by US manufacturers. Requests will be received from foreign manufacturers if signed by both such manufacturers and their agents residing in the United States. Certification of a color by an official agency of a foreign country cannot, under the provisions of the Federal Food, Drug, and Cosmetic Act, be accepted as a substitute for certification by the Food and Drug Administration.

Copies of regulations governing the listing, certification, and use of colors in foods, drugs, devices, and cosmetics shipped in interstate commerce or offered for entry into the United States, or answers to questions concerning them, are available from the Food and Drug Administration. Recommendations on submission of chemical and technological data are provided in the FDA's online publication (http://vm.cfsan.fda.gov/~dms/opa-col1.html).

3.7 Pesticide residues

Pesticide chemical residues are not considered 'food additives'. Therefore they must adhere to different requirements. Section 408 of the FDCA authorizes the Environmental Protection Agency (EPA) to establish a tolerance for the maximum amount of a pesticide residue that may be legally present in or on a raw agricultural commodity. This section also authorizes the EPA to exempt a pesticide residue in a raw agricultural commodity from the requirement of a tolerance. The requirements for pesticide chemical residues depend on whether they are present on raw agricultural commodities or processed foods. Pesticide residue limits for specific foods are provided by the EPA and can be accessed at http://www.epa.gov/pesticides/food/viewtols.htm.

Raw agricultural commodities are foods that are still in their 'natural state' such as an apple or head of lettuce. Any fruit, vegetable, nut or grain that has not been processed is considered a raw agricultural commodity. Washing, waxing, or coloring any raw, unpeeled commodity is not considered 'processing.' For these commodities, any pesticide residue present must be at a level lower than the tolerance set by the EPA or it will be adulterated. If no tolerance has been set for a particular chemical and the EPA has not exempted it, then the food is adulterated. Products of this kind containing pesticide residues are in violation of the Federal Food, Drug, and Cosmetic Act unless: (1) the pesticide chemical has been exempted from the requirement of a residue tolerance; or (2) a tolerance has been established for the particular pesticide on the specific food and the residue does not exceed the tolerance (Sec. 408).

Processed foods may not contain more than the published tolerance allowed for the raw commodity. Tolerances and exemptions from tolerances established by the EPA for pesticide residues in raw agricultural commodities are listed in 40 CFR Part 180. Food additive regulations issued by the EPA for pesticide residues in processed food and feed appear in 21 CFR Part 193 and in 21 CFR Part 561, respectively.

Tolerances for pesticide residues on many raw agricultural commodities have been established under Section 408 of the Federal Food, Drug, and Cosmetic Act. Tolerances are established, revoked or changed, as the facts warrant such action, by the Environmental Protection Agency. Firms considering offering for entry into the United States foods that may contain pesticide residues should contact the Communication Services Branch, Office of Pesticide Programs, Environmental Protection Agency, 401 M Street, SW, Washington, DC 20460. The Agency maintains a database of pesticide residue tolerances on the Internet at http://www.epa.gov/opprd001/tolerance/tsinfo/. The Office of Pesticide Programs also has information on the Internet at http://www.epa.gov/pesticides/ and sponsors a National Pesticide Telecommunications Network, which can be reached at 1-800-858-7378 or on the Internet at http://ace.ace.orst.edu/info/nptn/ (FDA website).

For protection of public health, there must be some control of the experimental use of pesticide chemicals in obtaining data necessary for registration of a marketable product. Experimental use permits are issued under provisions, at 40 USC 136c, of the Federal Insecticide, Fungicide, and Rodenticide Act. Either temporary tolerance or an exemption can be granted with a permit. When the EPA takes such action, a notice is published in the *Federal Register* (Schultz, 1981).

3.8 Setting tolerance levels

Federal agencies such as the FDA and EPA require a battery of toxicity tests in laboratory animals to determine an additive's or a pesticide's potential for causing adverse health effects, such as cancer, birth defects, and adverse effects on the nervous system or other organs. Tests are conducted for both short-term (acute) and long-term (chronic) toxicity. For chronic effects other than cancer, laboratory animals are exposed to different doses to determine the level at which no adverse effects occur. This level is divided by an uncertainty or 'safety' factor (usually 100) to account for the uncertainty of extrapolating from laboratory animals to humans and for individual human differences in sensitivity. The resulting figure, termed the Reference Dose, is the level of exposure that the EPA judges an individual could be exposed to on a daily basis for a lifetime with minimal probability of experiencing any adverse effect. For cancer risks, the EPA evaluates multi-year tests of laboratory animals to estimate levels unlikely to pose more than a negligible risk. Tolerances are only approved if the expected exposure is below these health concern levels.

Several of the types of studies are designed specifically to assess risks to infants and children. These include developmental toxicity studies, which examine risks to developing fetuses that result from exposure of the mother to additives or pesticides during pregnancy; developmental neurotoxicity studies, which specifically examine the risks to the developing nervous system; and two-generation reproduction studies, which provide information about the possible health effects on both the individual and its offspring resulting from exposure.

The FDA and EPA recognize that the diets of infants and children may differ substantially from those of adults and that they consume more food for their size than adults. As a result, they may be exposed to proportionately more residues. The FDA and EPA address these differences by combining survey information on food consumption by nursing infants, non-nursing infants, and children with data on residues to estimate their dietary exposure. The FDA and EPA also use this process to estimate exposure for other age groups, as well as several different ethnic groups and regional populations.

Information about exposure to infants, children, and other subgroups is then combined with toxicity information to determine the potential risks posed. If risks are unacceptable, the FDA and EPA will not approve the tolerances. Some of the EPA's recent important regulatory decisions, for example, were based on concerns about childhood or infant exposures, such as the elimination of the use of aldicarb on bananas and EBDCs on a number of fruits and vegetables.

The FDA and EPA believe that the tolerance process is protective of human health because it is based on extensive testing and on a combination of conservative assumptions and risk assessment practices developed using current scientific knowledge.

3.9 The approval process

There are a number of FDA websites (many are listed under the reference section of this chapter) which provide the most current information regarding the approval process. Excerpts related to the approval process (at the time of publication) from several of the FDA websites are provided below.

3.9.1 Guidance for submitting petitions and notifications (http://www.cfsan.fda.gov/~dms/opa-toc.html)

This document has been prepared in response to the many requests from industry for guidance in the preparation of petitions for submission to the Center for Food

Safety and Applied Nutrition. These guidelines are not requirements of the FDA, but rather represent a suggested format that can be followed by a petition if so desired. Legal requirements for GRAS, color additive, and food additive regulations can be found in 21 CFR Parts 71, 170 and 171. It includes information such as number of copies, pagination, binding, labeling, etc.

3.9.2 Recommendations for submission of chemical and technological data for direct food additive and GRAS food ingredient petitions, 1993 (http://vm.cfsan.fda.gov/~dms/opa-cg4.html)

This document describes the types of chemical and technological data that the FDA's Office of Premarket Approval in the Center for Food Safety and Applied Nutrition considers necessary for the evaluation of petitions seeking regulation of the safe use of direct food additives or for the affirmation of the use of food ingredients as generally recognized as safe (GRAS).

As prescribed by the Federal Food, Drug, and Cosmetic Act (the Act) and Title 21, Code of Federal Regulations (21 CFR) 170.3(h)(i), food additives must be shown to be safe (i.e., to have a reasonable certainty of no harm) under their intended conditions of use before they can be intentionally added to food. Food additives generally fall into two broad categories: (1) those that are added directly to food, as codified in 21 CFR Part 172, and (2) those that are added indirectly to food through contact of the food with packaging materials, processing equipment, or other food-contact materials, as codified in 21 CFR Parts 174–178. In addition, there is another class of food additives that does not neatly fit into either the direct or indirect food additive categories. These substances are called secondary direct food additives, and are codified in 21 CFR Part 173. These are substances whose functionality is required during the manufacture or processing of food and are ordinarily removed from the final food; although residuals might carry over to the final product, these residuals are not expected to exhibit any technical effect in food. Examples of secondary direct additives include enzyme immobilizing agents, ion exchange resins, boiler water additives and other processing aids.

Section 201(s) of the Act exempts substances that are GRAS from the definition of a food additive. There are two broad categories of GRAS substances codified in 21 CFR: (1) those listed as GRAS in 21 CFR Part 182 during 1958–1962 without having been subjected to a detailed scientific review of all information available on the safety of these substances; and (2) those in Part 184 that have been affirmed as GRAS by the Agency since the Presidential Directive of 1969. This Directive required the Agency to initiate a safety review of the status of all ingredients that were on the GRAS list, with the intention of affirming their use as GRAS, determining that they were subject to prior sanction, or concluding that they should be regulated as food additives. (Substances subject to prior sanction are listed in 21 CFR Part 181.) To affirm the use of a food ingredient as GRAS, the ingredient must be generally recognized among experts qualified by training and experience to evaluate its safety. Safety, under the conditions of intended use, may be shown through (1) scientific procedures or (2) experience based on common use in food prior to January 1, 1958.

Direct food additive petitions
Section 409(b)(2) of the Act describes the statutory requirements for food additive petitions. Briefly, these encompass five general areas of information:

(1) the identity of the additive;
(2) the proposed use of the additive;
(3) the intended technical effect of the additive;
(4) a method of analysis for the additive in food; and
(5) full reports of all safety investigations with respect to the additive.

In addition, the petitioner shall furnish, upon request, a complete description of the methods, facilities, and controls used in or for the production of the additive (21 CFR 409(b)(3)) and/or samples of the additive and of food in which the additive will be used (21 CFR 409(b)(4)).

21 CFR 171.1(c) describes in greater detail the data requirements for food additive petitions, including the five basic areas of information and scientific data noted above, as well as other administrative information and environmental assessment requirements. This document addresses chemistry-related issues only. As described in 21 CFR 171.1(h), certain data and information contained in food additive petitions are available for public disclosure, while other data are not. Questions in this regard should be directed to the Office of Premarket Approval.

GRAS affirmation petitions
The Act does not provide specific statutory requirements for GRAS affirmation petitions. As noted above, Section 201(s) of the Act exempts GRAS substances from the meaning of the term 'food additive' and, therefore, from the statutory requirements for food additive petitions. Section 201(s) also provides that general recognition of safety must be established either through (1) scientific procedures or (2) experience based upon common use in food prior to January 1, 1958. Under 21 CFR 170.35, the Agency has established a voluntary GRAS affirmation petition process in order to minimize controversy over whether the Agency agrees that the use of a substance is GRAS.

The eligibility requirements for classification of a substance as GRAS are described in 21 CFR 170.30. For a substance to be eligible for GRAS affirmation based upon its history of use in food prior to 1958, certain criteria must be met. If the substance was commonly used in food in the United States prior to 1958, then information documenting this use must be generally available (21 CFR 170.30(c)(1)). If the substance was only used in food outside the United States prior to 1958, then published documents, or other information, that shall be corroborated by information from a second, independent source that confirms the history and conditions of use, must be readily available in the country in which the history of use occurred, as well as to experts in the United States (21 CFR 170.30(c)(2)). If the substance did not have a common history of use in food prior to 1958, then the substance can be considered for affirmation as GRAS based only upon scientific procedures, as set forth in 21 CFR 170.30(b).

The regulatory requirements for GRAS affirmation petitions differ from food additive petitions in terms of the format, data requirements, and administrative requirements. In addition, there are distinctions in data requirements for GRAS petitions based upon history of use and those based upon scientific procedures. The data requirements for GRAS affirmation of a substance are set forth in 21 CFR 170.30(b) and (c). GRAS affirmation based upon scientific procedures requires 'the same quantity and quality of scientific evidence as is required to obtain approval of a food additive' (21 CFR 170.30(b)); thus, in preparing a petition for GRAS affirmation based upon scientific procedures, consideration shall also be given to the requirements set forth in 21 CFR 171.1(c) for food additive petitions, as well as those in 21 CFR 170.35(c)(1). The data requirements for a petition for GRAS affirmation based upon history of use are set forth in 21 CFR 170.35(c)(1); these petitions do not require the same quantity and quality of scientific evidence that is required for approval of a food additive regulation (21 CFR 170.30(c)(1)).

Under 21 CFR 170.35(c)(3), all GRAS affirmation petitions are placed on view in the office of the Dockets Management Branch, where they are available for public inspection; thus, any trade secret information in a GRAS affirmation petition would not be protected under 21 CFR 20.61. Hence, trade secret information should not be included in GRAS affirmation petitions. Additionally, scientific information essential to a GRAS affirmation petition must be available to the public in the open literature. For further information visit http://vm.cfsan.fda.gov/~dms/opa-cg4.html

3.9.3 Recommendations for chemistry data for indirect food additive petitions, June, 1995
(http://vm.cfsan.fda.gov/~dms/opa-cg5.html)
Section 409(a) of the Federal Food, Drug, and Cosmetic Act (the Act) states that use of a food additive shall conform to a regulation prescribing the conditions under which the additive may safely be used. The definition of a food additive (Section 201(s)) includes substances used in the processing, packaging, holding, and transporting of food that have no functional effect in the food but which may reasonably be expected to become components of food. These latter substances are known as indirect food additives. Specific regulations are established to cover the safe use of indirect food additives. These regulations are set forth in Title 21 of the Code of Federal Regulations (21 CFR) Parts 175–179. In addition to the specific regulations, 21 CFR 174.5 lays out general safety requirements for all indirect food additives.

Anyone intending to use an additive that does not conform to an existing regulation must file a petition proposing the issuance of a new regulation.

54   Food chemical safety

Section 409(b)(2) of the Act sets forth the statutory requirements for such a petition. These requirements are described in greater detail in 21 CFR 171.1. That section also specifies the format of the petition. Alternatively, in some cases, an exemption to regulation as a food additive may be pursued.

These recommendations are intended to amplify and explain the statutory chemistry requirements for indirect food additive petitions. The science and technology of food-packaging and food-contact articles as well as the scientific basis for evaluating exposure to indirect food additives are continually evolving. Therefore, for the most current information go directly to http://vm.cfsan.fda.gov/~dms/opa-cg5.html

3.9.4   Enzyme preparations: chemistry recommendations for food additive and GRAS affirmation petitions
(http://vm.cfsan.fda.gov/~dms/opa-cg7.html)

Enzyme preparations are produced in varying degrees of purity from animal, plant, and microbial sources. They may consist of whole killed cells, parts of cells, or cell-free extracts. They may also contain carriers, solvents, preservatives, and antioxidants. The enzyme preparations may be formulated as liquid, semi-liquid, or dry solid preparations. Food enzyme preparations have traditionally been added directly to food during processing. For many applications, the components of the preparation remain in the processed food product. In recent years, enzymes immobilized on solid supports have gained importance. Immobilized enzyme preparations may range from those that contain a highly specific, purified enzyme, to those that contain whole killed cells or structurally intact viable cells. For some enzymatic processes, co-immobilization of enzymes and cells may be advantageous. Immobilized enzyme preparations are not intended to become food components.

Enzyme preparations are regulated either as 'secondary' direct food additives under Title 21 of the Code of Federal Regulations (CFR), Part 173, or are affirmed as GRAS substances in 21 CFR Part 184. The regulatory status of food additives or substances affirmed as GRAS is established through the petition process. According to Section 409(b)(1) of the Federal Food, Drug, and Cosmetic Act (the Act), anyone may file a petition proposing the issuance of a regulation. Section 409(b)(2) of the Act prescribes the statutory requirements for food additive petitions. The requirements for food additive petitions are discussed in greater detail in 21 CFR 171.1. The Act does not provide specific statutory requirements for GRAS affirmation petitions. Although it exempts substances generally recognized as safe from the term 'food additive,' it states that general recognition of safety must be demonstrated through scientific procedures or common use of the ingredient in food prior to January 1, 1958 (Section 201(s)). The eligibility requirements for classification of a substance as GRAS are described in 21 CFR 170.30 and the requirements for GRAS affirmation petitions are set forward in 21 CFR 170.35. The recommendations contained in this document are intended to aid petitioners in assembling the chemical and technological data currently considered appropriate for a food additive or GRAS affirmation petition for an enzyme preparation. They cover data needs in the following areas: identity, manufacturing process, purity, use, analytical methodologies, technical effects, and probable human exposure. The recommendations do not address other data needs, such as those pertaining to microbiological, toxicological and environmental considerations. The recommendations will be updated as needed to reflect new developments in the manufacture and use of food enzyme preparations.

Not all petitions will need to be supported by the same quantity and quality of data. The data and information needs outlined on the website should not be regarded as absolute. Questions regarding the tailoring of the data package for specific cases may be discussed with agency personnel prior to submitting a petition.

3.9.5 Estimating exposure to direct food additives and chemical contaminants in the diet
(http://vm.cfsan.fda.gov/~dms/opa-cg8.html)

The US Food and Drug Administration (FDA) regulates substances either intentionally added to food to accomplish a proven technical effect or inadvertently added through contamination, processing, or packaging. The premarket approval process for food additives requires an estimate of the probable consumer exposure to the additive to determine whether its use or presence in a food at a given concentration is safe. The intent of this document is to provide an understanding of the databases and methodologies used by the Office of Premarket Approval (OPA) in the FDA's Center for Food Safety and Applied Nutrition to estimate exposure to food additives and other substances (e.g., chemical contaminants) found in the diet. This document is primarily directed at petitioners for food additive regulations. Illustrative examples of the calculations that the technical reviewers of OPA's Chemistry Review Branch (CRB) perform to obtain the legally-required estimate of probable exposure for substances added to food are included. It should also be useful to other parties interested in the means by which exposures to food additives and chemical contaminants can be estimated.

The key determinant in the safety evaluation of a substance found in or added to the diet is the relation of its probable human exposure to the level at which adverse effects are observed in toxicological studies. Simply, 'the dose makes the poison.' The implications of this adage as it pertains to food can be illustrated with two clear examples. While 'pure' water can be viewed as the safest of foods, excessive intake can lead to a potentially fatal electrolyte imbalance. Conversely, pure concentrated sulfuric acid destroys human tissue, but it is affirmed as 'Generally Recognized as Safe' (GRAS) by the FDA in the Code of Federal Regulations (CFR) when used to control pH during the processing of alcoholic beverages or cheeses. Clearly, the use and the dose (i.e., exposure) are overriding considerations when discussing the safety of a component of food.

Direct vs. indirect food additives

The definition of a food additive in Section 201(s) of the Federal Food, Drug, and Cosmetic Act, as amended ('the Act') refers to substances whose intended use results directly or indirectly in the substance becoming a component of food. The FDA refers to direct food additives as those added to a food to accomplish an intended effect. Indirect additives are those that unintentionally, though predictably, become components of food. Components of plastic packaging materials that can migrate to food are indirect additives.

The differences between direct and indirect additives are such that different methodologies are necessary to prepare estimates of probable human exposure. It is relatively straightforward to determine how much of a direct additive could be present in any given food. Indirect additives, on the other hand, cannot be treated in a parallel manner due to the great variety of packaging materials (glass, paper, coated papers, plastics, laminates, and adjuvants used in their manufacture), and the great variability in the use of packaged foods. Also, the fact that these additives are not intentionally added to food (which is to say that 'use levels' in food cannot be defined), makes direct comparisons with methods used for estimating exposures to direct additives inappropriate.

This document deals only with estimating exposure to direct additives and chemical contaminants. The procedures used to estimate exposure to chemical contaminants in food (including naturally occurring toxicants, such as mycotoxins) are essentially the same as those used for direct additives. Thus, contaminants will be considered in the discussion of direct additive exposure estimation. The procedures discussed herein are equally applicable to color additives, GRAS substances, prior-sanctioned ingredients, and pesticide residues.

Direct additives and chemical contaminants

Direct food additives are regulated in 21 CFR Part 172 (GRAS ingredients are regulated in 21 CFR Parts 182 and 184). Secondary direct additives, a sub-class of direct additives, are primarily processing aids. These materials, which are used to accomplish a technical effect during the processing of food but are not intended to serve a technical function in the finished food, are regulated in 21 CFR part 173. Secondary direct food additives are like indirect additives in that only residues of the additive, or its components, are typically found in the food. The estimate of daily intake (EDI) for a secondary direct additive, however, is generally calculated in the same manner as those for direct additives and will be considered in this document.

Chemical contaminants are substances that are present in food or food additives either unavoidably or unintentionally. Typically, there are two types of chemical contaminants that are encountered:

(1) those present in food additives (generally, from manufacture), and
(2) those present in food itself (due to manufacture, natural, or environmental contamination, e.g., lead, aflatoxin, or polychlorinated biphenyls).

The practical difference between estimating exposure for contaminants and additives is the derivation of the concentration of the substance. Contaminant concentrations are usually determined experimentally, while additive concentrations or use levels are proposed by a petitioner for specified uses of a food additive.

For more information visit http://vm.cfsan.fda.gov/~dms/opa-cg8.html

3.10 References

SCHULTZ, H.W. 1981. *Food Law Handbook*. The AVI Publishing Company, Westport, Connecticut.

TAYLOR, MICHAEL. 1984. Food and Color Additives: Recurring Issues in Safety Assessment and Regulation. In Seventy-Fifth Anniversary Commemorative Volume of *Food and Drug Law*, Food and Drug Law Institute Series, Food and Drug Law Institute, Washington, D.C.

3.10.1 Website references

Environmental Protection Agency http://www.epa.gov/
  To search for pesticide residue limits on specific foods
  http://www.epa.gov/pesticides/food/viewtols.htm

Food and Drug Administration http://www.fda.gov/
  Pesticide Analytical Manual (PAM) http://www.cfsan.fda.gov/~frf/pami1.html
    Contains the procedures and methods used in FDA labs for regulatory examination of food and feed samples to determine compliance with the FD&C Act
  Requirements of Laws and Regulations Enforced by the U.S. Food and Drug Administration http://www.fda.gov/opacom/morechoices/smallbusiness/blubook.htm
    Summarizes the principal requirements of laws enforced by FDA
  Listing and Certification of Color Additives for Foods, Drugs, Devices, and Cosmetics Chapter VII – General Authority, Subchapter B – Colors
    http://www.fda.gov/opacom/laws/fdcact/fdcact7b.htm
  Guidance for Submitting Petitions and Notifications
    http://www.cfsan.fda.gov/~dms/opa-toc.html
  Recommendations for Submission of Chemical and Technological Data for Direct Food Additive and GRAS Food Ingredient Petitions, 1993
    http://vm.cfsan.fda.gov/~dms/opa-cg4.html
  Recommendations for Chemistry Data for Indirect Food Additive Petitions, June, 1995
    http://vm.cfsan.fda.gov/~dms/opa-cg5.html
  Enzyme Preparations: Chemistry Recommendations for Food Additive and GRAS Affirmation Petitions

http://vm.cfsan.fda.gov/~dms/opa-cg7.html
Estimating Exposure to Direct Food Additives and Chemical Contaminants in the Diet
http://vm.cfsan.fda.gov/~dms/opa-cg8.html
Toxicological Testing of Food Additives http://vm.cfsan.fda.gov/~dms/opa-tg1.html

Part II

Analysing additives

4

Risk analysis of food additives

D. R. Tennant, Consultant, UK

4.1 Introduction

Additives are used in foods to perform a variety of functions, many of which are described elsewhere in this book. The use of additives is intended to provide some benefit to the consumer such as improved shelf life, taste or texture. However, where additives are used in foods, the public is entitled to expect that they will not be exposed to unacceptable risks should they consume such foods. This chapter is devoted to describing risk assessment methods that are applied at national and international levels.

Any reliable system for assessing and controlling chemical risks must contain six key elements whose relationships are described in Fig. 4.1:

1. *Hazard identification.* It is necessary to be aware of what chemicals might be used as additives in a particular foodstuff and the nature of the harmful consequences to human health that might be associated with them.
2. *Dose-response characterisation.* Different chemicals will be associated with different toxicological end-points and the risk of any individual experiencing toxicity is related to the dose that they receive. Very often it is possible to identify a dose level below which the probability of anyone experiencing an adverse effect is very low or zero. For additives this is usually referred to as the Acceptable Daily Intake (ADI).
3. *Exposure analysis.* The amount of any chemical that an individual is exposed to will depend upon the levels that occur in food and the amounts of those foods that are consumed. Different population groups will often have different levels of exposure and it is therefore necessary to identify such sub-groups. The exposure level for additives is frequently referred to as the Estimated Daily Intake (EDI).

Fig. 4.1 Food chemical risk assessment and risk management.

4. *Risk evaluation.* If an ADI has been identified then it is necessary to identify any population sub-groups whose exposure (EDI) might exceed that level. Risk evaluation should aim to quantify the level of risk that any such populations may be exposed to.
5. *Risk management.* If any population sub-group has been identified as being potentially at risk then measures to control the risk must be assessed and introduced. Any benefits associated with the foods affected must be taken into account and the costs associated with alternative methods of control evaluated.
6. *Risk communication.* Where there are potential risks associated with food additives other interested parties, in any sub-groups particularly affected, must be informed.

The remaining parts of this chapter will provide detailed information about each of these six steps. However, specific strategies for risk assessment and management may need additional elements, depending on the nature of the hazard, the foods in which it may occur and other specific conditions. For example, some additives may also occur as natural constituents in the diet or may be chemically altered when they come into contact with food components. For such substances a unique approach may need to be developed.

4.2 Hazard identification in the food chain

Food additives can be used in foods to perform a variety of functions. Since the use of food additives is regulated by national or supra-national legislation, the identification of additives that may be present in foods is usually a relatively straightforward step. In the EU, for example, many approved additives are listed in the annexes to Directives 94/35/EC,[1] 94/36/EC[2] and 95/2/EC[3] on sweeteners, colours and other food additives, that apply in all EU countries. For certain classes of additives, such as flavours and processing aids, no EU legislation exists and so these are controlled at a national level. A safety evaluation is part of the approval process for all food additives and this should reveal any toxicological hazards that relate to a particular additive. For most food additives no definite toxicological end-point can be defined other than a general failure to gain weight. This is usually attributed to the amount in the diet being so high as to render the food unpalatable and so suppressing the appetite. If such high levels were necessary to achieve a technological effect, then the additive would not be practicable. For such substances no specific hazard can be identified.

4.3 Dose-response characterisation

The severity of any adverse effect associated with a chemical used as a food additive is usually directly related to the dose. Severity can be measured as either the degree of damage to an individual or the probability of an individual being affected, or a combination of these effects. For substances that can cause tissue damage, increasing the dose will tend to increase the degree of damage. For carcinogens, where just one molecule has the theoretical ability to induce a tumour, increasing the dose increases the probability that an individual will contract a tumour. In either case there may be some threshold level below which no effects are observed.

4.3.1 Thresholded end-points

For many substances the body's own mechanisms for de-toxification and repair mean that low doses of some chemicals can be tolerated without experiencing any adverse effects. However, once a certain threshold has been exceeded then the degree of adverse effect is related to the dose. The highest dose at which no adverse effects are observed in the most susceptible animal species is identified as the No Observed Adverse Effect Level (NOAEL). The NOAEL is used as the basis for setting human safety standards for food additive Acceptable Daily Intakes (ADIs).[4]

The ADI was defined by the Joint WHO/FAO Expert Committee on Food Additives and Contaminants (JECFA)[5] as:

> an estimate of the amount of a food additive, expressed on a body weight basis, that can be ingested daily over a lifetime without appreciable health risk.

It is related to the NOAEL so that:

$$\text{ADI (mg/kg bw/day)} = \text{NOAEL (mg/kg bw/day)} \times UF_1 \times UF_2$$

where: UF_1 = uncertainty factor to allow for extrapolation from animal species to humans.
UF_2 = uncertainty factor to allow for inter-individual variability in humans.

Uncertainty factors usually have a default value of 100 so that the ADI is usually equal to the NOAEL $\times$ 100. If human data are available then UF_1 is usually taken to be one. Note that the NOAEL and ADI are corrected for bodyweight.

This is to allow for the fact that smaller individuals (e.g. children) can tolerate relatively lower doses before any adverse effect is experienced.

Intakes that exceed the ADI will not necessarily result in any adverse effect because the uncertainty factors are designed to be conservative. In practice it is probable that most people could exceed the ADI by a considerable margin before suffering any harm. Nevertheless, the probability that an individual will suffer harm (risk) increases once the ADI is exceeded and so this must be balanced against the costs of control. Conversely, the level of risk below the ADI is never quite zero because there is always a residual risk that relates to the lack of absolute certainty in the methods used for toxicological testing. In some cases no adverse end-point can be identified, such as for many naturally-occurring compounds that are widespread in foods. In such cases an ADI Not Specified (ADI NS) is allocated.

4.3.2 Non-thresholded end-points

Some chemicals are believed to have no threshold above which toxic effects are observed. In other words, a single molecule has the potential to induce an adverse effect. The most common group of hazards in this respect are genotoxic carcinogens. Chemical carcinogens are not normally approved as food additives because an acceptable daily intake cannot be established.

4.4 Exposure analysis

Having identified information about the relationship between dose of a food additive and any toxicological response, and determined an ADI, it is necessary to investigate the levels of actual doses in the human population. Exposure analysis is used to find out if any individuals have potential intakes that might exceed the ADI for a particular additive and if so, by how much. Two pieces of information are vital for this:

1. *Usage data*. The concentrations of the food additive in foods plus, if available, the patterns of use.
2. *Food consumption data*. The amounts of the affected foods eaten including, if necessary, consumption by sub-groups.

The Estimated Daily Intake (EDI) can then be calculated using a relatively simple equation:

$$\text{EDI (mg/kg bw/week)} = \frac{\text{usage (mg/kg)} \times \text{consumption (kg/week)}}{\text{bodyweight (kg)}}$$

The EDI is corrected for bodyweight so that a direct comparison with the ADI can be made.

Many factors can influence the accuracy of intake estimates and it is of primary importance to ensure that the assumptions made and data used are relevant to the specific risk analysis.[6] The selection of inappropriate data and

Risk analysis of food additives 65

methods can easily lead to estimates of intake that are orders of magnitude greater or less than real levels. A particular question is the selection of statistics to represent food consumption by a particular population. In the past a population average figure would have typically been used, but this approach could very easily under-estimate intakes of consumers at the upper end of the range of possible intake levels. The use of single 'worst-case' figures to describe the levels present in food and the amounts of food consumed will also not provide an accurate representation of true intakes and are only useful as a tool for screening.

Modern guidelines demand that particular sub-groups (such as children) are considered, including the intakes at the upper end of the intake range (commonly 90th or 97.5th percentiles). This means that the statistical distributions of food additive concentrations in food and food consumption patterns must be taken into account.

4.4.1 Usage data

The EU Directives 94/35/EC, 94/36/EC and 95/2/EC on sweeteners, colours and food additives other than colours and sweeteners, limit the amounts of certain food additives that can be used and the range of foods in which they are permitted. Similarly, the Codex Committee on Food Additives and Contaminants (CCFAC) has published its General Standard on Food Additives (GSFA), which lists the maximum use levels recorded world-wide. Care should be taken when using data from the EU Directive annexes or the GSFA because the figures represent the maximum permitted in each food group. In practice, use levels may need to be much lower to achieve the desired technical effect, particularly if used in combination with other additives intended for the same purpose. Furthermore, the additive is unlikely to be used in all foods in which it is permitted because other additives compete for the same function in the marketplace.

Although all foods must be labelled with the additives they contain, few comprehensive data are available about which specific, branded foods contain which additives. The Irish National Food Ingredient Database[7] (INFID) provides data about food additives present in specific food products available in the Irish market. Although many of the branded foods are the same as those available in other markets, the exact composition of foods may vary from country to country, depending on local regulations, product availability, national preferences, etc. Extrapolation from the INFID must therefore be done with caution.

The actual levels at which additives are used in foods may differ from those listed in the EU Directives since these are intended to achieve the maximum technological effect. In practice use levels might be much lower. For example, it is not always necessary to create very intense colours and the amount of colourant needed will also depend on the natural colour of the matrix. Similarly, many sweeteners are used in combinations in order to control costs whilst avoiding unpleasant side-flavours. This means that in order to gain an accurate impression of use levels in food it is necessary to either approach manufacturers for data or to conduct analytical surveys. The former task is made difficult because additive producers will not necessarily know what levels food manufacturers are using and it would therefore be necessary to obtain data from every manufacturer of each food that might contain a particular additive. Analytical surveys also present problems because of the high costs of chemical analysis and the large number of different food products that would need to be analysed.

4.4.2 Food consumption data

There are many sources of data on food consumption although not all are necessarily appropriate for risk assessment. The most readily available data are Food Balance Sheets (FBSs) that are prepared globally every year by the UN Food and Agriculture Organisation (FAO).[8] These list the domestic production, imports, exports and non-food uses for major raw food commodities for each country together with the calculated *per capita* annual consumption. Such data are invaluable for making comparisons between national diets since they provide a good indication of the types of food being consumed in each country. They are of limited value for risk assessment since they provide only the mean consumption and give no indication of the *range* of food consumption patterns within each country. Furthermore, since the data sheets relate to raw agricultural commodities, they are of limited value when additives in manufactured foods are being considered.

Food consumption surveys conducted at the household level provide more information about the distribution of consumption levels. Household budget surveys gather data on the expenditure by households on particular types of goods. The EC Eurostat Population and Social Conditions databases summarise this information but the food categories used are very restricted (food and non-alcoholic beverages: bread and cereals; meat; fish; milk, cheese and eggs; oils and fat; fruit; vegetables including potatoes and other tubers; sugar, including honey, syrups, chocolate and confectionery). Since the data relate to expenditure on the classes of foods it is very difficult to relate this to quantities of food consumed.

With data on the amounts of food consumed in households and the composition of the household, modelling can be used to develop some basic information about consumption by individuals.[9] However, for reliable estimates of food consumption by individuals the weighed diary method is probably the best. In this type of survey respondents are asked to weigh and record everything that they eat for the period of the survey. Subjects are usually selected from geographical regions and at different times of the year so that the survey is as representative as possible. The principal disadvantage of this type of survey is that it can only cover a few consecutive days and so food consumption over longer time-scales cannot be determined without a supplementary questionnaire about frequency of consumption.

4.4.3 Estimating intakes

Methods for calculating food additive intakes range from the very crude to the statistically sophisticated. The crude methods are generally quick and simple, but give little valid information about true intake levels. Sophisticated methods are more demanding in resources, particularly for detailed data about additive usage patterns and food consumption, but aim to provide more accurate results. In most cases a tiered approach is used, making simple yet conservative estimates in the first instance, to identify those additives for which no further action is required because intakes will always be less than the ADI. For those additives that present the potential to exceed ADIs in simple screens, more sophisticated methods can be used to investigate further.

Simple intake estimates

Production statistics can be used to provide very simple estimates of *per capita* intake if the size of the market is known:

$$per\ capita\ \text{additive intake} = \frac{\text{gross production} + \text{imports} - \text{exports}}{\text{population}}$$

Such estimates are useful for providing comparisons between additives and between countries but unless the entire population consumes the additive in exactly identical amounts, they provide little information about real intakes, particularly for those individuals at the upper end of the intake range.

Budget methods are based on the fact that there is a physiological upper limit to the amount of food and drink that can be consumed on any day. If the amount of additive present in that food is also limited (e.g. by national legislation) then there is an absolute maximum that can be ingested on any day. The assumptions in the budget method are extended to allow for the fact that only a proportion of the diet is likely to contain additives (Table 4.1).[10] A conversion factor is produced which is used to derive the maximum use level from the ADI:

$$\text{maximum level in drinks (mg/kg)} = \text{ADI (mg/kg bw/day)} \times 40$$

$$\text{maximum level in food (mg/kg)} = \text{ADI (mg/kg bw/day)} \times 160$$

The method can also be used to provide a simple estimate of intake if maximum usage levels are known (the 'reverse budget' method):

Table 4.1 Figures most commonly used in the standard budget method

	Upper limit of daily consumption	Proportion likely to contain additives	Resulting factor
Beverages	100 ml/kg bw	25%	40
Food	25 g/kg bw (=50 kcal/kg bw)	25%	160

$$\text{potential intake} = \frac{\text{maximum level in drinks}}{40} + \frac{\text{maximum level in food}}{160}$$

The budget method has been generally accepted as a conservative screening method for additives unless they are used in a wide range of foods. However, the method has been criticised because it does not take into account sufficiently potential intakes by children who have higher energy intakes in relation to their bodyweights. A European intergovernmental working group has evaluated the budget method[10] and recommended lowering the drinks 'factor' from 40 to 10 to allow for the relatively high level of consumption of soft drinks by small children.

The SCOOP Report 4.2[10] lists those additives that have potential intakes that exceed the ADI when screened using the budget method (Table 4.2). Out of 58 additives examined, 36 had intakes greater than the ADI for adults and 48 had intakes greater than the ADI for children. It was recommended that all of these additives required further examination. Because the budget method is such a crude tool, intakes that appear to exceed the ADI may seldom do so in reality. For example, the additive with highest intakes in relation to its ADI in the SCOOP study was adipic acid and its salts (E355–7 and E359) with intakes 57.5 times the ADI for adults and 207.5 times the ADI for children. However, Directive 95/2/EC restricts the use of adipic acid and its salts to fillings and toppings for fine bakery wares, dry powdered dessert mixes, gel-like desserts, fruit-flavoured desserts, and powders for home preparation of drinks. Given such a narrow range of uses it is unlikely that an individual would have a high level of intake for any sustained period of time. Nevertheless, any indication of the possibility of high level intakes should be followed up with intake calculations based on actual food consumption data.

Intake calculations based on individual food consumption data
To calculate estimates of intake it is necessary to multiply additive levels present in food by the amounts of food consumed. In the budget method very crude assumptions about food consumption and additive levels are used. In reality, individuals will select different foods containing different additive levels and so a distribution of intakes will be produced. Those individuals who consume the greatest amounts of foods containing particular additives at the highest levels will have highest intakes. Several countries, including the USA, Germany, and the UK have high-quality data on the food consumption patterns of their respective populations. These are often collected by asking volunteers (or their carers) to keep a weighed record of all food consumed during a set period – usually 3 days to one week. Such surveys are limited by the relatively short time span and the number of participants. Nevertheless they can provide a rich source of information if used with care.

A histogram of the consumption of carbonated soft drinks by UK adults and pre-school children is provided in Fig. 4.2. This shows a typical distribution with a wide range of values. In this case only about one half of respondents reported soft drink consumption during the survey and those that did consume show a

Table 4.2 Food additives requiring further attention for adults or children following budget method screening[10]

E Number	Name	TMDI/ADI Adults	TMDI/ADI Children	E Number	Name	TMDI/ADI Adults	TMDI/ADI Children
E 102	Tartrazine	0.8	1.4	E 416	Karaya gum	2.5	2.5
E 104	Quinoline yellow	0.6	1.1	E 432–6	Polysorbates	1.9	1.6
E 110	Sunset yellow FCF	2.5	2.1	E 442	Ammonium phosphatides	2.1	2.1
E 120	Carmines	1.3	2.1	E 444	Sucrose acetate isobutyrate	0.8	3
E 122	Azorubine	1.6	1.3	E 320	Butylated hydroxyanisole	2.5	2.5
E 124	Ponceau 4R	1.6	1.3	E 321	Butylated hydroxytoluene	12.5	12.5
E 127	Erythrosine BS	12.5	12.5	E 338–43, E 450–52	Phosphorus	0.7	1.5
E 128	Red 2G	1.3	1.3	E 355–7 E 359	Adipic acid + salts	57.5	207.5
E 129	Allura red AC	0.9	1.5	E 473–4	Sucrose esters/sucroglycerides	7.8	26.6
E 131	Patent blue V	0.4	1.4	E 475	Polyglycerol esters of fatty acids	2.5	2.5
E 132	Indigotine	1.3	2.1	E 476	Polyglycerol esters of polycondensed fatty acids of castor oil	4.2	4.2
E 133	Brilliant blue FCF	0.6	1.1	E 479b	Thermally oxidised soya bean oil with mono/diglycerides of fatty acids	1.3	1.3
E 142	Green S	1.3	2.1	E 481–2	Steroyl lactates	1.6	1.6
E 151	Black PN	1.3	2.1	E 483	Stearyl tartrate	1.6	1.6
E 155	Brown HT	2.1	1.8	E 491–2 E 495	Sorbitan esters (1)	1.3	1.3
E 160b	Annatto extracts	2.4	2.4	E 493–4	Sorbitan esters (2)	6.3	6.3
E 161g	Canthaxanthine	1.9	1.9	E 520–3 E 541 E 554–9 E 573	Aluminium	13.1	13.1
E 200 E 202–3	Sorbic acid + salts	0.8	1.7	E 535–6	Ferrocyanides	5	5
E 210–3	Benzoic acid + salts	2	4.3	E 950	Acesulfame-K	1.5	4.4
E 220–4 E 226–8	Sulphites	11.6	33	E 951	Aspartame	0.5	1.7
E 249 E250	Nitrites	3.1	3.1	E 952	Cyclamic acid + salts	1.4	4.1
E 297 E 365–7	Fumaric acid + salts	8.3	20.8	E 954	Saccharin	0.8	2.3
E 310–2	Gallates	2.5	2.5	E 959	Neohesperidine	0.4	1.1
E 315–6	Erythrobic acid	1	1	E 999	Quilliai extract	1	4

70 Food chemical safety

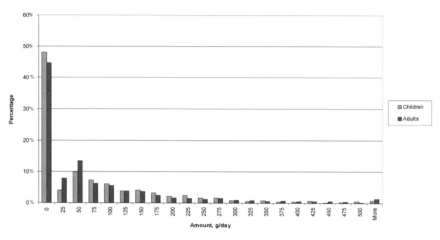

Fig. 4.2 Consumption of carbonated drinks by UK adults and pre-school children.

very skewed log-normal type of distribution. This means that the majority of consumers are consuming relatively small amounts of the product whilst a small proportion is consuming significantly larger amounts. This distribution reflects the fact that carbonated drinks are generally regarded as a leisure product rather than a serious food item. Dietary staples, such as bread, tend to have a higher proportion of consumers and distributions that begin to take on more of the characteristics of the normal distribution (Fig. 4.3).

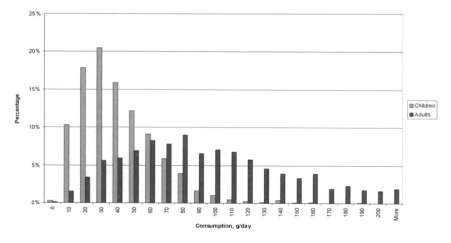

Fig. 4.3 Consumption of bread by UK adults and pre-school children.

Risk analysis of food additives 71

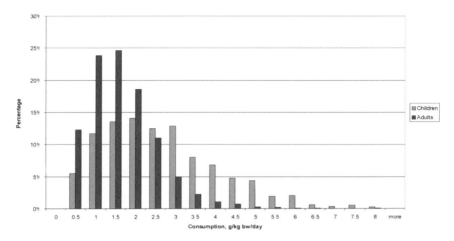

Fig. 4.4 Consumption of bread by UK adults and pre-school children – corrected for individual bodyweight.

Note that in Figs 4.2 and 4.3 adults generally have slightly higher consumption of both carbonated drinks and bread when expressed on a simple gram per person per day basis although for carbonated drinks, average intakes are about 70 g per day for both adults and children. However, to make comparisons with the ADI, intake estimates need to be based on the consumption per kilogram bodyweight per day. Bread consumption, corrected for bodyweight, is shown in Fig. 4.4 and it can be seen that children's consumption now exceeds that of adults. This is because children have much higher energy, and thus food, demands than adults. The effect is more pronounced for foods that are consumed preferentially by children, such as carbonated soft drinks (Fig. 4.5). After correction for bodyweight, children's average consumption of carbonated drinks is five times that of adults. These differences between adults and children can have very significant effects on predicted intakes of additives.

In order to use the food consumption data to predict intakes of food additives it is necessary to combine them with information about the levels in foods. Data on usage have been described above and at the simplest level it is usual to use the maximum permitted levels from national or supra-national legislation.

Butylated hydroxytoluene (BHT – E 321) is permitted for use as an antioxidant in:

- fats and oils for the professional manufacture of heat-treated foodstuffs (100 mg/kg in fat)
- frying oil and frying fat, excluding olive, pomace oil (100 mg/kg in fat)
- lard; fish oil; beef, poultry and sheep fat (100 mg/kg in fat)

72   Food chemical safety

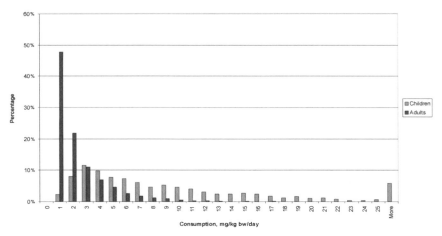

Fig. 4.5 Consumption of carbonated drinks by UK adults and pre-school children – corrected for individual bodyweight.

- chewing gum (400 mg/kg; total gallates + BHT + BHA in fat)
- dietary supplements (400 mg/kg; total gallates + BHT + BHA in fat).

In practice this means that BHT can be present at up to 100 mg/kg in the fat fraction of manufactured foods such as baked goods and fried foodstuffs such as potato crisps or fried chicken. Although higher levels are permitted in chewing gum and dietary supplements, these might be expected to have a lesser impact on intakes because of the relatively small amounts of such foods consumed.

Intakes are calculated on the basis of the fat content of each food with the exception of chewing gum. Table 4.3 provides data on the mean, 90th percentile and 97.5th percentile consumption of food, fat consumption for each food group, daily BHT intake and daily BHT intake corrected for bodyweight. For chewing gum and dietary supplements the number of consumers was too small to make reliable predictions of intake across the population. The results of such analyses should therefore be interpreted with great caution.

Distributional analysis generates estimates of intakes of BHT that exceed the ADI for adults and children for a small number of consumers (Fig. 4.6). In this case children's intakes are lower than adults, reflecting the fact that children eat less of these foods in relation to their bodyweight (with the exception of chewing gum). Although the estimates are less conservative than those generated by the budget method, they are still probably over-estimating true intakes. This is because it is assumed that on every occasion that fats and oils are used in the food manufacturing applications identified above, BHT is present at its maximum permitted level as an antioxidant. In reality BHT competes with other antioxidants in the additives market and so this scenario is unlikely to prevail in the real world. In practice BHT will be used in only a fraction of these

Table 4.3 Consumption of foods and their fat content and intakes of BHT by UK adults and pre-school children

	% Consuming	Food consumption (g/day)			Fat consumption (g/day)			BHT intake (mg/day)			BHT intake (mg/kg bw/day)		% ADI
		Mean	90th %ile	97.5th %ile	Mean	90th %ile	97.5th %ile	Mean	90th %ile	97.5th %ile	Mean	90th %ile	97.5th %ile	
Adults														
Baked goods	89	37	75	111	7	14	21	0.69	1.44	2.06	0.010	0.021	0.031	62
Fried foods	90	56	123	234	10	20	33	0.98	2.03	3.29	0.014	0.029	0.049	98
Chewing gum	0													
Supplements	1	9	26	56				0.12	0.16	0.81	0.002	0.002	0.013	27
All foods	100	84	161	255	15	28	41	1.49	2.79	4.06	0.022	0.041	0.062	124
Pre-school children														
Baked goods	91	23	46	65	5	9	13	0.07	0.13	0.19	0.005	0.009	0.013	26
Fried foods	95	31	64	101	7	15	21	0.10	0.21	0.30	0.007	0.014	0.021	43
Chewing gum	2	1	2	2				0.40	0.60	0.92	0.029	0.041	0.063	126
Supplements	5	1	2	2				0.02	0.07	0.28	0.002	0.004	0.016	32
All foods	100	50	93	119	11	20	26	0.16	0.30	0.42	0.012	0.020	0.030	59

ADI = 0.05 mg/kg bw

74 Food chemical safety

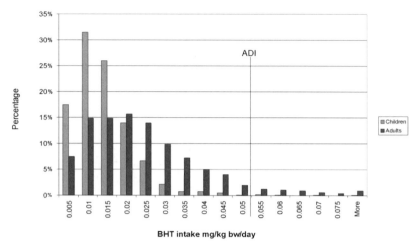

Fig. 4.6 Intakes of BHT by UK adults and pre-school children.

applications. If it was known what particular foods contained BHT then it would be possible to generate a more accurate estimate of intake. In certain circumstances, data sources such as the Irish National Food Ingredient Database could be useful to identify which branded foods contained a particular additive and then to eliminate all other foods from the analysis. The case of BHT is somewhat different because the fats and oils in which it is used are ingredients of other foods and additives used in ingredients that need not be declared on food labels. The presence of BHT in fats and oils may therefore not be recorded.

Probabilistic intake modelling
The approaches for estimating intakes described thus far have relied on being able to extract a single figure to represent the concentrations of additives in foods. The calculations assume that the additive is present in all foods in which it is permitted at the maximum permitted level. In reality additive concentrations will show a distribution of values and this distribution tends to be overlooked using conventional methods. Probabilistic (also known as Monte Carlo) modelling provides an opportunity for making use of all of the data from the distribution of additive levels and food consumption to predict the probability of a given level of intake occurring.

Probabilistic modelling works by taking a random sample from the distribution of additive levels for a given food and combining it with a random sample from the distribution of food consumption for that food. This sampling is repeated several thousand times until a smooth intake distribution curve is produced.

Probabilistic modelling has been widely applied for determining acute intakes of pesticide residues. The method works well because for any randomly selected individual, the level of pesticide residue in a given food item is a relatively random event for most consumers. This is because of the vast distribution networks for agricultural commodities which mean that a given item of food could have come from almost any source. The situation is different for many additives because food selection is not such a random event. For example, additives tend to be used in branded foods and brand loyalties often mean that a consumer will always select a particular brand and so the additive level will be unaltered. Random sampling in current Monte Carlo methods would only be valid if shoppers shopped 'blind' so that the probability of choosing any one brand was the same as that for any other.

4.5 Risk evaluation

Risk evaluation is the apparently simple task of comparing an estimate of intake with the ADI. If intakes are below the ADI then there is virtually no risk whereas if they exceed the ADI then some risk management action may be required. In practice risk evaluation is a far less certain science.

A vital and often over-looked aspect of risk evaluation is ensuring that the estimate of intake corresponds to the ADI so that like is being compared with like. For example, toxicological end-points are frequently time-related. The toxicological end-points relating to many additives are non-specific and probably relate to intakes averaged over the long term. Some additive end-points can relate to short-term gastro-intestinal effects such as stomach irritation or diarrhoea. In such cases it is the size of the most recent dose that is probably critical rather than average amounts over the longer term. This is important because for foods that are not consumed very frequently, the amount consumed on a day when the food is eaten can be very much greater than when that amount is averaged over a longer period of time.

4.6 Methods for risk management

The outputs from risk assessment will normally include information about the relationship between dose and risk, and estimates of levels of doses and thus risks in the population. This information needs to be balanced with information about the possible benefits associated with a food additive in making judgements about optimal risk management. For many additives, benefits are restricted to improving consumer appeal for a food, such as colour, texture or flavour. For other additives their presence may prevent the growth of microorganisms or chemical oxidation and so they would be associated with potential safety benefits. Other additives may bring benefits to only a few. For example, for the general population artificial sweeteners bring few benefits apart from lowering the cost of certain foods. However, for those who cannot consume sucrose, either for medical reasons such as diabetes or because of the need to lose weight, artificial sweeteners can make foods both palatable and safe to eat.

Decisions about the risk management of additives must take into account both risks and benefits. The hypothetical risk associated with occasional exceedance of the ADI may be more acceptable for one additive, if it has tangible benefits associated with it, than for another.

4.6.1 Consumer perceptions of risk

In making risk management decisions it is important to take into account non-technical factors in addition to scientific and economic information. Recent crises in the food industry have indicated that consumers' perceptions about risks are driven by factors that would not be considered in conventional risk assessments. Research has shown that factors such as whether sub-groups (particularly children) might be affected, whether the hazard is familiar, if there are effects on the environment or if risks and benefits are equitably shared can determine consumers' reactions to an issue.

Risk managers must be aware that in the event of a crisis consumers' perceptions about risks can have as great or greater impact on the outcome than the real food safety issues.

Recent trends have seen more open acknowledgement of the need to balance scientific and social factors. For example, in the UK the Food Advisory Committee that advises the Food Standards Agency on food safety issues is comprised of a wide range of expertise including toxicologists, chemists, food technologists, economists, and representatives of consumer organisations and the food industry. The Committee is thus able to provide a balanced view that takes all interests into account.

4.7 Risk communication

An important part of the risk management process involves informing consumers, industry and other stakeholders of the decisions made by regulatory authorities. However, this is a narrow view of risk communication that does not take into account the potential for dialogue between interested parties that can result in better decision-making. Understanding how consumers view the potential risks associated with food additives can help to avoid either under- or over-regulation.

4.8 Future trends

Risk analysis, like most scientific disciplines is subject to continual evolution. This means that methods and concepts that are in common use today may well become discredited and obsolete in the future. This presents a problem for risk managers because it creates the impression that everything done before was somehow 'wrong'. In fact, most changes are gradual and tend to take effect at the margins rather than overturning all previous assessments. An example of this is the introduction of the separate consideration of children in the risk assessment process. Some toxicologists argue that the way that the ADI is established means that it applies to the entire life-time and so only average life-time intakes need to be considered in risk assessment. Others take the view that intakes should not be allowed to exceed the ADI on any occasion during a life-time. However, the consensus view is that occasional exceedances of the ADI are tolerable since the probability of harm is extremely small. For most consumers, intakes of any additive will vary from day to day and so the occasional exceedance should be balanced by many other days well below the ADI.

Two factors mean that children may differ from the majority of consumers in their intakes of additives: firstly, they have higher energy and thus food needs in relation to their lower bodyweights, than adults. This means that if their diet contains foods that contain additives then they may have consistently raised intakes of additives. Secondly, children tend to have preferences for particular foods and for many children these foods are sweetened, commercially-prepared foods such as soft drinks and confectionery. Therefore, for such foods, children may have much higher intakes of additives than the energy intake : bodyweight ratio would cause alone. Most leading food regulatory authorities now take separate account of intakes by children.

4.9 Conclusion

There is considerable scope for producing a wide variety of intake estimates for any given scenario, depending on the method chosen for the analysis and the data available. It is therefore vital that the underlying toxicological concerns and the nature of food consumption and additive usage data are fully understood if an intake estimate that is relevant to the particular situation is to be provided. Given the complexity of the available methods and data it is likely that any estimate of additive intake will contain errors. It is therefore vital that information about uncertainties are provided alongside intake estimates so that the correct interpretation can be applied.

4.10 Sources of further information and advice

World-Wide-Web resources

The EU Commission Directorate General on Health and Consumer Protection	http://europa.eu.int/_omm./dgs/health_consumer/index_en.htm
World Health Organisation	http://www.who.int/
UK Pesticides Safety Directorate	http://www.pesticides.gov.uk/
UK Food Standards Agency	http://www.foodstandards.gov.uk/

Joint FAO/WHO Expert Committee on Food Additives and Contaminants http://www.inchem.org/aboutjecfa.html

Institute of Food Science and Technology (IFST) http://www.ifst.org/

International Society of Exposure Analysis (ISEA) http://www.ISEAweb.org

Publications

CASWELL, J. A. (1991). *The Economics of Food Safety*. Elsevier Applied Science Publishers Ltd. ISBN 0-444-01614-7.

TENNANT, D. R. (1997). *Food Chemical Risk Analysis*. Blackie Academic and Professional, Chapman and Hall, London. ISBN 0 412 723107.

4.11 References

1. EU Directive 94/35/EC on sweeteners for use in foodstuffs. OJ L237/3 10.9.94.
2. EU Directive 94/36/EC on colours for use in foodstuffs. OJ L237/13 10.9.94.
3. EU Directive 95/2/EC on food additives other than colours and sweeteners. OJ L61/1 18.3.95.
4. RUBERY E. D., BARLOW S. M. and STEADMAN J. H. (1990). Criteria for setting quantitative estimates of acceptable intakes of chemicals in food in the UK. *Food Additives and Contaminants* 7/3, 287–302.
5. WHO (1987). *Principles for the Safety Assessment of Food Additives and Contaminants in Food*. Environmental Health Criteria 70. World Health Organisation, Geneva.
6. DOUGLAS, J. S. and TENNANT, D. R. (1997). 'Estimations of dietary intake of food chemicals,' pp. 195–215 in Tennant, D. R. (Ed.) *Food Chemical Risk Analysis*. Blackie Academic and Professional, Chapman and Hall, London.
7. INSTITUTE OF EUROPEAN FOOD STUDIES (1999). The Irish National Food Ingredient Database User's Manual. Trinity College, Dublin, Ireland.
8. Can be down-loaded from the FAO website at: http://apps.fao.org/lim500/nph-wrap.pl?FoodBalanceSheet&Domain=FoodBalanceSheet
9. COMBRIS, P., BERTAIL, P. and BOIZOT, C. (1998). La Consommation Alimentaire en 1995: Distribution des Quantitiés Consommées à Domicile. INRA Laboratoire de Recherche sur la Consommation.
10. EU SCIENTIFIC CO-OPERATION PROGRAMME (1997). Report on methodologies for the monitoring of food additive intake across the European Union. SCOOP Task 4.2 Report. Ministry of Agriculture, Fisheries and Food, UK.

5

Analytical methods: quality control and selection

R. Wood, Food Standards Agency, Norwich

5.1 Introduction

It is now universally recognised as being essential that a laboratory produces and reports data that are fit-for-purpose. For a laboratory to produce consistently reliable data it must implement an appropriate programme of quality assurance measures; such measures are now required by virtue of legislation for food control work or, in the case of the UK Food Standards Agency (FSA), in their requirements for contractors undertaking surveillance work. Thus customers now demand of providers of analytical data that their data meet established quality requirements. These are further described below. The significance of the measures identified are then described and some indications are given as to the future of analytical methodology within the food laboratory. These are then discussed.

Methods of analysis have been prescribed by legislation for a number of foodstuffs since the UK acceded to the European Community in 1972. However, the Community now recognises that the quality of results from a laboratory is equally as important as the method used to obtain the results. This is best illustrated by consideration of the Council Directive on the Official Control of Foodstuffs (OCF) which was adopted by the Community in June, 1989.[1] This, and the similar Codex Alimentarius Commission requirements, are described below. As a result of this general recognition there is a general move away from the need to prescribe all analytical methodology in detail towards the prescription of the general quality systems within which the laboratory must operate. This allows greater flexibility to the laboratory without detracting from the quality of results that it will produce.

5.2 Legislative requirements

5.2.1 The European Union

For analytical laboratories in the food sector there are legislative requirements regarding analytical data which have been adopted by the European Union. In particular, methods of analysis have been prescribed by legislation for a number of foodstuffs since the UK acceded to the European Community in 1972. However, the Union now recognises that the competency of a laboratory (i.e. how well it can use a method) is equally as important as the 'quality' of the method used to obtain results. The Council Directive on the Official Control of Foodstuffs which was adopted by the Community in 1989[1] looks forward to the establishment of laboratory quality standards, by stating that 'In order to ensure that the application of this Directive is uniform throughout the Member States, the Commission shall, within one year of its adoption, make a report to the European Parliament and to the Council on the possibility of establishing Community quality standards for all laboratories involved in inspection and sampling under this Directive' (Article 13).

Following that, in September 1990, the Commission produced a Report which recommended establishing Community quality standards for all laboratories involved in inspections and sampling under the OCF Directive. Proposals on this have now been adopted by the Community in the Directive on Additional Measures Concerning the Food Control of Foodstuffs (AMFC).[2] In Article 3 of the AMFC Directive it states:

1. Member States shall take all measures necessary to ensure that the laboratories referred to in Article 7 of Directive 89/397/EEC[1] comply with the general criteria for the operation of testing laboratories laid down in European standard EN 45001[3] supplemented by Standard Operating Procedures and the random audit of their compliance by quality assurance personnel, in accordance with the OECD principles Nos. 2 and 7 of good laboratory practice as set out in Section II of Annex 2 of the Decision of the Council of the OECD of 12 Mar 1981 concerning the mutual acceptance of data in the assessment of chemicals.[4]
2. In assessing the laboratories referred to in Article 7 of Directive 89/397/EEC Member States shall:
   (a) apply the criteria laid down in European standard EN 45002;[5] and
   (b) require the use of proficiency testing schemes as far as appropriate.

   Laboratories meeting the assessment criteria shall be presumed to fulfil the criteria referred to in paragraph 1. Laboratories which do not meet the assessment criteria shall not be considered as laboratories referred to in Article 7 of the said Directive.
3. Member States shall designate bodies responsible for the assessment of laboratories as referred to in Article 7 of Directive 89/397/EEC. These bodies shall comply with the general criteria for laboratory accreditation bodies laid down in European Standard EN 45003.[6]

4. The accreditation and assessment of testing laboratories referred to in this article may relate to individual tests or groups of tests. Any appropriate deviation in the way in which the standards referred to in paragraphs 1, 2 and 3 are applied shall be adopted in accordance with the procedure laid down in Article 8.

In Article 4, it states:

> Member States shall ensure that the validation of methods of analysis used within the context of official control of foodstuffs by the laboratories referred to in Article 7 of Directive 89/397/EEC comply whenever possible with the provisions of paragraphs 1 and 2 of the Annex to Council Directive 85/591/EEC of 23 December 1985 concerning the introduction of Community methods of sampling and analysis for the monitoring of foodstuffs intended for human consumption.[7]

As a result of the adoption of the above directives legislation is now in place to ensure that there is confidence not only in national laboratories but also those of the other Member States. As one of the objectives of the EU is to promote the concept of mutual recognition, this is being achieved in the laboratory area by the adoption of the AMFC directive. The effect of the AMFC Directive is that organisations must consider the following aspects within the laboratory: its organisation, how well it actually carries out analyses, and the methods of analysis used in the laboratory. All these aspects are inter-related, but in simple terms may be thought of as:

- becoming accredited to an internationally recognised standard; such accreditation is aided by the use of internal quality control procedures
- participating in proficiency schemes, and
- using validated methods.

In addition it is important that there is dialogue and co-operation by the laboratory with its customers. This is also required by virtue of the EN 45001 Standard at paragraph 6, and will be emphasised even more in future with the adoption of ISO/IEC Guide 17025.[8]

The AMFC Directive requires that food control laboratories should be accredited to the EN 45000 series of standards as supplemented by some of the OECD GLP principles. In the UK, government departments have nominated the United Kingdom Accreditation Service (UKAS) to carry out the accreditation of official food control laboratories for all the aspects prescribed in the Directive. However, as the accreditation agency will also be required to comply with the EN 45003 Standard and to carry out assessments in accordance with the EN 45002 Standard, all accreditation agencies that are members of the European Co-operation for Accreditation of Laboratories (EA) may be asked to carry out the accreditation of a food control laboratory within the UK. Similar procedures will be followed in the other Member States, all having or developing equivalent organisations to UKAS. Details of the UK requirements for food control laboratories are described later in this chapter.

5.2.2 Codex Alimentarius Commission: guidelines for the assessment of the competence of testing laboratories involved in the import and export control of food

The decisions of the Codex Alimentarius Commission are becoming of increasing importance because of the acceptance of Codex Standards in the World Trade Organisation agreements. They may be regarded as being semi-legal in status. Thus, on a world-wide level, the establishment of the World Trade Organisation (WTO) and the formal acceptance of the Agreements on the Application of Sanitary and Phytosanitary Measures (SPS Agreement) and Technical Barriers to Trade (TBT Agreement) have dramatically increased the status of Codex as a body. As a result, Codex Standards are now seen as *de facto* international standards and are increasingly being adopted by reference into the food law of both developed and developing countries.

Because of the status of the CAC described above, the work that it has carried out in the area of laboratory quality assurance must be carefully considered. One of the CAC Committees, the Codex Committee on Methods of Analysis and Sampling (CCMAS), has developed criteria for assessing the competence of testing laboratories involved in the official import and export control of foods. These were recommended by the Committee at its 21st Session in March 1997[9] and adopted by the Codex Alimentarius Commission at its 22nd Session in June 1997;[10] they mirror the EU recommendations for laboratory quality standards and methods of analysis. The guidelines provide a framework for the implementation of quality assurance measures to ensure the competence of testing laboratories involved in the import and export control of foods. They are intended to assist countries in their fair trade in foodstuffs and to protect consumers.

The criteria for laboratories involved in the import and export control of foods, now adopted by the Codex Alimentarius Commission, are:

- to comply with the general criteria for testing laboratories laid down in ISO/IEC Guide 25: 1990 'General requirements for the competence of calibration and testing laboratories'[8] (i.e. effectively accreditation)
- to participate in appropriate proficiency testing schemes for food analysis which conform to the requirements laid down in 'The International Harmonised Protocol for the Proficiency Testing of (Chemical) Analytical Laboratories'[11] (already adopted for Codex purposes by the CAC at its 21st Session in July 1995)
- to use, whenever available, methods of analysis that have been validated according to the principles laid down by the CAC
- to use internal quality control procedures, such as those described in the 'Harmonised Guidelines for Internal Quality Control in Analytical Chemistry Laboratories'.[12]

In addition, the bodies assessing the laboratories should comply with the general criteria for laboratory accreditation, such as those laid down in the ISO/IEC Guide 58:1993: 'Calibration and testing laboratory accreditation systems – General requirements for operation and recognition'.[13]

Thus, as for the European Union, the requirements are based on accreditation, proficiency testing, the use of validated methods of analysis and, in addition, the formal requirement to use internal quality control procedures which comply with the Harmonised Guidelines. Although the EU and Codex Alimentarius Commission refer to different sets of accreditation standards, the ISO/IEC Guide 25: 1990 and EN 45000 series of standards are similar in intent. It is only through these measures that international trade will be facilitated and the requirements to allow mutual recognition to be fulfilled will be achieved.

5.3 FSA surveillance requirements

The Food Standards Agency undertakes food chemical surveillance exercises. It has developed information for potential contractors on the analytical quality assurance requirements for food chemical surveillance exercises. These requirements are describe below but reproduced as an appendix to this chapter; they emphasise the need for a laboratory to produce and report data of appropriate quality. The requirements are divided into three parts dealing with:

- Part A: quality assurance requirements for surveillance projects provided by potential contractors at the time that tender documents are completed and when commissioning a survey. Here information is sought on:
  - the formal quality system in the laboratory if third-party assessed (i.e. if UKAS accredited or GLP compliant)
  - the quality system if not accredited
  - proficiency testing
  - internal quality control
  - method validation.

- Part B: information to be defined by the FSA customer once the contract has been awarded to a contractor, e.g. the sample storage conditions to be used, the methods to be used, the IQC procedures to be used, the required measurement limits (e.g. limit of detection (LOD), limit of determination/quantification (LOQ), and the reporting limits)
- Part C: information to be provided by the contractor on an on-going basis once contract is awarded – to be agreed with the customer to ensure that the contractor remains in 'analytical control'.

5.4 Laboratory accreditation and quality control

Although the legislative requirements apply only to food-control laboratories, the effect of their adoption is that other food laboratories will be advised to achieve the same standard in order for their results to be recognised as equivalent and accepted for 'due diligence' purposes. In addition, the Codex requirements affect all organisations involved in international trade and thus provide an important 'quality umbrella'.

As shown above, these include a laboratory to be third-party assessed to international accreditation standards, to demonstrate that it is in statistical control by using appropriate internal quality control procedures, to participate in proficiency testing schemes which provide an objective means of assessing and documenting the reliability of the data it is producing and to use methods of analysis that are 'fit-for-purpose'. These requirements are summarised below and then described in greater detail later in this chapter.

5.4.1 Accreditation

The AMFC Directive requires that food-control laboratories should be accredited to the EN 45000 series of standards as supplemented by some of the OECD GLP principles. In the UK, government departments will nominate the United Kingdom Accreditation Service (UKAS) to carry out the accreditation of official food-control laboratories for all the aspects prescribed in the Directive. However, as the accreditation agency will also be required to comply to EN 45003 Standard and to carry out assessments in accordance with the EN 45002 Standard, any other accreditation agencies that are members of the European Co-operation for Accreditation of Laboratories (EA) may also be nominated to carry out the accreditation. Similar procedures will be followed in the other Member States, all having or developing equivalent organisations to UKAS.

It has been the normal practice for UKAS to accredit the scope of laboratories on a method-by-method basis. In the case of official food-control laboratories undertaking non-routine or investigative chemical analysis it is accepted that it is not practical to use an accredited fully documented method in the conventional sense, which specifies each sample type and analyte. In these cases a laboratory must have a protocol defining the approach to be adopted which includes the requirements for validation and quality control. Full details of procedures used, including instrumental parameters, must be recorded at the time of each analysis in order to enable the procedure to be repeated in the same manner at a later date. It is therefore recommended that for official food-control laboratories undertaking analysis, appropriate methods are accredited on a generic basis with such generic accreditation being underpinned where necessary by specific method accreditation.

Food-control laboratories seeking to be accredited for the purposes of the Directive should include, as a minimum, the following techniques in generic protocols: HPLC, GC, atomic absorption and/or ICP (and microscopy). A further protocol on sample preparation procedures (including digestion and solvent dissolution procedures) should also be developed. Other protocols for generic methods which are acceptable to UKAS may also be developed. Proximate analyses should be addressed as a series of specific methods including moisture, fat, protein and ash determinations.

Where specific Regulations are in force then the methods associated with the Regulations shall be accredited if the control laboratory wishes to offer enforcement of the Regulations to customers. Examples of these are methods of analysis for aflatoxins and methods of analysis for specific and overall migration for food contact materials.

By using the combination of specific method accreditation and generic accreditation it will be possible for laboratories to be accredited for all the analyses of which they are capable and competent to undertake. Method performance validation data demonstrating that the method was fit-for-purpose shall be demonstrated before the test result is released and method performance shall be monitored by on-going quality-control techniques where applicable. It will be necessary for laboratories to be able to demonstrate quality-control procedures to ensure compliance with the EN 45001 Standard,[3] an example of which would be compliance with the ISO/AOAC/IUPAC Guidelines on Internal Quality Control in Analytical Chemistry Laboratories.[12]

5.4.2 Internal quality control

Internal quality control (IQC) is one of a number of concerted measures that analytical chemists can take to ensure that the data produced in the laboratory are of known quality and uncertainty. In practice this is determined by comparing the results achieved in the laboratory at a given time with a standard. IQC therefore comprises the routine practical procedures that enable the analyst to accept a result or group of results or reject the results and repeat the analysis. IQC is undertaken by the inclusion of particular reference materials, 'control materials', into the analytical sequence and by duplicate analysis.

ISO, IUPAC and AOAC INTERNATIONAL have co-operated to produce agreed protocols on the 'Design, Conduct and Interpretation of Collaborative Studies'[14] and on the 'Proficiency Testing of [Chemical] Analytical Laboratories'.[11] The Working Group that produced these protocols has prepared a further protocol on the internal quality control of data produced in analytical laboratories. The document was finalised in 1994 and published in 1995 as the 'Harmonised Guidelines For Internal Quality Control In Analytical Chemistry Laboratories'.[12] The use of the procedures outlined in the Protocol should aid compliance with the accreditation requirements specified above.

Basic concepts
The protocol sets out guidelines for the implementation of internal quality control (IQC) in analytical laboratories. IQC is one of a number of concerted measures that analytical chemists can take to ensure that the data produced in the laboratory are fit for their intended purpose. In practice, fitness for purpose is determined by a comparison of the accuracy achieved in a laboratory at a given time with a required level of accuracy. Internal quality control therefore comprises the routine practical procedures that enable the analytical chemist to accept a result or group of results as fit-for-purpose, or

reject the results and repeat the analysis. As such, IQC is an important determinant of the quality of analytical data, and is recognised as such by accreditation agencies.

Internal quality control is undertaken by the inclusion of particular reference materials, called 'control materials', into the analytical sequence and by duplicate analysis. The control materials should, wherever possible, be representative of the test materials under consideration in respect of matrix composition, the state of physical preparation and the concentration range of the analyte. As the control materials are treated in exactly the same way as the test materials, they are regarded as surrogates that can be used to characterise the performance of the analytical system, both at a specific time and over longer intervals. Internal quality control is a final check of the correct execution of all of the procedures (including calibration) that are prescribed in the analytical protocol and all of the other quality assurance measures that underlie good analytical practice. IQC is therefore necessarily retrospective. It is also required to be as far as possible independent of the analytical protocol, especially the calibration, that it is designed to test.

Ideally both the control materials and those used to create the calibration should be traceable to appropriate certified reference materials or a recognised empirical reference method. When this is not possible, control materials should be traceable at least to a material of guaranteed purity or other well characterised material. However, the two paths of traceability must not become coincident at too late a stage in the analytical process. For instance, if control materials and calibration standards were prepared from a single stock solution of analyte, IQC would not detect any inaccuracy stemming from the incorrect preparation of the stock solution.

In a typical analytical situation several, or perhaps many, similar test materials will be analysed together, and control materials will be included in the group. Often determinations will be duplicated by the analysis of separate test portions of the same material. Such a group of materials is referred to as an analytical 'run'. (The words 'set', 'series' and 'batch' have also been used as synonyms for 'run'.) Runs are regarded as being analysed under effectively constant conditions. The batches of reagents, the instrument settings, the analyst, and the laboratory environment will, under ideal conditions, remain unchanged during analysis of a run. Systematic errors should therefore remain constant during a run, as should the values of the parameters that describe random errors. As the monitoring of these errors is of concern, the run is the basic operational unit of IQC.

A run is therefore regarded as being carried out under repeatability conditions, i.e. the random measurement errors are of a magnitude that would be encountered in a 'short' period of time. In practice the analysis of a run may occupy sufficient time for small systematic changes to occur. For example, reagents may degrade, instruments may drift, minor adjustments to instrumental settings may be called for, or the laboratory temperature may rise. However, these systematic effects are, for the purposes of IQC, subsumed into the repeatability variations. Sorting the materials making up a run into a randomised order converts the effects of drift into random errors.

Scope of the guidelines
The guidelines are a harmonisation of IQC procedures that have evolved in various fields of analysis, notably clinical biochemistry, geochemistry, environmental studies, occupational hygiene and food analysis. There is much common ground in the procedures from these various fields. However, analytical chemistry comprises an even wider range of activities, and the basic principles of IQC should be able to encompass all of these. The guidelines will be applicable in the great majority of instances although there are a number of IQC practices that are restricted to individual sectors of the analytical community and so not included in the guidelines.

In order to achieve harmonisation and provide basic guidance on IQC, some types of analytical activity have been excluded from the guidelines. Issues specifically excluded are as follows:

- *Quality control of sampling.* While it is recognised that the quality of the analytical result can be no better than that of the sample, quality control of sampling is a separate subject and in many areas not yet fully developed. Moreover, in many instances analytical laboratories have no control over sampling practice and quality.
- *In-line analysis and continuous monitoring.* In this style of analysis there is no possibility of repeating the measurement, so the concept of IQC as used in the guidelines is inapplicable.
- *Multivariate IQC.* Multivariate methods in IQC are still the subject of research and cannot be regarded as sufficiently established for inclusion in the guidelines. The current document regards multianalyte data as requiring a series of univariate IQC tests. Caution is necessary in the interpretation of this type of data to avoid inappropriately frequent rejection of data.
- *Statutory and contractual requirements.*
- *Quality assurance measures* such as pre-analytical checks on instrumental stability, wavelength calibration, balance calibration, tests on resolution of chromatography columns, and problem diagnostics are not included. For present purposes they are regarded as part of the analytical protocol, and IQC tests their effectiveness together with the other aspects of the methodology.

Recommendations
The following recommendations represent integrated approaches to IQC that are suitable for many types of analysis and applications areas. Managers of laboratory quality systems will have to adapt the recommendations to the demands of their own particular requirements. Such adoption could be implemented, for example, by adjusting the number of duplicates and control material inserted into a run, or by the inclusion of any additional measures favoured in the particular application area. The procedure finally chosen and its accompanying decision rules must be

codified in an IQC protocol that is separate from the analytical system protocol. The practical approach to quality control is determined by the frequency with which the measurement is carried out and the size and nature of each run. The following recommendations are therefore made. (The use of control charts and decision rules are covered in Appendix 1 to the guidelines.)

In all of the following the order in the run in which the various materials are analysed should be randomised if possible. A failure to randomise may result in an underestimation of various components of error.

Short (e.g. $n < 20$) frequent runs of similar materials
Here the concentration range of the analyte in the run is relatively small, so a common value of standard deviation can be assumed. Insert a control material at least once per run. Plot either the individual values obtained, or the mean value, on an appropriate control chart. Analyse in duplicate at least half of the test materials, selected at random. Insert at least one blank determination.

Longer (e.g. $n > 20$) frequent runs of similar materials
Again a common level of standard deviation is assumed. Insert the control material at an approximate frequency of one per ten test materials. If the run size is likely to vary from run to run it is easier to standardise on a fixed number of insertions per run and plot the mean value on a control chart of means. Otherwise plot individual values. Analyse in duplicate a minimum of five test materials selected at random. Insert one blank determination per ten test materials.

Frequent runs containing similar materials but with a wide range of analyte concentration
Here we cannot assume that a single value of standard deviation is applicable. Insert control materials in total numbers approximately as recommended above. However, there should be at least two levels of analyte represented, one close to the median level of typical test materials, and the other approximately at the upper or lower decile as appropriate. Enter values for the two control materials on separate control charts. Duplicate a minimum of five test materials, and insert one procedural blank per ten test materials.

Ad hoc analysis
Here the concept of statistical control is not applicable. It is assumed, however, that the materials in the run are of a single type. Carry out duplicate analysis on all of the test materials. Carry out spiking or recovery tests or use a formulated control material, with an appropriate number of insertions (see above), and with different concentrations of analyte if appropriate. Carry out blank determinations. As no control limits are available, compare the bias and precision with fitness-for-purpose limits or other established criteria.

By following the above recommendations laboratories would introduce internal quality control measures which are an essential aspect of ensuring that data released from a laboratory are fit-for-purpose. If properly executed, quality control methods can monitor the various aspects of data quality on a run-by-run basis. In runs where performance falls outside acceptable limits, the data produced can be rejected and, after remedial action on the analytical system, the analysis can be repeated.

The guidelines stress, however, that internal quality control is not foolproof even when properly executed. Obviously it is subject to 'errors of both kinds', i.e. runs that are in control will occasionally be rejected and runs that are out of control occasionally accepted. Of more importance, IQC cannot usually identify sporadic gross errors or short-term disturbances in the analytical system that affect the results for individual test materials. Moreover, inferences based on IQC results are applicable only to test materials that fall within the scope of the analytical method validation. Despite these limitations, which professional experience and diligence can alleviate to a degree, internal quality control is the principal recourse available for ensuring that only data of appropriate quality are released from a laboratory. When properly executed it is very successful.

The guidelines also stress that the perfunctory execution of any quality system will not guarantee the production of data of adequate quality. The correct procedures for feedback, remedial action and staff motivation must also be documented and acted upon. In other words, there must be a genuine commitment to quality within a laboratory for an internal quality control programme to succeed, i.e. the IQC must be part of a complete quality management system.

5.5 Proficiency testing

Participation in proficiency testing schemes provides laboratories with an objective means of assessing and documenting the reliability of the data they are producing. Although there are several types of proficiency testing schemes they all share a common feature: test results obtained by one laboratory are compared with those obtained by one or more testing laboratories. The proficiency testing schemes must provide a transparent interpretation and assessment of results. Laboratories wishing to demonstrate their proficiency should seek and participate in proficiency testing schemes relevant to their area of work.

The need for laboratories carrying out analytical determinations to demonstrate that they are doing so competently has become paramount. It may well be necessary for such laboratories not only to become accredited and to use fully validated methods but also to participate successfully in proficiency testing schemes. Thus, proficiency testing has assumed a far greater importance than previously.

5.5.1 What is proficiency testing?
A proficiency testing scheme is defined as a system for objectively checking laboratory results by an external agency. It includes comparison of a

laboratory's results at intervals with those of other laboratories, the main object being the establishment of trueness. In addition, although various protocols for proficiency testing schemes have been produced the need now is for a harmonised protocol that will be universally accepted; the progress towards the preparation and adoption of an internationally recognised protocol is described below. Various terms have been used to describe schemes conforming to the protocol (e.g. external quality assessment, performance schemes, etc.), but the preferred term is 'proficiency testing'.

Proficiency testing schemes are based on the regular circulation of homogeneous samples by a co-ordinator, analysis of samples (normally by the laboratory's method of choice) and an assessment of the results. However, although many organisations carry out such schemes, there has been no international agreement on how this should be done – in contrast to the collaborative trial situation. In order to rectify this, the same international group that drew up collaborative trial protocols was invited to prepare one for proficiency schemes (the first meeting to do so was held in April 1989). Other organisations, such as CEN, are also expected to address the problem.

5.5.2 Why proficiency testing is important

Participation in proficiency testing schemes provides laboratories with a means of objectively assessing, and demonstrating, the reliability of the data they produce. Although there are several types of schemes, they all share a common feature of comparing test results obtained by one testing laboratory with those obtained by other testing laboratories. Schemes may be 'open' to any laboratory or participation may be invited. Schemes may set out to assess the competence of laboratories undertaking a very specific analysis (e.g. lead in blood) or more general analysis (e.g. food analysis). Although accreditation and proficiency testing are separate exercises, it is anticipated that accreditation assessments will increasingly use proficiency testing data.

5.5.3 Accreditation agencies

It is now recommended by ISO Guide 25,[8] the prime standard to which accreditation agencies operate, that such agencies require laboratories seeking accreditation to participate in an appropriate proficiency testing scheme before accreditation is gained. It should be recognised that in practice the ISO/IEC Guide 25 has been superseded by the ISO/IEC 17025:2000 Standard, even though it is the former Standard that is frequently referred to in official documents. There is now an internationally recognised protocol to which proficiency testing schemes should comply; this is the IUPAC/AOAC/ISO Harmonised Protocol described below.

5.5.4 ISO/IUPAC/AOAC International Harmonised Protocol For Proficiency Testing of (Chemical) Analytical Laboratories

The International Standardising Organisations, AOAC, ISO and IUPAC, have co-operated to produce an agreed 'International Harmonised Protocol For Proficiency Testing of (Chemical) Analytical Laboratories'.[11] That protocol is recognised within the food sector of the European Community and also by the Codex Alimentarius Commission. The protocol makes the following recommendations about the organisation of proficiency testing, all of which are important in the food sector.

Framework
Samples must be distributed regularly to participants who are to return results within a given time. The results will be statistically analysed by the organiser and participants will be notified of their performance. Advice will be available to poor performers and participants will be kept fully informed of the scheme's progress. Participants will be identified by code only, to preserve confidentiality. The scheme's structure for any one analyte or round in a series should be:

- samples prepared
- samples distributed regularly
- participants analyse samples and report results
- results analysed and performance assessed
- participants notified of their performance
- advice available for poor performers, on request
- co-ordinator reviews performance of scheme
- next round commences.

Organisation
The running of the scheme will be the responsibility of a co-ordinating laboratory/organisation. Sample preparation will either be contracted out or undertaken in house. The co-ordinating laboratory must be of high reputation in the type of analysis being tested. Overall management of the scheme should be in the hands of a small steering committee (Advisory Panel) having representatives from the co-ordinating laboratory (who should be practising laboratory scientists), contract laboratories (if any), appropriate professional bodies and ordinary participants.

Samples
The samples to be distributed must be generally similar in matrix to the unknown samples that are routinely analysed (in respect of matrix composition and analyte concentration range). It is essential they are of acceptable homogeneity and stability. The bulk material prepared must be effectively homogeneous so that all laboratories will receive samples that do not differ significantly in analyte concentration. The co-ordinating laboratory should also show the bulk sample is sufficiently stable to ensure it will not undergo

significant change throughout the duration of the proficiency test. Thus, prior to sample distribution, matrix and analyte stability must be determined by analysis after appropriate storage. Ideally, the quality checks on samples referred should be performed by a different laboratory from that which prepared the sample, although it is recognised that this would probably cause considerable difficulty to the co-ordinating laboratory. The number of samples to be distributed per round for each analyte should be no more than five.

Frequency of sample distribution
Sample distribution frequency in any one series should not be more than every two weeks and not less than every four months. A frequency greater than once every two weeks could lead to problems in turn-round of samples and results. If the period between distributions extends much beyond four months, there will be unacceptable delays in identifying analytical problems and the impact of the scheme on participants will be small. The frequency also relates to the field of application and amount of internal quality control that is required for that field. Thus, although the frequency range stated above should be adhered to, there may be circumstances where it is acceptable for a longer time scale between sample distribution, e.g. if sample throughput per annum is very low. Advice on this respect would be a function of the Advisory Panel.

Estimating the assigned value (the 'true' result)
There are a number of possible approaches to determining the nominally 'true' result for a sample but only three are normally considered. The result may be established from the amount of analyte added to the samples by the laboratory preparing the sample; alternatively, a 'reference' laboratory (or group of such expert laboratories) may be asked to measure the concentration of the analyte using definitive methods or thirdly, the results obtained by the participating laboratories (or a substantial sub-group of these) may be used as the basis for the nominal 'true' result. The organisers of the scheme should provide the participants with a clear statement giving the basis for the assignment of reference values which should take into account the views of the Advisory Panel.

Choice of analytical method
Participants can use the analytical method of their choice except when otherwise instructed to adopt a specified method. It is recommended that all methods should be properly validated before use. In situations where the analytical result is method-dependent the true value will be assessed using those results obtained using a defined procedure. If participants use a method that is not 'equivalent' to the defining method, then an automatic bias in result will occur when their performance is assessed.

Performance criteria
For each analyte in a round a criterion for the performance score may be set, against which the score obtained by a laboratory can be judged. A 'running

score' could be calculated to give an assessment of performance spread over a longer period of time.

Reporting results
Reports issued to participants should include data on the results from all laboratories together with participants' own performance score. The original results should be presented to enable participants to check correct data entry. Reports should be made available before the next sample distribution. Although all results should be reported, it may not be possible to do this in very extensive schemes (e.g. 800 participants determining 15 analytes in a round). Participants should, therefore, receive at least a clear report with the results of all laboratories in histogram form.

Liaison with participants
Participants should be provided with a detailed information pack on joining the scheme. Communication with participants should be by newsletter or annual report together with a periodic open meeting; participants should be advised of changes in scheme design. Advice should be available to poor performers. Feedback from laboratories should be encouraged so participants contribute to the scheme's development. Participants should view it as their scheme rather than one imposed by a distant bureaucracy.

Collusion and falsification of results
Collusion might take place between laboratories so that independent data are not submitted. Proficiency testing schemes should be designed to ensure that there is as little collusion and falsification as possible. For example, alternative samples could be distributed within a round. Also instructions should make it clear that collusion is contrary to professional scientific conduct and serves only to nullify the benefits of proficiency testing.

5.5.5 Statistical procedure for the analysis of results

The first stage in producing a score from a result x (a single measurement of analyte concentration in a test material) is to obtain an estimate of the bias, thus:

$$\text{bias} = x - X$$

where X is the true concentration or amount of analyte.

The efficacy of any proficiency test depends on using a reliable value for X. Several methods are available for establishing a working estimate of $\hat{X}$ (i.e. the assigned value).

Formation of a z-score
Most proficiency testing schemes compare bias with a standard error. An obvious approach is to form the z-score given by:

$$z = (x - \hat{X})/\sigma$$

where σ is a standard deviation. σ could be either an estimate of the actual variation encountered in a particular round ($\tilde{s}$) estimated from the laboratories' results after outlier elimination or a target representing the maximum allowed variation consistent with valid data.

A fixed target value for σ is preferable and can be arrived at in several ways. It could be fixed arbitrarily, with a value based on a perception of how laboratories should perform. It could be an estimate of the precision required for a specific task of data interpretation. σ could be derived from a model of precision, such as the 'Horwitz Curve'.[15] However, while this model provides a general picture of reproducibility, substantial deviation from it may be experienced for particular methods.

Interpretation of z-scores
If $\hat{X}$ and σ are good estimates of the population mean and standard deviation then z will be approximately normally distributed with a mean of zero and unit standard deviation. An analytical result is described as 'well behaved' when it complies with this condition.

An absolute value of z ($|z|$) greater than three suggests poor performance in terms of accuracy. This judgement depends on the assumption of the normal distribution, which, outliers apart, seems to be justified in practice.

As z is standardised, it is comparable for all analytes and methods. Thus values of z can be combined to give a composite score for a laboratory in one round of a proficiency test. The z-scores can therefore be interpreted as follows:

$|z| < 2$ 'Satisfactory': will occur in 95% of cases produced by 'well behaved results'

$2 < |z| < 3$ 'Questionable': but will occur in $\approx$5% of cases produced by 'well behaved results'

$|z| > 3$ 'Unsatisfactory': will only occur in $\approx$0.1% of cases produced by 'well behaved results'.

Combination of results within a round of the trial
There are several methods of combining the z-scores produced by a laboratory in one round of the proficiency test described in the Protocol. They are:

The sum of scores, $SZ = \Sigma z$

The sum of squared scores, $SSZ = \Sigma z^2$

The sum of absolute values of the scores, $SAZ = \Sigma |z|$

All should be used with caution, however. It is the individual z-scores that are the critical consideration when considering the proficiency of a laboratory.

Calculation of running scores
Similar considerations apply for running scores as apply to combination scores above.

5.6 Analytical methods

Analytical methods should be validated as fit-for-purpose before use by a laboratory. Laboratories should ensure that, as a minimum, the methods they use are fully documented, laboratory staff trained in their use and control mechanisms established to ensure that the procedures are under statistical control.

The development of methods of analysis for incorporation into International Standards or into foodstuff legislation was, until comparatively recently, not systematic. However, the EU and Codex have requirements regarding methods of analysis and these are outlined below. They are followed by other International Standardising Organisations (e.g. AOAC International (AOACI) and the European Committee for Standardization (CEN)).

5.6.1 Codex Alimentarius Commission

This was the first international organisation working at the government level in the food sector that laid down principles for the establishment of its methods. That it was necessary for such guidelines and principles to be laid down reflects the confused and unsatisfactory situation in the development of legislative methods of analysis that existed until the early 1980s in the food sector.

The 'Principles for the Establishment of Codex Methods of Analysis'[16] are given below; other organisations which subsequently laid down procedures for the development of methods of analysis in their particular sector followed these principles to a significant degree. They require that preference should be given to methods of analysis the reliability of which have been established in respect of the following criteria, selected as appropriate:

- specificity
- accuracy
- precision; repeatability intra-laboratory (within laboratory), reproducibility inter-laboratory (within laboratory and between laboratories)
- limit of detection
- sensitivity
- practicability and applicability under normal laboratory conditions
- other criteria which may be selected as required.

5.6.2 The European Union

The Union is attempting to harmonise sampling and analysis procedures in an attempt to meet the current demands of the national and international enforcement agencies and the likely increased problems that the open market will bring. To aid this the Union issued a Directive on Sampling and Methods of Analysis.[7] The Directive contains a technical annex, in which the need to carry out a collaborative trial before it can be adopted by the Community is emphasised.

The criteria to which Community methods of analysis for foodstuffs should now conform are as stringent as those recommended by any international organisation following adoption of the Directive. The requirements follow those described for Codex above, and are given in the Annex to the Directive. They are:

1. Methods of analysis which are to be considered for adoption under the provisions of the Directive shall be examined with respect to the following criteria:
   (i) specificity
   (ii) accuracy
   (iii) precision; repeatability intra-laboratory (within laboratory), reproducibility inter-laboratory (within laboratory and between laboratories)
   (iv) limit of detection
   (v) sensitivity
   (vi) practicability and applicability under normal laboratory conditions
   (vii) other criteria which may be selected as required.
2. The precision values referred to in 1 (iii) shall be obtained from a collaborative trial which has been conducted in accordance with an internationally recognised protocol on collaborative trials (e.g. International Organisation of Standardization 'Precision of Test Methods').[17] The repeatability and reproducibility values shall be expressed in an internationally recognised form (e.g. the 95% confidence intervals as defined by ISO 5725/1981). The results from the collaborative trial shall be published or be freely available.
3. Methods of analysis which are applicable uniformly to various groups of commodities should be given preference over methods which apply to individual commodities.
4. Methods of analysis adopted under this Directive should be edited in the standard layout for methods of analysis recommended by the International Organisations for Standardization.

5.6.3 Other organisations

There are other international standardising organisations, most notably the European Committee for Standardization (CEN) and AOACI, which follow similar requirements. Although CEN methods are not prescribed by legislation, the European Commission places considerable importance on the work that CEN carries out in the development of specific methods in the food sector; CEN has been given direct mandates by the Commission to publish particular methods, e.g. those for the detection of food irradiation. Because of this some of the methods in the food sector being developed by CEN are described below. CEN, like the other organisations described above, has adopted a set of guidelines to which its Methods Technical Committees should conform when developing a method of analysis. The guidelines are:

> Details of the interlaboratory test on the precision of the method are to be summarised in an annex to the method. It is to be stated that the values derived from the interlaboratory test may not be applicable to analyte concentration ranges and matrices other than given in annex.
> The precision clauses shall be worded as follows:

Repeatability: The absolute difference between two single test results found on identical test materials by one operator using the same apparatus within the shortest feasible time interval will exceed the repeatability value r in not more than 5% of the cases.
The value(s) is (are): ...

Reproducibility: The absolute difference between two single test results on identical test material reported by two laboratories will exceed the reproducibility, R, in not more than 5% of the cases.
   The value(s) is (are): ...

> There shall be minimum requirements regarding the information to be given in an Informative Annex, this being:

> Year of interlaboratory test and reference to the test report (if available)
> Number of samples
> Number of laboratories retained after eliminating outliers
> Number of outliers (laboratories)
> Number of accepted results
> Mean value (with the respective unit)
> Repeatability standard deviation (s_r) (with the respective unit)
> Repeatability relative standard deviation (RSD_r) (%)
> Repeatability limit (r)w (with the respective units)
> Reproducibility relative standard deviation (s_R) (with the respective unit)
> Reproducibility relative standard deviation (RSD_R) (%)
> Reproducibility (R) (with the respective unit)
> Sample types clearly described
> Notes if further information is to be given.

98   Food chemical safety

5.6.4   Requirements of official bodies
Consideration of the above requirements confirms that in future all methods must be fully validated if at all possible, i.e. have been subjected to a collaborative trial conforming to an internationally recognised protocol. In addition this, as described above, is now a legislative requirement in the food sector of the European Union. The concept of the valid analytical method in the food sector, and its requirements, is described below.

5.6.5   Requirements for valid methods of analysis
It would be simple to say that any new method should be fully tested for the criteria given above. However, the most 'difficult' of these is obtaining the accuracy and precision performance criteria.

Accuracy
Accuracy is defined as the closeness of the agreement between the result of a measurement and a true value of the measureand.[18] It may be assessed with the use of reference materials. However, in food analysis, there is a particular problem.

In many instances, though not normally for food additives and contaminants, the numerical value of a characteristic (or criterion) in a Standard is dependent on the procedures used to ascertain its value. This illustrates the need for the (sampling and) analysis provisions in a Standard to be developed at the same time as the numerical value of the characteristics in the Standard are negotiated to ensure that the characteristics are related to the methodological procedures prescribed.

Precision
Precision is defined as the closeness of agreement between independent test results obtained under prescribed conditions.[19] In a standard method the precision characteristics are obtained from a properly organised collaborative trial, i.e. a trial conforming to the requirements of an International Standard (the AOAC/ISO/IUPAC Harmonised Protocol or the ISO 5725 Standard). Because of the importance of collaborative trials, and the resource that is now being devoted to the assessment of precision characteristics of analytical methods before their acceptance, they are described in detail below.

Collaborative trials
As seen above, all 'official' methods of analysis are required to include precision data. These may be obtained by subjecting the method to a collaborative trial conforming to an internationally agreed protocol. A collaborative trial is a procedure whereby the precision of a method of analysis may be assessed and quantified. The precision of a method is usually expressed in terms of repeatability and reproducibility values. Accuracy is not the objective.

Recently there has been progress towards a universal acceptance of collaboratively tested methods and collaborative trial results and methods, no matter by whom these trials are organised. This has been aided by the publication of the IUPAC/ISO/AOAC Harmonisation Protocol on Collaborative Studies.[14] That Protocol was developed under the auspices of the International Union of Pure and Applied Chemists (IUPAC) aided by representatives from the major organisations interested in conducting collaborative studies. In particular, from the food sector, the AOAC International, the International Organisation for Standardisation (ISO), the International Dairy Federation (IDF), the Collaborative International Analytical Council for Pesticides (CIPAC), the Nordic Analytical Committee (NMKL), the Codex Committee on Methods of Analysis and Sampling and the International Office of Cocoa and Chocolate were involved. The Protocol gives a series of 11 recommendations dealing with:

- the components that make up a collaborative trial
- participants
- sample type
- sample homogeneity
- sample plan
- the method(s) to be tested
- pilot study/pre-trial
- the trial proper.

5.6.6 Statistical analysis

It is important to appreciate that the statistical significance of the results is wholly dependent on the quality of the data obtained from the trial. Data that contain obvious gross errors should be removed prior to statistical analysis. It is essential that participants inform the trial co-ordinator of any gross error that they know has occurred during the analysis and also if any deviation from the method as written has taken place. The statistical parameters calculated and the outlier tests performed are those used in the internationally agreed Protocol for the Design, Conduct and Interpretation of Collaborative Studies.[14]

5.7 Standardised methods of analysis for additives

There are many organisations that publish standardised methods of analysis for additives, such methods normally having been validated through a collaborative trial organised to conform to one of the internationally accepted protocols described previously. Such organisations will include AOACI, the European Organisation for Standardisation (CEN) and the Nordic Committee for Food Analysis (NMKL). Within Europe, the most important of these international standardising organisations is probably CEN. CEN has a technical committee dealing with horizontal methods of analysis in which both additive and contaminant methods of analysis are discussed (TC 275). The methods of analysis for additives within its work programme are outlined below. This is

100 Food chemical safety

given by Working Group. The titles under the Working Group heading refer to the work item (topic area) of that Working Group. The Working Groups not listed (e.g. 3, 4 etc.) are concerned with contaminant methods of analysis and have been discussed previously.[20]

Work programme of CEN TC 275 Working Group 1: Sulfites
Published standards
1988-1:1998   Determination of sulfite – Part 1: Optimized Monier Williams procedure
1988-2:1998   Determination of sulfite – Part 2: Enzymatic method

Work programme of CEN TC 275 Working Group 2: Intense Sweeteners
Published standards
1376:1996   Determination of saccharin in table top preparations – Spectrometric method
1377:1996   Determination of acesulfame-K in table top preparations – Spectrometric method
1378:1996   Determination of aspartame in table top preparations – High performance liquid chromatographic method
1379:1996   Determination of cyclamate and saccharin in table top preparations – High performance liquid chromatographic method
12856:1999 Determination of acesulfame-K, aspartame and saccharin – High performance liquid chromatographic method
12857:1999 Determination of cyclamate – High performance liquid chromatographic method

On-going activities
Determination of neohesperidine DHC by HPLC
Determination of sugar alcohols

Work programme of CEN TC 275 Working Group 7: Nitrates
Published standards
12014-1:1997/A1:1999   Determination of nitrate and/or nitrite content – Part 1: General considerations
12014-2:1997   Part 2: HPLC/IC method for the determination of nitrate content of vegetables and vegetable products
12014-3:1998 ENV   Part 3: Spectrometric determination of nitrate and nitrite and nitrite content of meat products after enzymatic reduction of nitrate or nitrite
12014-4:1998 ENV   Part 4: IC method for the determination of nitrate and nitrite of meat and meat products
12014-5:1998   Part 5: Enzymatic determination of nitrate content of vegetable-containing food for babies and infants
12014-7:1998   Part 7: Continuous flow method for the

Analytical methods: quality control and selection 101

determination of nitrate content of vegetables and vegetable products after cadmium reduction

Work programme of CEN TC 275 Working Group 8: Food irradiation
Published standards
1786:1996	Detection of irradiated food containing bone – Method by ESR-spectroscopy
1787:2000	Detection of irradiated food containing cellulose – Method by ESR-spectroscopy
1788:1996	Detection of irradiated food from which silicate minerals can be isolated – Method by thermoluminescence
1784:1996	Detection of irradiated food containing fat – Gas chromatographic analysis of hydrocarbons
1785:1996	Detection of irradiated food containing fat – Gas chromatographic/mass spectrometric analysis of 2 alkyl cyclobutanones

On-going activities
Detection of irradiated food containing crystalline sugar by ESR-spectroscopy
Thermoluminescence detection of irradiated food from which silicate minerals can be isolated
DNA Comet assay screening for the detection of irradiated foodstuffs
Detection of irradiated food using photostimulated luminescence
Microbiological screening for irradiated food using LAL/GNB procedures
Microbiological screening for irradiated food using DEFT/APC

Work programme of CEN TC 275 Working Group 9: Vitamins
Published standards
12823-1:2000	Determination of vitamin A by HPLC: Part 1: Measurement of all-trans and 13 cis retinol
12823-2:2000	Determination of vitamin A by HPLC: Part 2: Measurement of β-carotene
12821:2000	Determination of vitamin D by HPLC – Measurement of cholecalciferol (D_3) and ergocalciferol (D_2)
12822:2000	Determination of vitamin E by HPLC – Measurement of α-, β-, γ- and δ-tocopherols

On-going activities
Determination of vitamin B_1
Determination of vitamin B_2
Determination of vitamin B_6 by HPLC
Determination of vitamin B_6 by microbiological assay
Determination of folates
Determination of vitamin C
Determination of vitamin K_1

5.8 The future direction for methods of analysis

There is current discussion on an international basis whereby the present procedure by which specific methods of analysis are incorporated into legislation are replaced by one in which method performance characteristics are specified. This is because by specifying a single method:

- the analyst is denied freedom of choice and thus may be required to use an inappropriate method in some situations
- the procedure inhibits the use of automation
- it is administratively difficult to change a method found to be unsatisfactory or inferior to another currently available.

As a result the use of an alternative approach whereby a defined set of criteria to which methods should comply without specifically endorsing specific methods is being considered and slowly adopted in some sectors of food analysis. This approach will have a considerable impact on the food analytical laboratory. There are a number of issues that are of concern to the food analytical community of which analysts should be aware. These are outlined briefly below.

5.8.1 Measurement uncertainty

It is increasingly being recognised both by laboratories and the customers of laboratories that any reported analytical result is an estimate only and the 'true value' will lie within a range around the reported result. The extent of the range for any analytical result may be derived in a number of different ways, e.g. using the results from method validation studies or determining the inherent variation through different components within the method, i.e. estimating these variances as standard deviations and developing an overall standard deviation for the method. There is some concern within the food analytical community as to the most appropriate way to estimate this variability.

5.8.2 Single laboratory method validation

There is concern in the food analytical community that although methods should ideally be validated by a collaborative trial, this is not always feasible for economic or practical reasons. As a result, IUPAC guidelines are being developed for single laboratory method validation to give information to analysts on the acceptable procedure in this area. These guidelines should be finalised by the end of 2001.

5.8.3 Recovery

It is possible to determine the recovery that is obtained during an analytical run. Internationally harmonised guidelines have been prepared which indicate how recovery information should be handled. This is a contentious area amongst analytical chemists because some countries of the organisations require analytical methods to be corrected for recovery, whereas others do not. Food analysts should recognise that this issue has been addressed on an international basis.

5.9 References

1. EUROPEAN UNION, *Council Directive 89/397/EEC on the Official Control of Foodstuffs*, O.J. L186 of 30.6.1989.
2. EUROPEAN UNION, *Council Directive 93/99/EEC on the Subject of Additional Measures Concerning the Official Control of Foodstuffs*, O.J. L290 of 24.11.1993.
3. EUROPEAN COMMITTEE FOR STANDARDIZATION, *General Criteria for the Operation of Testing Laboratories – European Standard EN 45001*, Brussels, CEN/CENELEC, 1989.
4. ORGANISATION FOR ECONOMIC CO-OPERATION AND DEVELOPMENT, *Decision of the Council of the OECD of 12 Mar 1981 Concerning the Mutual Acceptance of Data in the Assessment of Chemicals*, Paris, OECD, 1981.
5. EUROPEAN COMMITTEE FOR STANDARDIZATION, *General Criteria for the Assessment of Testing Laboratories – European Standard EN45002*, Brussels, CEN/CENELEC, 1989.
6. EUROPEAN COMMITTEE FOR STANDARDIZATION, *General Criteria for Laboratory Accreditation Bodies – European Standard EN45003*, Brussels, CEN/CENELEC, 1989.
7. EURPEAN UNION, *Council Directive 85/591/EEC Concerning the Introduction of Community Methods of Sampling and Analysis for the Monitoring of Foodstuffs Intended for Human Consumption*, O.J. L372 of 31.12.1985.
8. INTERNATIONAL ORGANIZATION FOR STANDARDIZATION, *General Requirements for the Competence of Calibration and Testing Laboratories ISO/IEC 17025*, Geneva, ISO, 1999.
9. CODEX ALIMENTARIUS COMMISSION, *Report of the 21st Session of the Codex Committee on Methods of Analysis and Sampling – ALINORM 97/23A*, Rome, FAO, 1997.
10. CODEX ALIMENTARIUS COMMISSION, *Report of the 22nd Session of the Codex Alimentarius Commission – ALINORM 97/37*, Rome, FAO, 1997.
11. INTERNATIONAL UNION OF PURE AND APPLIED CHEMISTRY, *The International Harmonised Protocol for the Proficiency Testing of (Chemical) Analytical Laboratories*, ed. Thompson M and Wood R, *Pure Appl. Chem.*, 1993 65 2123–2144 (also published in *J. AOAC International*, 1993 76 926–940).
12. INTERNATIONAL UNION OF PURE AND APPLIED CHEMISTRY, *Guidelines on Internal Quality Control in Analytical Chemistry Laboratories*, ed. Thompson M and Wood R, *Pure Appl. Chem.*, 1995 67 649–666.

13. INTERNATIONAL ORGANIZATION FOR STANDARDIZATION, *Calibration and Testing Laboratory Accreditation Systems – General Requirements for Operation and Recognition – ISO/IEC Guide 58*, Geneva, ISO, 1993.
14. HORWITZ W, 'Protocol for the Design, Conduct and Interpretation of Method Performance Studies', *Pure Appl. Chem*, 1988 60 855–864 (revision published 1995).
15. HORWITZ W, 'Evaluation of Analytical Methods for Regulation of Foods and Drugs', *Anal. Chem.*, 1982 54 67A–76A.
16. CODEX ALIMENTARIUS COMMISSION, *Procedural Manual of the Codex Alimentarius Commission – Tenth Edition,* Rome, FAO, 1997.
17. INTERNATIONAL ORGANIZATION FOR STANDARDIZATION, *Precision of Test Methods – Standard 5725*, Geneva, ISO, 1981 (revised 1986 with further revision in preparation).
18. INTERNATIONAL ORGANIZATION FOR STANDARDIZATION, *International Vocabulary for Basic and General Terms in Metrology – 2nd Edition*, Geneva, ISO, 1993.
19. INTERNATIONAL ORGANIZATION FOR STANDARDIZATION, *Terms and Definitions used in Connections with Reference Materials – ISO Guide 30*, Geneva, ISO, 1992.
20. WOOD R, 'Analytical methods: quality control and selection', Chapter 3 in Watson, D.H. (ed.) *Food Chemical Safety, Volume 1: Contaminants*, Cambridge, Woodhead Publishing, pp. 57–61.

Appendix: Information for potential contractors on the analytical quality assurance requirements for food chemical surveillance exercises

Introduction

The FSA undertakes surveillance exercises, the data for which are acquired from analytical determinations. The Agency will take measures to ensure that the analytical data produced by contractors are sufficient with respect to analytical quality, i.e. that the results obtained meet predetermined analytical quality requirements such as fitness-for-purpose, accuracy and reliability. Thus when inviting tenders FSA will ask potential contractors to provide information regarding the performance requirements of the methods to be used in the exercise, e.g. limit of detection, accuracy, precision etc., and the quality assurance measures used in their laboratories. When presenting tenders, laboratories should confirm how they comply with these specifications and give the principles of the methods to be used. These requirements extend both to the laboratory as a whole and to the specific analytical determinations being required in the surveillance exercise. The requirements are described in three parts, namely:

Analytical methods: quality control and selection

- Part A: Quality assurance requirements for surveillance projects provided by potential contractors at the time ROAMEs are completed and when commissioning a survey
- Part B: Information to be defined by the FSA customer once the contract has been awarded – to be agreed with contractor
- Part C: Information to be provided by the contractor on an on-going basis once contract is awarded – to be agreed with the customer.

Each of these considerations is addressed in detail below. Potential contractors are asked to provide the information requested in Part A of this document when submitting ROAME forms in order to aid the assessment of the relative merits of each project from the analytical/data quality point of view. This information is best supplied in tabular form, for example that outlined in Part A, but may be provided in another format if thought appropriate. The tables should be expanded as necessary. Parts B and C should not be completed when submitting completed ROAME forms.

Explanation of Parts A, B and C of Document
Part A
Part A describes the information that is to be provided by potential contractors at the time that the ROAME Bs are completed for submission to the Group. Provision of this information will permit any FSA 'Analytical Group' and customers to make an informed assessment and comparison of the analytical quality of the results that will be obtained from the potential contractors bidding for the project. Previously potential contractors have not been given defined guidance on the analytical quality assurance information required of them and this has made comparison between potential contractors difficult. Part A is supplied to potential contractors at the same time as further information about the project is supplied.

The list has been constructed on the premise that contractors will use methods of analysis that are appropriate and accredited by a third party (normally UKAS), participate in and achieve satisfactory results in proficiency testing schemes and use formal internal quality control procedures. In addition, Parts B and C are made available to the potential contractors so that they are aware of what other demands will be made of them and can build the costs of providing the information into their bids.

Part B
This section defines the analytical considerations that must be addressed by both the customer and contractor before the exercise commences. Not all aspects may be relevant for all surveys, but each should be considered for relevancy. Agreement will signify a considerable understanding of both the analytical quality required and the significance of the results

obtained.

Part C
This section outlines the information that must be provided by the contractor to a customer on an on-going basis throughout the project. The most critical aspect is the provision of Internal Quality Control (IQC) control charts thus ensuring that the customer has confidence that the contractor is in 'analytical control'.

By following the above the FSA customers will have confidence that the systems are in place in contractors with respect to analytical control and that they are being respected. It is appreciated that not all aspects outlined in Parts A, B and C will be appropriate for every contract but all should be at least considered as to their appropriateness.

Contents of Parts A, B and C of document
Part A
Potential contractors should provide the information requested below. Please provide the information requested either in section 1 or in section 2 and then that in sections 3 to 5.

Section 1: Formal quality system in the laboratory if third party assessed (i.e. if UKAS accredited or GLP compliant)
Please describe the quality system in your laboratory by addressing the following aspects:

- To which scheme is your laboratory accredited or GLP compliant?
- Please describe the scope of accreditation, by addressing:
    1. the area that is accredited
    2. for which matrices, and
    3. for which analytes
  or supply a copy of your accreditation schedule.
- Do you foresee any situation whereby you will lose accreditation status due to matters outside your immediate control, e.g. closure of the laboratory?

Section 2: Quality system if not accredited
Please describe the quality system in your laboratory by addressing the following aspects:

- Laboratory Organisation:
    1. Management/supervision
    2. Structure and organisational chart
    3. Job descriptions if appropriate.
- Staff:

1. Qualifications
2. Training records
3. Monitoring of the analytical competency of individual staff members.
- Documentation
  1. General lab procedures
  2. Methods to be used (adequate/detailed enough to control consistent approach).
- Sample Preparation
  1. Location
  2. Documented procedures
  3. Homogenisation
  4. Sub-sampling
  5. Sample identification
  6. Cross-contamination risk
  7. Special requirements.
- Equipment Calibration
  1. Frequency
  2. Who
  3. Records
  4. Marking.
- Traceability
  1. Who did what/when
  2. Equipment – balances etc.
  3. Sample storage/temperature
  4. Calibration solutions: how prepared and stored.
- Results/Reports
  1. Calculation checks
  2. Typographic checks
  3. Security/confidentiality of data
  4. Software usage/control
  5. Job title of approved signatory.
- Laboratory Information Management System

Please outline the system employed.

- Internal Audits
  1. Audit plan
  2. Frequency
  3. Who carries out the audit?
  4. Are internal audit reports available?
  5. What are the non-compliance follow-up procedures?
- Sub-contracting
  1. In what circumstances is sub-contracting carried out?
  2. How is such sub-contracting controlled and audited?

Section 3: Proficiency testing

Please describe the arrangements for external proficiency testing in your laboratory by addressing the following aspects:

- Do you participate in proficiency testing schemes? If so, which schemes?
- Which analyte/matrices of the above schemes do you participate in?
- What are your individual proficiency scores and their classification, (e.g. z-scores or equivalent), over the past two years, for the analyte/matrices of relevance to this proposal?
- What remedial action do you take if you should get unsatisfactory results?

Section 4: Internal quality control
Please describe the IQC measures adopted in your laboratory by addressing the following:

- What control samples do you use in an analytical run?
- Do you follow the Harmonised Guidelines?[1]
- What IQC procedures are in place?
- Do you use Certified Reference Materials (CRMs), and if so, how? For example, specify the concentration(s) matrix type(s) etc.
- Which appropriate CRMs do you use?
- Do you use In-House Reference Materials (IHRM) and how are they obtained? For example, specify the concentration(s) matrix type(s).
- Are they traceable? For example, to CRM, a reference method, inter-laboratory comparison, or other.
- What criteria do you have regarding reagent blanks?
- What action/warning limits are applied for control charts?
- What action do you take if the limits are exceeded?
- Do you check new control materials and calibration standards? If so, how?
- Can we see the audit of previous results – what actions have been taken or trends observed?
- Do you make use of duplicate data as an IQC procedure?
- How frequently are control materials (CRMs, blanks, IHRM etc.) incorporated in the analytical run?
- Do you randomise your samples in an analytical run? (including duplicates).

Section 5: Method validation
Please describe the characteristics of the method of analysis you propose to use in the survey by addressing the following:

- What methods do you have to cover the matrix and analyte combinations required?
- Do you routinely use the method?
- Is the method accredited?

---

[1] 'Guidelines on Internal Quality Control in Analytical Chemistry Laboratories', ed. M. Thompson and R. Wood, *Pure Appl. Chem.*, 1995, **67**, 649–66.

- Has the method been validated by collaborative trial (i.e. externally)?
- Has the method been validated through any In-House Protocol?
- Is it a Standard (i.e. published in the literature or by a Standards Organisation) Method?
- Please identify the performance characteristics of the methods, i.e.
    1. LOQ
    2. LOD
    3. Blanks
    4. Precision values over the relevant concentration range expressed as relative standard deviations
    5. Bias and recovery characteristics including relevant information on traceability.
- Do you estimate measurement uncertainty/reliability?
- Do you normally give a measurement uncertainty/reliability when reporting results to your customer?

Part B

The FSA customer is to consider and then define the following in consultation with the contractor:

1. What analysis is required for what matrices.
2. The sample storage conditions to be used. Are stability checks for specific analytes undertaken?
3. The methods to be used and a copy of Standard Operating Procedures (SOPs) where accredited, including any sampling and sample preparation protocols, to be supplied to the customer.
4. The IQC procedures to be used. In particular the following should be considered:
    - the use of the International Harmonised Guidelines for IQC
    - the use of control charts
    - randomisation within the run
    - the composition of the analytical run (e.g. the number of control samples, and in particular the number of blanks, spikes, IHRMs etc.)
    - the reference materials to be used
    - the determination of recoveries on each batch using procedures as described in the International Harmonised Guidelines with all results to be corrected for recovery except where otherwise specified (i.e. for pesticides) and for the recovery data quoted to be reported.
5. The measurement limits (i.e. limit of detection (LOD): limit of determination/quantification (LOQ), and reporting limits, etc.).
6. The maximum acceptable measurement reliability (also known as measurement uncertainty) for each analytical result.
7. The treatment of individual results with respect to uncertainty, reliability, i.e. as
    (a) $x \pm y\,\mu g/kg$ where y is the measurement reliability (i.e. as if the sample

were to be a 'historic' surveillance result), or
(b) not less/more than $x\,\mu g/kg$ where x is the analytical result determined less the measurement reliability (i.e. as if the sample were to be an 'enforcement style' result) when assessing compliance with a (maximum) limit.
8. Whether there are to be action limits whereby the customer is immediately notified of 'abnormal' results.
9. The procedures to be used for confirmation of 'abnormal' results, e.g. those that exceed any defined statutory limit. The procedures to be used if qualitative analysis is to be undertaken.
10. The consistent way of expressing results, e.g. (a) on a wet (as is) basis, on a dry weight basis or on a fat weight basis, and (b) the reporting units for specific analytes to be used throughout survey (i.e. mg/kg etc.).
11. The time interval for customer visits (e.g. once every three months, or as otherwise appropriate) and for submission of control charts.
12. Whether there are any possibilities of developing integrated databases between customer and major contractors. If not, the customer to provide reporting guidelines.
13. The procedure for logging in of samples and traceability of sample in the laboratory.
14. The security of samples within the laboratory.

Part C

The following are to be provided by the contractor on an on-going basis throughout the contract to confirm that the contractor remains in 'analytical control'.
1. Copies of the control charts and duplicate value control charts or other agreed measures to monitor IQC.
2. Records of action taken to remedy out-of-control situations to be provided at the same time with control charts.
3. Where action limits have been identified in Part B (see para. 8), the results of samples that exceed the action limits are to be sent to the customer as soon as available.
4. Any relevant proficiency testing scheme results obtained during the course of the survey.

6
New methods in detecting food additives
C. J. Blake, Nestlé Research Centre, Lausanne

6.1 Introduction

Food additives include a multitude of compounds for stabilising, colouring, sweetening, thickening, emulsifying, preserving, and for adjusting pH or buffering food products and as antioxidants. Salt, glutamates and other nucleotides are added to improve the flavour of food products. Mineral salts, vitamins and related compounds are used as nutrients to fortify food products, but in some cases are also added as food additives, for example riboflavin and carotenes may be used as food colours. The food analyst faces a complex task to identify and quantify this enormous range of compounds. An overview is presented of current methods of analysis in a rapidly changing field with some pointers to emerging techniques for the future.

In the food industry analytical methods can be divided into two classes:

1. *Reference and research methods.* These are generally more sophisticated procedures that are used by central quality control laboratories or by government agencies with qualified personnel. These methods are often used to verify results obtained by the 'rapid' methods described below or to calibrate the instruments. They may also be used to verify additives declared on the label of a food product or to check for the use of non-permitted additives. This aspect is of increasing importance since legislation may vary between different countries, and food products are frequently subject to inter-market 'exports'. The most important characteristics of these analytical procedures are:

    - high selectivity and specificity
    - good precision and accuracy

- good limits of quantification
- robust
- multi-analyte if possible.

Many of the more established techniques have been validated through collaborative studies which becomes of greater importance as laboratories seek to become accredited via ISO, EN or related systems where the use of official or well validated methods is mandatory. New instrumental techniques are constantly being reported in the literature but it often requires many years before procedures are introduced, validated and then applied within the food industry. Recent techniques that can be included in this category are capillary electrophoresis and liquid chromatography-mass spectrometry (LC-MS). In time procedures based on these techniques will also become accepted as routine methods and are likely to be adopted by some of the official international bodies like the AOAC International, CEN, ISO, etc.

2. *Rapid or alternative methods.* These techniques are used increasingly to control concentrations of additives in raw materials and finished products and should have some of the following characteristics:

- preferably utilisable by non-skilled operator
- reasonably accurate results
- reduced measurement time
- lower method complexity
- reduction of human influence (errors) on results
- low cost equipment or low-operating cost
- potential for on-line or in-line automation.

6.2 Reference or research methods

The analytical techniques used for additives analysis are reviewed below. They are mainly chromatographic but enzymatic, flow injection analysis, inductively coupled plasma–atomic emission spectrometry and atomic absorption methods are also used.

6.2.1 High performance liquid chromatography

High performance liquid chromatography (HPLC) is the most widely used chromatographic method for analysis of additives in food products. Liquid chromatography-mass spectrometry (LC-MS) and LC-MS/MS are also beginning to be used for analysis of additives. Numerous books and reviews cover the principles of these techniques, for example Nollet (1996a,b), Nielsen (1998) and Nolle (2000). In recent years column separation technology, particularly for reversed phase HPLC columns, has markedly improved enabling more reproducible and robust separations. The range of detection methods has also been improved with the development of diode-array-detection (DAD), post-column reaction with fluorescence detection, evaporative light scattering detection (ELSD) and biosensors. HPLC is usually applied to the analysis of classes of additives and sometimes for several types of additives in certain food products. Applications of HPLC are described below according to the type of additive.

The analysis of synthetic and natural colours was reviewed by Maslowska (1996) and Cscerháti and Forgács (1999), while Glória (2000) has reviewed comprehensively the determination of synthetic colours. With respect to synthetic colours or coal-tar dyes, little harmonisation of legislation has occurred and different colours are permitted in individual countries. Several hundred dyes are known and problems may arise in identifying non-permitted and sometimes unknown compounds on a chromatogram. Therefore methods capable of extracting, identifying and quantifying many colours simultaneously are necessary to verify compliance with local regulations. The choice of sample preparation for extraction of synthetic colours depends on the type of food sample. Liquid samples such as beverages can be diluted in water and filtered. Carbonated beverages are degassed. Water-soluble products like dehydrated drinks, jellies, jams and sweets are dissolved in warm water and then filtered. Incomplete extraction may occur for some food colours due to the binding of the dyes to proteins, lipids and carbohydrates (Glória, 2000). Various pretreatments using enzymes such as papain, lipase, amyloglucosidase, cellulase, pectinase and phospholipase can be used to release synthetic colours from the food matrix. Petroleum ether can be used to remove lipids. Synthetic dyes may be isolated or purified by column chromatography with polyamide, ion-pair or solvent extraction or by solid phase extraction (SPE) with C_{18} cartridges. The main difficulty in applying these procedures is to ensure a complete extraction without degradation of the various colours. Neier and Matissek (1998) reported recoveries of 60–100% for various blue, yellow and red colours from spiked gum sweets and marzipan. Separation of the various colours is normally made by reversed phase chromatography with ion-pairing (Berzas et al., 1997; NMKL, 1989) or with acetonitrile/phosphate buffer pH 5.5 (Neier and Matissek, 1998). UV detection or DAD set at the optimum wavelength for each colour can be used. Certain colours like Quinoline Yellow or Brilliant Blue FCF may contain several components (Fig. 6.1) which complicates interpretation of the chromatograms obtained. A validated method for the determination of some synthetic colours by ion-pairing HPLC has been published by the NMKL (1989). It includes the following colours: Tartrazine, Quinoline Yellow, Sunset Yellow FCF, Amaranth, Ponceau 4R, Erythrosine, Patent Blue V, Indigotine, Brilliant Black PN, Black 7984, Fast Green FCF and Blue VRS.

Natural colours include: annatto, anthocyanins, beetroot red (betalaines), caramel, carotenoids, cochineal and lac pigments, flavanoids, chlorophylls and tumeric. There is a trend towards encapsulating natural colours for food use, but this is not yet reflected in the extraction techniques described in the published analytical methods. Lancaster and Lawrence (1996) described the extraction and

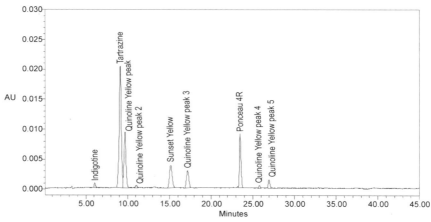

Fig. 6.1 Separation of colours extracted from a dry dog food (poultry, cereals, green vegetable and carrots).

chromatographic separation of the active principles of cochineal (carminic acid) and annatto (α-bixin, β-bixin, α-norbixin and β-norbixin) from fruit beverages, candy and yoghurt. Coffrey et al. (1997) reported a method for estimating the concentration of Caramel type III added to foods. The use of carotenoids as natural colorants is permitted within the European Union and the United States. Sample preparation methods may involve saponification and solvent extraction (Schüep and Schierle, 1997) or direct solvent extraction with acetone/petroleum ether or hexane/ethanol/acetone, or supercritical fluid extraction (Eder, 2000). Reversed-phase chromatography using C_{18} columns has become the method of choice, although C_{30} stationary phases are useful for separation of carotene isomers (Sander et al., 2000). A review of the chromatographic determination of carotenoids in foods was made by Oliver and Palou (2000). Chlorophylls and their derivatives (Mg-chlorophyll and Cu-chlorophyll) may be used as food colorants to restore the colour of freshly harvested crops as they are permitted food additives. These compounds are exceptionally labile and all manipulations during analysis must be carried out in the dark and rapidly to prevent photodestruction or photo-isomerisation. Cold extractions with acetone/methanol are commonly used, followed by cleanup on C_{18} SPE cartridges. Reversed-phase HPLC with absorption or fluorimetric detection is the method of choice (Eder, 2000) for quantitative analysis of chlorophylls in food. Analysis of anthocyanins and betalins was reviewed by Eder (2000) but a more up-to-date review of anthocyanin analysis by HPLC and LC-MS has been reported by Da Costa et al. (2000).

Sweeteners can be roughly divided into two groups: bulk and intense sweeteners. Prodolliet (1996) and Glória (2000) reviewed thoroughly the analysis and properties of intense sweeteners: acesulfame-K, alitame, cyclamate, aspartame, glycyrrhizin, neohesperidin DC, saccharin, stevioside, sucralose and thaumatin. They are generally used in low calorie products such as diet beverages, dairy products and confectionaries. The European Committee for Standardisation (CEN) (1999) recently published a validated method for the determination of acesulfame-K, aspartame and saccharin in a wide variety of food products. It also allows the determination of caffeine, sorbic acid and benzoic acid. In this procedure the sample is extracted or diluted with water. If necessary, the extracted sample solution is purified on a SPE column or with Carrez reagents. The intense sweeteners in the sample test solution are separated on a reversed-phase chromatography column and determined by UV detection at 220 nm. A similar method was applied to the determination of aspartame in beverages, milk products and sweets (Gromes et al., 1995). Separation was carried out on a C_{18} column with a KH_2PO_4/methanol (70:30, v/v) mobile phase. A comparison was made between results by HPLC and an enzymatic method; similar values were obtained with recoveries of aspartame in the range 93–102%. The NMKL (1997a) has published a validated reference method for determination of saccharin in beverages and sweets. Galletti et al. (1996) described the analysis of aspartame by means of LC-electrospray mass spectrometry using selected ion monitoring at the (M+H)+ peak at m/z 295. The method was applied to soft drinks. A new method for determination of cyclamate in low-calorie foods by HPLC with indirect visible photometry was reported by Choi et al. (2000).

Sugars may be added to foods as bulk sweeteners or for their preservative action (osmotic pressure) or for their crystallinity while polysaccharides provide foods with texture, body and colloidal properties. Peris-Tortajada (2000) reviewed the analysis of sugars and separation of mono- and disaccharides is often performed with an amino-bonded silica column with acetonitrile/water eluent and refractive index (RI) detection. Sugar alcohols like sorbitol and xylitol (often used in dietetic hard candy and chewing gum) as well as nutritive sweeteners like sucralose and kestose may also be separated by this technique. The use of the $-NH_2$ stationary phase also permits the separation of glucose oligosaccharides up to DP 30–35. These carbohydrates are often extracted from food products with water, aqueous-ethanol or aqueous-acetonitrile at high temperature (reflux) or at room temperature. The use of a high temperature may cause degradation of labile carbohydrates thus it is often better to perform the extractions at room temperature. Further cleanup is often necessary with clarifying agents like Carrez solutions (zinc acetate and potassium ferricyanide) to precipitate proteins and fatty acids. Other cleanup methods include ion-exchange, solvent precipitation, SPE, and HPLC guard columns (Peris-Tortajada, 2000). A recent innovation is to use a dialysis unit (automated liquid-liquid extraction procedure) which can be coupled directly with an automatic injector prior to HPLC analysis. One of the major problems in using HPLC for sugar analysis is the lack of specificity and sensitivity of the RI detector. In recent years more sensitive RI detectors have been developed and some have been adapted for gradient elution. Other possibilities are the ELSD, which is more sensitive than the RI detector and allows gradient elution, or biosensor detection (Sprenger et al., 1999).

The analysis of synthetic antioxidants has been reviewed by Diaz and Cabanillas (1996) and Karovičová and Šimco (2000a,b). Primary antioxidants are mainly phenolic substances that inhibit the process of fat oxidation by interrupting the chain of free radicals and include natural and synthetic tocopherols, alkyl gallates, butylated hydroxyanisole (BHA), butylated hydroxytoluene (BHT) ter-butylhydroquinone (TBHQ) and nordihydroguaiaretic acid (NDGA). Ethoxyquin is not generally allowed for human consumption but may be used in petfoods and in animal feeds and fruits as an antiscald agent. It is also used in the spice industry to prevent carotenoid loss during post-harvest handling. Another class of antioxidants, which operate as oxygen consumers, includes ascorbic acid, ascorbyl palmitate and erythorbic acid. A further class of antioxidants includes tocopherols and tocotrienols. The main difficulty is to extract quantitatively the various phenolic antioxidants present in low concentrations from the food matrix and numerous techniques have been described: steam distillation, Soxhlet extraction with petroleum ether or hexane as well as direct extraction with solvents (for example, dry methanol, dimethylsulphoxide, acetonitrile, pentane, cyclohexane). Some methods describe a double extraction with methanol and petroleum-ether (Diaz and Cabanillas, 1996; Karovičová and Šimco, 2000a,b). Care has to be taken during solvent evaporation to reduce pressure to prevent losses of BHT or oxidation of TBHQ (Karovičová and Šimco, 2000b). Gradient elution is often used with a C_{18} column with detection at 280 nm for most phenolic antioxidants but at 360 nm for ethoxyquin (Diaz and Cabanillas, 1996). A reference method for determination of phenolic antioxidants in oils, fats and butter oil is described by the AOAC Int. (2000) and by the AOCS (1998). Dieffenbacher (1998) reported the results of collaborative studies organised by the antioxidants working group of the AIIBP on determination of BHA and BHT in oils, fats and dry soup powders. A modified IUPAC method was tested for soup powders in which methanol-methanolic, pH 3.5 KH_2PO_4 buffer (1:1) gave rapid elution of co-extracted substances and reduced interferences with the antioxidant peaks. Abidi (2000) has reviewed the separation of tocopherols and tocotrienols by normal and reversed phase chromatography by HPLC and LC-MS techniques. A reference method for determination of tocopherols in vegetable oils and fats by HPLC with fluorescence detection has been published by the AOCS (1998). Ethoxyquin was determined in various meals and extruded petfoods (Schreier and Greene, 1997). Extraction was made with acetonitrile on a series of sixteen samples that contained ethoxyquin in the range 0.25–289 ppm. Separation was made on a C_{18} HPLC column with a mobile phase of acetonitrile/ammonium acetate solution. An interesting procedure was reported by Chen and Fu (1995) for the simultaneous determination of TBHQ, BHA as well as some preservatives and sweeteners. Products including soy sauce, dried roast beef and sugared fruit were mixed with α-hydroxyisobutyric acid solution and hexadecyltrimethylammonium bromide as ion-pair reagent, followed by absorption onto a C_{18} SPE cartridge. The additives were desorbed in two fractions which were subsequently combined and filtered. Fifteen additives were separated by ion-pair reversed-phase HPLC analysis with UV detection at 233 nm. A new electrochemical detection technique was described by McCabe and Acworth (1998). Samples were mixed with hexane and the phenolic antioxidants were partitioned into acetonitrile. After separation on a Supelcosil® C_{18} column with gradient elution, detection was made with a high pressure Coul Array system with a series of electrochemical sensors. The method was applied to the simultaneous detection of PG, THBP, TBHQ, NDGA, BHA, Ionox 100, OG, BHT and DG in butter, margarine, shortening, lard and hand cream. The advantages of this technique with respect to the AOAC method were claimed to be simpler sample preparation, better limits of detection, better linearity, higher sample throughput and lower costs. Abidi (2000) noted the emergence of supercritical fluid chromatography for the analysis of tocopherols and tocotrienols in foods including vegetable oils.

Preservatives are added to food products to improve their shelf life by preventing or inhibiting the activity and growth of microorganisms. The most commonly used preservatives are benzoic, sorbic and propionic acids along with sulphites and the methyl-, ethyl- and propyl esters of p-hydroxybenzoic acid (parabens). The analysis of preservatives by HPLC has been reviewed by Karovičová and Šimco (1996, 2000a). Extraction from food products is usually made by one of the following methods: solvent extraction, SPE, ion-pair extraction and steam distillation. Reversed phase HPLC is commonly used for separation of the various preservatives with UV detection at around 230 nm, except for sulphites which may be analysed by indirect photometry-HPLC (Pizzoferrato et al., 1998). An AOAC interlaboratory study was performed for benzoic acid in orange juice (Lee, 1995). Another validated reference method has been issued by the NMKL (1997b) for determination of benzoic acid, sorbic acid and p-hydroxybenzoic acid esters which is applicable to a range of foods. The main drawback of this method is that one set of chromatographic conditions is required to separate sorbic and benzoic acids and another for the parabens. However, all five preservatives can be readily separated in a single run on a reversed phase C_{18} column with a mobile phase of 5 mM ammonium acetate (pH 4.4) and acetonitrile (7+3 v/v) gradient elution (Schulte, 1995). A typical separation obtained with a similar procedure is shown in Fig. 6.2. Certain common food components like vanillin, ethylvanillin, benzaldehyde and eugenol may interfere and so each preservative peak should be identified by comparing its diode-array spectrum with that of the standard. Patz et al. (1997) determined sulfite in fruit juices by HPLC-biosensor coupling.

Glutamic acid and its salts (especially the sodium salt) enhance the flavour of many convenience foods. Bejaars et al. (1996) described the determination of free glutamic acid in soups, meat products and Chinese food. The method involves hot water extraction of test portions followed by filtration and dilution. The extracts were treated with N,N-dimethyl-2-mercapto-ethyl-ammonium chloride and o-phtaldialdehyde to convert glutamic acid into a stable fluorescent, 1-alkylthio-2-alkylisoindole. Homocysteic acid was used as the internal standard. Separation was made on a C_{18} column, eluted with

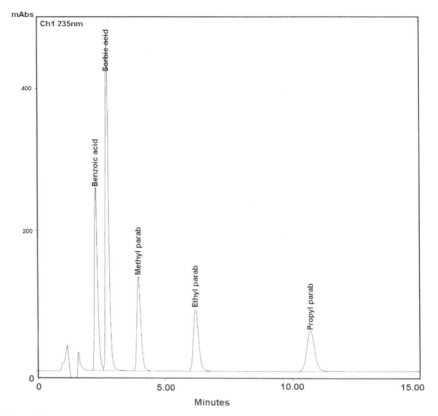

Fig. 6.2 Separation of benzoic acid, sorbic acid and p-hydroxybenzoic acid esters by HPLC.

acetonitrile/phosphate buffer (pH7)/water with fluorescence detection. The method was studied collaboratively and adopted as a first action AOAC method.

Potassium bromate is a widely used dough conditioner. However, if it is used in excessive quantities in bread products then appreciable residues (> 1 mg/kg) can remain which is of concern since it is a cancer suspect agent. Its routine analysis is laborious, time-consuming and difficult by HPLC, and Cunningham and Warner (2000) described the development of an instrumental neutron activation method for determination of bromine while HPLC was used to determine bromate in selected samples.

Vitamins occur naturally in many foods and raw materials. However the natural contents are often supplemented in many food products to ensure an adequate intake, for example in infant formulae, breakfast cereals and clinical nutrition products. Vitamins are usually added as nutrients and thus not covered in this chapter but may also be added as food colours (riboflavin, carotenes). The reader should refer to the following references for recent developments in

New methods in detecting food additives 119

vitamins analysis by HPLC or LC-MS (De Leenheer et al., 2000; Ball, 2000; Russel, 2000; Song et al., 2000). The analysis of tocopherols, added as antioxidants, has been reviewed by Abidi (2000). In foods, vitamin C exists as two vitamers: L-ascorbic acid (AA) and its oxidation product, dehydro-L-ascorbic acid (DHAA) which is unstable. Russel (2000) and Nyssönen et al. (2000) reviewed methods of ascorbic acid analysis. HPLC is often used with UV (Ali and Phillippo, 1996), electrochemical detection (Hidiroglou et al., 1998) or post-column derivatisation with fluorescence detection (Bognár and Daood, 2000). A stereoisomer, isoascorbic acid (IAA) is often added to food as an antioxidant but IAA and its oxidation product, dehydroisoascorbic acid (DHIAA) have only five per cent of the vitamin C activity of AA and DHAA. Ali and Phillippo (1996) and Hidiroglou et al. (1998) described HPLC methods to separate AA and IAA in meat-based food products.

Emulsifiers are an important class of food additives and include:

- lecithins
- mono- and diglycerides
- mono- and diglyceride esters
- other speciality emulsifiers (e.g. glycol alginate esters, ammonium phosphatides, sucrose esters, sorbitan esters, polyglycerol esters and Ca and Na steroyl-2-lactylate).

A useful review of food emulsifiers was reported by Hasenhuettl and Hartel (1997) that included the classical methods of analysis. Some of the most recent chromatographic methods are described below. The analysis of the phospholipid composition is of importance mainly in the certification and quality control of lecithins (Van der Meeren and Vanderleelen, 2000). HPLC separation with ELSD has become the method of choice for this analysis and a typical chromatogram of a commercial soybean lecithin is shown in Fig. 6.3. Caboni et al. (1996) described the separation and analysis of phospholipids in a range of foods. Total lipids were extracted by the Folch method followed by SPE on C_8 cartridges. The recovery of phospholipids from a series of spiked food products was in the range 93–100%. The extracted phospholipid fraction containing L-phosphatidyl-DL-glycerol, phosphatidylethanolamine, phosphatidylinositol, phosphatidylcholine and sphingomyelin was then analysed by HPLC with ELSD. Monoglycerides and diglycerides are among the major emulsifiers used in food products and Liu et al. (1993) and AOCS (1998) described a quantitative method for their determination by HPLC with ELSD.

Gordon and Krishnakumar (1999) observed that the traditional emulsifiers market within Europe has been hit by concerns about BSE (use of animal fat) and by the use of genetically modified soybean, and observed that speciality emulsifiers may gradually replace lecithin and the mono- and diglycerides within Europe. A typical example is the sucrose ester group. Sucrose polyester with six of the eight sucrose hydroxyl groups esterified with fatty acids, has physical and organoleptic properties similar to those of cooking and frying fats. However, because of its high molecular weight and resistance to lipolysis, it

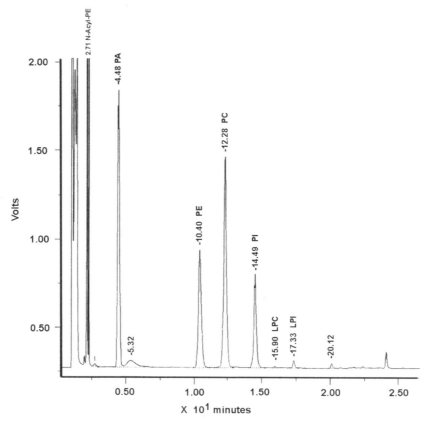

Fig. 6.3 Separation of lecithin phospholipids by HPLC with evaporative light scattering detector.

cannot be readily metabolised in the body. It is often used as a fat substitute in snack foods. Sucrose polyester in salad dressings has been determined by HPLC (Chase et al., 1995). Extraction was performed with methylene chloride/ isopropanol/anhydrous $MgSO_4$. Acylglycerols and sucrose polyester were collected as a fraction from a gel permeation column and then analysed by HPLC with ELSD. Sucrose octaacetate was added as an internal standard. Moh et al. (2000) described an improved separation of sucrose ester isomers using gradient HPLC with ELSD. Another class of emulsifiers, the fatty acid esters of sorbitol and its anhydrides are approved as food additives in most countries. A new method for determination of sorbitan tristearate in vegetable oils and fats was reported by Thyssen and Andersen (1998). This method is based on the isolation of sorbitan tristearate compounds on a silica cartridge and their in-situ hydrolysis with sulphuric acid at 90°C/24 h. The polyalcohols formed (sorbitans, isosorbide and sorbitol) were eluted from the silica cartridge followed by separation on a Shodex® SUGAR SC1011 column with RI detection.

The detailed analysis of the polysaccharide hydrocolloids added to foods as thickeners or stabilisers is a great challenge for the analyst since they are often added at low levels (0.01–1.0%) and other hydrocolloids or food components may interfere in the analysis. Roberts and Quemener (1999) reviewed the analysis of carrageenans by HPLC and other techniques. A common method of analysis is to depolymerise hydrocolloids to their constituent sugars or anhydrosugars followed by analysis by HPLC or IC. A useful approach was recently reported by Quemener et al. (2000a) who described a procedure for quantitative determination of kappa and iota carrageenans, in complex food systems. This involved three main steps: freeze drying of sample, mild methanolysis of carrageenans, and reverse phase HPLC analysis of the 3,6 anhydrogalactose dimethylacetal released. Recoveries were generally higher than 80% with a precision of 3.2% and a limit of quantification of 0.03%. As a second step, Quemener et al. (2000b) described procedures for quantitative determination of pectins, alginates and xanthan in complex food systems. The three main steps were: freeze drying of sample, methanolysis to release uronic acids and neutral sugars, and reverse phase separation of the constituent sugars. Recoveries were generally higher than 80% with limits of quantification around 0.03%.

6.2.2 Ion chromatography

Ion chromatography (IC) is used increasingly in the food industry and a useful reference book was written by Weiss (1995). Henshall (1997) reviewed applications for food and beverage analysis. Recent reviews have been published for anions and cations (Buldini et al., 1997; Danielson et al., 2000) and carbohydrates (Cataldi et al., 2000).

The separation of sugars with cation-exchange columns loaded with metal ions (including calcium, lead and silver) with water eluent and refractive index detection is a classical analytical technique in sugars analysis. It is described in an ISO (1998a) method. The main drawbacks of this approach are due to an early elution of oligosaccharides immediately after the void volume, often leading to poor resolution and the use of the non-specific and low sensitivity RI detector. The development of high performance anion-exchange chromatography (HPAEC) with post amperometric detection (PAD) has provided analytical chemists with an excellent procedure for separation of monosaccharides, disaccharides and oligosaccharides in a single run with very high sensitivity. A range of commercial anion-exchange columns (CarboPac®) are available offering different selectivities for different classes of saccharides and alditols. The mobile phase is normally composed of dilute NaOH solution, but better selectivity can be obtained when using eluents of NaOH and $Ba(OAc)_2$ (Cataldi et al., 2000). High sensitivity and system stability is provided by the PAD working electrode. Corradini et al. (1997) separated isomalt, maltitol, glycerol, xylitol, sorbitol, mannitol and fructose on a Carbopac MA-1® column. Alditols in sugar-free products and a low calorie sweetener containing sorbitol, mannitol and fructose can also be analysed by this method. Ito (1999) reported a method for the analysis of the non-nutritive sweetener sucralose in a variety of foods with aqueous extraction, cleanup on a SPE cartridge and determination by IC-PAD.

A novel method for determination of carrageenans and agar containing 3,6-anhydrogalactose was reported by Jol et al. (1999). The technique uses reductive hydrolysis with a methylmorpholine-borane complex in the presence of acid. This was followed by HPAEC analysis of alditols without derivatisation. The technique is said to be sensitive and accurate, and the results were comparable with those obtained by GC in about half the time. Koswig et al. (1997) described a HPAEC method for identification of the thickening gums in fruit juices (carrageenans, guar gum, gum arabic, gum tragacanth, xanthan, pectin from apple and pectin from orange). The optimum hydrolysis conditions were reaction at 120°C with 1 M sulphuric acid, before neutralisation at pH7 which were reported to give the most complete and reproducible hydrolysis across the range of hydrocolloids tested. Subsequent analysis by HPAEC demonstrated good resolution between all constituent monosaccharides and so evaluation of chromatograms was reported to be straightforward. It appears that this method cannot specifically detect carrageenan via the 3,6-anhydrogalactose residue as it is completely destroyed under the acidic conditions used.

Polyphosphates are used as food additives in cheese, fish paste, and various processed meat products and petfoods. Suppressed IC is often used for this difficult determination. Sekiguchi et al. (2000) described an improved IC method using an on-line hydroxide eluent generator which gave much less baseline shifts during the hydroxide gradients, greater retention time reproducibility, and better precision of results. A separation of phosphate, citrate and polyphosphates up to DP8 was obtained. The presence of calcium in the extracts of cheese caused peak distortion and retention time shortening for P4 and IP6. Thus calcium was removed by passage of the extract through a cation-exchange resin before IC analysis. A useful method for simultaneous determination of chloride, nitrite, nitrate and polyphosphates in meat and seafood products was reported by Kaufmann et al. (2000). The preservatives nitrate and nitrite were extracted from processed meat products with water and then separated by suppressed IC with UV detection at 225 nm (Siu and Henshall, 1998). Recoveries were 90–100% for both nitrate and nitrite. An alternative procedure was validated and issued by the CEN (1998). An aqueous extract of the meat product is treated with acetonitrile to remove any interfering substances and then separation of nitrite and nitrate is made on an ion-exchange column with UV detection at 205 nm in the unsuppressed mode. The main difficulty of this procedure is the long conditioning time necessary to stabilise the ion-exchange column before use with the mobile phase of acetonitrile/lithium borate gluconate buffer. IC techniques have considerable advantages over the traditional spectrophotometric method for this determination since they avoid the use of the troublesome cadmium reduction of nitrate to nitrite.

Sulphite which may be added to foods, beers and wines as an antioxidant or anti-browning agent can be determined by IC with an ion-exclusion column and amperometric detection (AOAC Int., 2000). The main disadvantage is due to the rapid fouling of the platinum working electrode, which results in a significant decrease in detector response over time. A more rugged procedure was reported by Wygant et al. (1997), in which the use of PAD was found to overcome these problems by applying cleaning potentials to the working electrode.

Chen et al. (1997a) analysed sodium saccharin in soft drinks, orange juice and lemon tea after filtration by injection into an ion-exclusion column with detection at 202 nm. Recoveries of 98–104% were obtained. They reported that common organic acids like citric and malic and other sweeteners did not interfere. Qu et al. (1999) determined aspartame in fruit juices, after degassing and dilution in water, by IC-PAD. The decomposition products of aspartame, aspartic acid and phenylanaline were separated and other sweeteners did not interfere. The recoveries of added aspartame were 77–94%. Chen et al. (1997b) separated and determined four artificial sweeteners and citric acid.

6.2.3 Thin-layer chromatography

This classical technique has been used for many years but there is a continuing interest particularly as a simple screening procedure for classes of additives. Several recent articles have described the identification of natural colours. Hirokado et al. (1999) extracted several natural colours (madder colour, cochineal extract, lac colour, carthamus yellow and carthamus red) from processed foods and then separated them on either a C_{18} reverse phase RP-18 F245S plate or cellulose high-performance plate. The method was applied to 30 kinds of processed foods such as jelly, cake and beverages. Hayashi et al. (1999) separated beta-carotene and paprika. Itakura et al. (1999) separated lac I and II and cochineal colours using reverse phase TLC scanning densitometry. Samples were homogenised with 80% aqueous methanol, cleaned up on an SPE cartridge, and then separated on an RP-18F245S plate. The method was applied to 122 foods including juices, jellies and candies. Brockmann (1998) organised a collaborative study for the qualitative analysis of added natural colours in raw sausage including red sandalwood, betanin, cochineal and angkak pigments. The various colours were successfully separated on cellulose and silica plates respectively. Many other applications for food additives were recently reviewed by Sherma (2000).

6.2.4 Capillary electrophoresis

Capillary electrophoresis (CE) is a modern analytical technique that allows the rapid and efficient separation of sample components based on differences in their electrophoretic mobilities as they migrate or move through narrow bore capillary tubes (Frazier et al., 2000a). While widely accepted in the pharmaceutical industry, the uptake of CE by food analysts has been slow due to the lack of literature dedicated to its application in food analysis and the absence of well-validated analytical procedures applicable to a broad range of food products.

Swedberg (1997), Fernandes and Flick (1998) and Frazier et al. (2000a) have described the principles and some applications of CE. A recent review by Kok (2000) describes capillary electrophoresis instrumentation and operation. The major formats of CE are:

- capillary zone electrophoresis (CZE)
- micellar electrokinetic chromatography (MEKC or MECC)
- capillary isoelectric focusing (CIEF)
- capillary gel electrophoresis (CGE)
- capillary isotachyphoresis (CITP).

The main advantages of CE are its ease of use, speed and efficiency, while disadvantages are matrix effects causing shifts in retention time and variations in peak area response, non-gaussian peak shapes for certain molecules which are difficult to integrate, and limited detector sensitivity.

Numerous CE separations have been published for synthetic colours, sweeteners and preservatives (Frazier et al., 2000a; Sádecká and Polonský, 2000; Frazier et al., 2000b). A rapid CZE separation with diode array detection for six common synthetic food dyes in beverages, jellies and syrups was described by Pérez-Urquiza and Beltrán (2000). Kuo et al. (1998) separated eight colours within 10 minutes using a pH 9.5 borax-NaOH buffer containing 5 mM β-cyclodextrin. This latter method was suitable for separation of synthetic food colours in ice-cream bars and fruit soda drinks with very limited sample preparation. However the procedure was not validated for quantitative analysis. A review of natural colours and pigments analysis was made by Watanabe and Terabe (2000). Da Costa et al. (2000) reviewed the analysis of anthocyanin colours by CE and HPLC but concluded that the latter technique is more robust and applicable to complex sample types. Caramel type IV in soft drinks was identified and quantified by CE (Royle et al., 1998).

A range of other applications has been described. Pérez-Ruiz et al. (2000) analysed glutamate in beverages, sauces and gravy by CE with laser-induced fluorescence detection of the fluorecein isothiocyanate derivative of glutamate. Recoveries of added glutamate were in the range 94–107%. Sulphite analysis in foods and beverages by CE was described by Trenerry (1996). The sulphite was converted to SO_2 and finally to sulphate using a Monier-Williams distillation. This method measures the sulphate levels free from other volatile compounds, which often interfere with the standard Monier-Williams distillation/titrimetric method in ingredients like onions, garlic and cabbages. Hall et al. (1994) described an MEKC technique for separation of the synthetic antioxidants PG, BHA, BHT and TBHQ in six minutes while Pant and Trennery (1995) extracted and separated sorbic and benzoic acids from foods and beverages. Zhao and Johnson (2000) described a rapid method for the analysis of sucralose by capillary electrochromatography. Separations of anion and cations were reviewed by Mopper (1996) and Sádecká and Polonský (1999). Such methods could be applied to analysis of mineral additives in food products. Certain polysaccharides like polydextrose, Pinefibre and guar gum may be used as soluble fibre in soft drinks. Kitada et al. (1997) extracted these polysaccharides by gel-permeation chromatography and then separated and determined them by CE.

There is a recent trend towards simultaneous CE separations of several classes of food additives. This has so far been applied to soft drinks and preserved fruits, but could also be used for other food products. An MEKC method was published (Lin et al., 2000) for simultaneous separation of intense sweeteners (dulcin, aspartame, saccharin and acesulfame K) and some preservatives (sorbic and benzoic acids, sodium dehydroacetate, methyl-, ethyl-, propyl- and isopropyl- p-hydroxybenzoates) in preserved fruits. Ion pair extraction and SPE cleanup were used prior to CE analysis. The average recovery of these various additives was 90% with good within-laboratory reproducibility of results. Another procedure was described by Frazier et al. (2000b) for separation of intense sweeteners, preservatives and colours as well as caffeine and caramel in soft drinks. Using the MEKC mode, separation was obtained in 15 min. The aqueous phase was 20 mM carbonate buffer at pH 9.5 and the micellar phase was 62 mM sodium dodecyl sulphate. A diode array detector was used for quantification in the range 190–600 nm, and limits of quantification of 0.01 mg/l per analyte were reported. The authors observed that their procedure requires further validation for quantitative analysis.

A major problem in CE analysis is the poor detection limits that are obtained with UV or DAD. Some potential exists for coupling biosensors to CE as a specific detection method (Bossi et al., 2000) although few published applications have appeared yet.

6.2.5 Gas chromatography

The fundamentals of gas chromatography (GC) are well covered in numerous publications (Nollet, 1996b; Paré and Bélanger, 1997). The main limitation of this technique is that it can only be used for components that are volatile at less than 300°C and so is not readily applicable to many additives unless a volatile derivative can be formed. Thus the application of modern GC techniques for additives analysis is less than that of HPLC or the rapidly emerging CE. Cscerháti and Forgács (1999) reviewed applications of capillary GC and GC-MS which include capillary GC-MS analysis of potassium bromate, a bread improver, and the preservatives propionic acid and propionates. Another application described by these authors is for the analysis of the artificial sweetener aspartame in soft drinks by pyrolysis GC-MS. Preuß and Their (1996) described the isolation of natural thickeners and gums from food and their identification by determination of their constituent sugars by capillary column gas chromatography. A partially automated capillary GC method with FID/MS detection was reported for simultaneous determination of phenolic antioxidants, tocopherols and preservatives in fatty foods without derivatisation (Gonzáles et al., 1999).

6.2.6 Flow-injection analysis

This technique has been established for many years particularly for water, soil and feeding-stuff analysis, where a large number of analyses are required for quality control or monitoring purposes. A number of applications have been published for food additives including aspartame (Fatibello et al., 1999), citric acid (Prodromidis et al., 1997), chloride, nitrite and nitrate (Ferreira et al., 1996), cyclamates (Cabero et al., 1999), sulphites (Huang et al., 1999; AOAC Int, 2000), and carbonate, sulphite and acetate (Shi et al., 1996). Yebra-Biurrun (2000) reviewed the determination of artificial sweeteners (saccharin, aspartame and cyclamate) by flow injection.

There is increasing interest in the use of specific sensor or biosensor detection systems with the FIA technique (Galensa, 1998). Tsafack et al. (2000) described an electrochemiluminescence-based fibre optic biosensor for choline with flow-injection analysis and Su et al. (1998) reported a flow-injection determination of sulphite in wines and fruit juices using a bulk acoustic wave impedance sensor coupled to a membrane separation technique. Prodromidis et al. (1997) also coupled a biosensor with an FIA system for analysis of citric acid in juices, fruits and sports beverages and Okawa et al. (1998) reported a procedure for the simultaneous determination of ascorbic acid and glucose in soft drinks with an electrochemical filter/biosensor FIA system.

6.2.7 Inductively coupled plasma atomic emission spectrometry and atomic absorption spectrophotometry

The analysis of cations like sodium, potassium, calcium and magnesium is commonly performed by ICP-AES which is gradually replacing flame atomic absorption spectrometry within the food industry. It is normally impossible to differentiate cations present naturally in the food product from those added as food additives. The most critical part of the analysis is to transform the food product into a solution suitable for ICP-AES. Sun et al. (2000) compared five methods of sample preparation but concluded that the classical dry-ashing method is still preferable to microwave acid digestions or wet-digestion with nitric/perchloric acids. A new method was recently reported for determination of the common food colour titanium dioxide in foods using ICP-AES (Lomer et al., 2000). Samples were digested with 18 M sulphuric acid for 1 h at 250°C with determination of titanium as a marker for titanium dioxide at 336.121 nm.

An atomic absorption method was published by AOAC Int. (2000) for determination of the anti-foaming agent polydimethylsiloxane in pineapple juice, that is based on extraction with 4-methyl-2-pentanone and aspiration into a nitrous oxide/acetylene flame. A silicone lamp was used for detection.

6.2.8 Enzymatic methods

The NMKL (1997c) described an enzymatic method for determination of starch and glucose in foods that was validated in a collaborative study. Starch was hydrolysed to glucose with Termamyl enzyme® at pH5 for 15 min. at 100°C. A similar method was published by ISO (1998b). Chemically modified starches are widely used within the food industry but little recent information is available for improved methods of analysis. ISO (1997) published an enzymatic procedure for analysis of added citric acid in cheese.

6.3 Rapid or alternative methods

This group of methods can be applied to routine quality control analyses or for process control of food additives. Many publications describe new developments but few validated procedures are available in the literature. Some applications used within the food industry remain unpublished but some details are given below. A wide variety of techniques are available including biosensors, enzymatic, pH differential methods, X-ray fluorescence and NIR.

6.3.1 Immunoassay

The area of immunoassay was reviewed by Dixon (1998) but has been applied infrequently to the area of food additives. One recent example is for carminic acid, the major component of cochineal dye which was screened and determined in foods by an ELISA technique (Yoshida, 1995). Monoclonal anticarminic acid antibody was obtained from A/J mice immunised with carminic and acid-human immunoglobulin G (IgG) conjugate. Carminic acid was extracted with water from milk products while meat and fish were digested with pronase, and then extracted with NaOH solution. Microtiter plates were coated with carminic acid-bovine serum albumin conjugate and goat anti-mouse IgG(H+L)-peroxidase complex was used as a second antibody, and 3,3',5,5'-tetramethylbenzidine is used as a substrate for the peroxidase. Recoveries of carminic acid from all products were greater than 85%. This ELISA system also responded to the structural analogue of carminic acid, laccaic acid.

6.3.2 Biosensors

A biosensor is an analytical device characterised by a biological sensing element intimately connected or integrated within a transducer, which converts the biological event into a response that can be further processed. The biological sensing element can be either catalytic (e.g. enzyme, microorganism) or non-catalytic, also denoted 'affinity sensing element' (e.g. antibody receptor). The development of biosensors for the food industry has historically been one of the most promising yet frustrating areas of development. A number of excellent reviews have been made in biosensor technology (Cunningham, 1998; Kress-Rogers, 1998; Ramsay, 1998; Scott, 1998). In spite of all this wealth of information, Scott (1998) concluded that 'the major reason for the poor progress in development is the lack of understanding found in the food industry

Table 6.1 Some recent publications describing applications of biosensors useful for additives analysis

Analyte	Biosensor type	Details	Reference
L-ascorbic acid	Potentiometric	Based on ascorbate oxidase of natural source immobilised on ethylene-vinylacetate membrane	Fernandes et al. (1999)
L-ascorbic acid	Amperometric	Enzyme-less biosensor using poly-L-histidine-copper complex as an alternative biocatalyst	Hasebe et al. (1998)
Antioxidants	Optical fibre chemiluminescence	Immobilised luminol/haematin reagent phase	Palaroan et al. (2000)
Choline	Biosensor	Rapid assay using microwave hydrolysis and a choline biosensor	Panfili et al. (2000)
Glutamate	Amperometric	Prussian blue-based 'artificial peroxidase' as a transducer for hydrogen peroxide	Karyakin et al. (2000)
Lecithin		A new biosensor operating directly in non-aqueous solvent	Campanella et al. (1998)
Starch	Enzymatic	Amyloglucosidase, mutarotase, glucose oxidase and catalase co-immobilised on either nylon net or activated collagen membrane	Vrbova et al. (1993)
Sulphite	Amperometric	Sulfite oxidase with cytochrome c, as electron acceptor, and a screen-printed transducer	Abass et al. (2000)

Table 6.2 Analytes measured by membrane biosensors (Yellow Springs Instruments)

Analyte	Sample type	Membrane
Glucose	Vegetables, ice-cream, cereals, peanut butter, etc.	Glucose oxidase
Sucrose	Vegetables, ice-cream, cereals, peanut butter, etc.	Glucose oxidase, mutarotase, invertase
Lactate	Lunch meat, cooked foods	Lactate oxidase
Glutamate (monosodium glutamate)	Food products	Glutamate oxidase
Choline	Infant formula	Choline oxidase

concerning biosensor technology and the lack of understanding in the biosensor community concerning the requirements and conditions of food analysis'. Numerous applications of biosensors have been published in the literature and a selection is given in Table 6.1. Enzyme biosensors are commercially available from several companies and Woodward et al. (1998) reviewed applications of enzyme membrane biosensors for food analysis (Table 6.2). The development and potential applications of chemiluminescence sensors for determination of a range of analytes was recently reviewed by Aboul-Enein et al. (2000).

Immunosensors have been developed commercially mostly for medical purposes but would appear to have considerable potential for food analysis. The Pharmacia company has developed an optical biosensor, which is a fully automated continuous-flow system which exploits the phenomenon of surface plasmon resonance (SPR) to detect and measure biomolecular interactions. The technique has been validated for determination of folic acid and biotin in fortified foods (Indyk, 2000; Bostrom and Lindeberg, 2000), and more recently for vitamin B_{12}. This type of technique has great potential for application to a wide range of food additives but its advance will be linked to the availability of specific antibodies or other receptors for the various additives. It should be possible to analyse a whole range of additives by multi-channel continuous flow systems with further miniaturisation.

6.3.3 X-ray fluorescence

This non-destructive technique is a very suitable tool for rapid in-line analysis of inorganic additives in food products (Price and Major, 1990; Anon, 1995). It can be readily used by non-skilled operators, and dry materials can be pressed into a pellet or simply poured into a sample cup. The principles of this technique related to food analysis are described by Pomeranz and Meloan (1994). A useful Internet site is http://www.xraysite.com, which includes information about different XRF instruments from various companies. Wavelength dispersive X-ray fluorescence (WD-XRF) or bench-top energy dispersive (ED-XRF) instruments are available. XRF is a comparative technique, thus a calibration curve needs to be established using food products of the same type as those to be analysed with known (previously analysed by a reference method) elemental content. Calibrations can be stored by the instrument software and restandardised on a daily basis with reference materials. This technique is not often described in scientific literature for food analysis, however typical applications include: determination of salt content in snack foods via analysis of chloride, and analysis of fortified minerals like iron and calcium in infant formula. Measurement time is usually about 5–10 minutes per sample.

6.3.4 Near infrared and Fourier transform infrared spectroscopy

The infrared technique has been described in numerous publications and recent reviews were published by Davies and Giangiacomo (2000), Ismail et al. (1997) and Wetzel (1998). Very few applications have been described for analysis of additives in food products. One interesting application is for controlling vitamin concentrations in vitamin premixes used for fortification of food products by attenuated total reflectance (ATR) accessory with Fourier transform infrared (FTIR) (Wojciechowski et al., 1998). Four vitamins were analysed – B_1 (thiamin), B_2 (riboflavin), B_6 (vitamin B_6 compounds) and Niacin (nicotinic acid) – in about 10 minutes. The partial least squares technique was used for calibration of the equipment. The precision of measurements was in the range 4–8%, similar to those obtained for the four vitamins by the reference HPLC method.

6.3.5 Various rapid enzymatic and test kit methods

Enzymatic assay techniques have been developed for several additives by Merck. BIOQUANT® kits are available for aspartame (intense sweetener) and nitrate (preservative). Gromes et al. (1995) applied the Bioquant kit to determination of aspartame in yoghurt, quark and confectionery. For low concentrations of aspartame a blank correction procedure was necessary. Recoveries of aspartame were in the range 93–102%.

The RQ flex test kit® (Merck) which uses specific test strips is useful for the semi-quantitative determination of several analytes. D(+) ascorbic acid can be determined in fortified food products with an accuracy of 85–115% (unpublished data), however the procedure cannot be applied to coloured food products. Added iron salts may be extracted from food products with dilute sulphuric acid and adjusted to pH2 with NaOH solution. Fe^{3+} is reduced to Fe^{2+} with ascorbic acid. Fe^{2+} reacts with Ferrospectral® to form a red-violet complex. An internal calibration is provided on a barcode which is read by the RQ-flex reflectometer prior to any measurements. This avoids the need to calibrate the instrument with standard solutions.

A colorimetric assay for lecithin and choline was described by Kotsira and Klonis (1998) using two enzymes (phospholipase and choline oxidase) and an indicator dye conjugate (bromothymol blue-glutathione) co-immobilised on a glutaraldehyde-activated polyacrylamide transparent gel. The change of the optical absorption of the biocatalytic gel is related linearly to the lecithin concentration in soya bean. The differential-pH technique shows some potential for analysis of various food additives including ascorbate, lactate, malate and citrate (Bucsis, 1999). However, validated procedures applicable to analysis of additives in food products have not yet been published.

6.3.6 Various photometric methods

Hermann (2000) described a rapid automated method involving generation of a known amount of free radicals and the detection of the excess by photochemi-luminescence. Test kits are available for determination of total water-soluble antioxidants, fat-soluble antioxidants and ascorbic acid. A luminometric method was developed for the determination of antioxidative activity and was subsequently applied to anthocyanin and betalaine colour concentrates (Kuchta et al., 1999). The method involved quantification of the interruption in luminescence from the hydrogen peroxide-horse radish peroxidase-luminol system in the presence of antioxidants.

6.4 Future trends

In terms of rapid methods the current trend in analytical chemistry is towards 'user-friendly' increasingly miniaturised instruments particularly for quality control applications. 'Laboratory-on-a-chip' applications in which an entire analytical procedure can be performed on a 'chip' has a promising future but still requires considerable development (Myers, 2000). The components are already available in research laboratories and a whole system including reagent pumps, detectors, CE capillaries, etc., can be housed in a 'chip' the size of a matchbox (Cefai, 2000). This is linked to the tremendous developments in palm-top PC technology that can be harnessed for calibration and data treatment. It may be some time before commercial applications are available but it is a fast-moving field and many companies and research groups are active in this area. Another major area is that of biosensors which still have unfulfilled potential. Since they are relatively easy to miniaturise there is some potential for their use as specific detectors in hand-held analysers for a wide variety of analytes. Immunoassay test methods are widely used in the food industry for detection of allergens, authenticity purposes or for detection of contaminants, but not yet for food additives. Future developments for food additives are likely to include the area of immunosensors as long as the specific receptors can be developed.

For reference methods, HPLC with various detectors has become the standard reference technique for analysis of food additives, but new developments in this area are mainly linked to detector technology. Diode array detectors have not totally met the expectations of food analysts in terms of their specificity and LC-MS is likely to fill the gap. Specific detection with biosensor chips may also have a future for certain analyses. The use of combined LC-MS/DAD systems is likely to become commonplace as user-friendly and cost-effective instruments are becoming available. This will enable the analysis of several classes of food additives simultaneously with greater specificity and better detection limits. However, this will also require better trained analytical chemists conversant with MS. Another area of great potential is that of CE which allows rapid and efficient separations but still suffers from detectors of low sensitivity and poor acceptance in the food industry due to the lack of validated applications. There is some potential in the future for coupled CE-MS systems or CE-biosensor chip detectors which could offer higher sensitivity and selectivity.

6.5 Sources of further information and advice

6.5.1 Key books
Some useful reference books are listed below:

AACC (2000), *Approved Methods of Analysis*. St. Paul, American Association of Cereal Chemists.

AOAC INT. (2000), *AOAC Official Methods of Analysis*. Gaithersburg, AOAC International.

AOCS (1998), *AOCS Official Methods*. American Oil Chemists Society.

CSCERHÁTI T and FORGÁCS E (1999), *Chromatography in Food Science and Technology*. Lancaster Technomic Publishing Co.

DALZEL J M (1998), *Ingredients Handbook: Fat Substitutes*. Leatherhead, Leatherhead Food RA.

DE LEENHEER A P, LAMBERT W E and VAN BOCXLAER J F (2000), *Modern Chromatographic Analysis of Vitamins*. New York, Marcel Dekker.

Food Chemicals Codex (1996) and supplements

FRAZIER R A, AMES J M and NURSTEN H E (2000a), *Capillary Electrophoresis for Food Analysis – Method Development*. Cambridge, Royal Society of Chemistry.

LMBG (2000), *Amtliche Sammlung von Untersuchungsverfahren*, nach § 35 LMBG, Berlin, Beuth Verlag GmbH.

NIELSEN S S (1998) *Food Analysis*. Gaithersburg, Aspen Publishers.

NOLLET L M L (1996a,b), *Handbook of Food Analysis*, 2 vols. New York, Marcel Dekker.

NOLLET L M L (2000), *Food Analysis by HPLC*. New York, Marcel Dekker.

PARÉ J R J and BÉLANGER J M R (1997), *Instrumental Methods in Food Analysis*. Amsterdam, Elsevier.

SCOTT A O (1998), *Biosensors for Food Analysis*. Cambridge, Royal Society of Chemistry.

TUNICK M H, PALUMBO S A and FRATAMICO P M (1998), *New Techniques in the Analysis of Foods*. New York, Kluwer Academic/Plenum.

WETZEL D L B and CHARALAMBOUS G (1998), *Instrumental Methods in Food and Beverage Analysis*. Amsterdam, Elsevier.

WEISS J (1995), *Ion Chromatography*. Weinheim, Germany, VCH.

6.5.2 Review articles

ABIDI S L (2000), 'Review: chromatographic analysis of tocol-derived lipid antioxidants', *J Chromatogr A*, 881, 197–216.

DA COSTA C T, HORTON D and MARGOLIS S A (2000), 'Review: Analysis of anthocyanins in foods by liquid chromatography, liquid chromatography-mass spectrometry and capillary electrophoresis', *J Chromatogr A*, 881, 403–10.

HENSHALL A (1997), 'Use of ion chromatography in food and beverage analysis', *Cereal Foods World*, 42, 414–19.

KAROVIČOVÁ J and ŠIMKO P (2000b), 'Determination of synthetic phenolic antioxidants in foods by HPLC', *J Chromatogr A*, 882, 271–81.

LOZANO A, ALEGRIA A, BARBERA R, FARRE R and LARGARDA M J (1998), 'Evaluation of methods for determining ascorbic acid, citric acid, lactose and maltodextrins in infant juices and formulas', *Alimentaria*, 298, 95–99.

OLIVER J and PALOU A (2000), 'Review: chromatographic determination of carotenenoids in foods'. *J Chromatogr A*, 881, 543–55.

ROBERTS M A and QUEMENER B (1999), 'Measurement of carrageenans in food: challenges, progress, and trends in analysis', *Trends in Food Sci and Technol*, 10, 169–81.

SÁDECKÁ J and POLONSKÝ J (1999), 'Determination of inorganic ions in foods and beverages by capillary electrophoresis', *J Chromatogr A*, 834, 401–17.

SÁDECKÁ J and POLONSKÝ J (2000), 'Review: electrophoretic methods in the analysis of beverages', *J Chromatogr A*, 880, 243–79.

SHERMA J (2000), 'Review: thin layer chromatography in food and agricultural analysis', *J Chromatogr A*, 880, 129–47.

WATANABE T and TERABE S (2000), 'Review: analysis of natural food pigments by capillary electrophoresis', *J Chromatogr A*, 880, 311–22.

YEBRA-BIURRUN M C (2000), 'Flow injection determination of artificial sweeteners: a review', *Food Addit Contam*, 17(9), 733–8.

6.5.3 Journals
Analyst
Analytical Chemistry
Food Chemistry
Food Additives and Contaminants
Fresenius' Journal of Analytical Chemistry
Journal of Agricultural and Food Chemistry
Journal of AOAC International
Journal of Capillary Electrophoresis
Journal of Chromatography A
Trends in Analytical Chemistry
Z. Lebensmittel Unters Forsch

134 Food chemical safety

6.5.4 Useful organisations for information on methods of analysis
Association of Official Analytical Chemists International (AOAC International)
Committee European Norms (CEN)
Food Chemicals Codex (FCC)
International Dairy Federation (IDF)
International Federation of Fruit Juice Producers
International Organisation for Standardisation (ISO)
Joint Executive Committee for Food Additives (JECFA)
LMBG (Germany)
Nordic Analytical Methods Committee (NMKL)

6.6 References

AACC (2000), *Approved Methods of Analysis*, 10th edn. St. Paul, American Association of Cereal Chemists.

ABAS A K, HART J P and COWELL D (2000), 'Development of an amperometric sulfite biosensor based on sulfite oxidase with cytochrome c, as electron acceptor, and screen-printed transducer', *Sens. Actuators B*, 62(2), 148–53.

ABIDI S L (2000), 'Review: chromatographic analysis of tocol-derived lipid antioxidants', *J Chromatogr A*, 881, 197–216.

ABOUL-ENEIN H Y, STEFAN R I, VAN STADEN J F, ZHANG X R, GARCIA-CAMPANA A M and BAEYENS W R G (2000), 'Recent developments and applications of chemiluminescence sensors', *Crit Rev Anal Chem*, 30(4), 271–89.

ALI M S and PHILLIPPO E T (1996), 'Simultaneous determination of ascorbic, dehydroascorbic, isoascorbic and dehydroisoascorbic acid in meat-based products by liquid chromatography with post-column fluorescence detection: a method extension', *J AOAC Int*, 79, 803–8.

ANON (1995), EDXRF. (The Lab-X 3000, an energy-dispersive X-ray fluorescence spectrometer from Oxford Instruments), *Lab News (April)*, 8.

AOAC INT. (2000), *AOAC Official Methods of Analysis*. 17th edn. Gaithersburg, AOAC International.

AOCS (1998), *AOCS Official Methods*, 5th edn. Champaign, American Oil Chemists Society.

BALL, G F M (2000), 'The fat soluble vitamins', in Nollet L M L, *Food Analysis by HPLC*, 2nd edn. New York, Marcel Dekker, pp. 321–402.

BEJAARS P R, VAN DIJK R, BISCHOP E and SPIEGELENBERG, W M (1996), 'Liquid chromatographic determination of free glutamic acid in soup, meat product and Chinese food: Interlaboratory study', *J AOAC Int*, 79(3), 697–702.

BERZAS N, GUIBERTEAU C and CONTENTO S (1997), 'Separation and determination of dyes by ion-pair chromatography', *J Liq Chromat Relat Technol*, 20, 3073–88.

BOGNÁR A and DAOOD H G (2000), 'Simple in-line postcolumn oxidation and derivatisation for the simultaneous analysis of ascorbic and dehydroascorbic acids in foods', *J Chromat Sci*, 38, 162–8.

BOSSI A, PILETSKY S A, RIGHETTI P G and TURNER A P F (2000), 'Review: capillary electrophoresis coupled to biosensor detection', *J Chromatogr A*, 892, 143–53.

BOSTROM C M and LINDEBERG J (2000), 'Biosensor-based determination of folic acid in fortified food', *Food Chem*, 70(4), 523–32.

BROCKMANN R (1998), 'Detection of natural colorants in raw sausage. Results of a collaborative test', *Fleischwirtschaft*, 78, 143–5.

BUCSIS L (1999), 'Schnelle enzymatische Analysen durch pH Anderungssenorik', *Labor Praxis*, 2, 48–50.

BULDINI P, CAVALLI S and TRIFIRO A (1997), 'State-of-the-art ion chromatographic determination of inorganic ions in food', *J Chromatogr A*, 789, 529–48.

CABERO C, SAURINA J and HERNANDEZ C (1999), 'Flow-injection spectrophotometric determination of cyclamate in sweetener products with sodium 1,2-naphthoquinone-4-sulfonate', *Anal Chim Acta*, 381, 307–13.

CABONI M F, MENOTTI S and LERCKER G (1996), 'Separation and analysis of phospholipids in different foods with a light scattering detector', *J AOCS*, 73(1), 1561–6.

CAMPENELLA L, PACIFICI F, SAMMARTINO M P and TOMASSETTI M (1998), 'Analysis of lecithin in pharmaceutical products and diet integrators using a new biosensor operating directly in non-aqueous solvent', *J Pharm Biomed Anal*, 18, 597–604.

CASELUNGHE M B and LINDEBERG J (2000), 'Biosensor-based determination of folic acid in fortified food', *Food Chem*, 70, 523–32.

CATALDI T R I, CAMPA C and DE BENEDETTO G E (2000), 'Carbohydrate analysis by high performance anion-exchange chromatography with pulsed amperometric detection: The potential is still growing', *Fresenius J Anal Chem*, 368, 739–58.

CEFAI J (2000), 'Complexity and miniaturisation', *LC GC Europe*, 13, 752–64.

CEN (1998), 'Foodstuffs – Determination of nitrate and/or nitrite content – Part 4: IC method for the determination of nitrate and nitrite content of meat products', European Committee for Food Standardisation, CEN/TC 275.

CEN (1999), 'Foodstuffs – Determination of acesulfame-K, aspartame and saccharin: HPLC method', European Committee for Food Standardisation, EN 12856:1999.

CHASE G W Jr., AKOH C C and EITENMILLER R L (1995), 'Liquid chromatographic analysis of sucrose polyester in salad dressings by evaporative light scattering detection', *J AOAC Int*, 78, 1324–7.

CHEN B H and FU S C (1995), 'Determination of preservatives, sweeteners and antioxidants in foods by paired-ion liquid chromatography', *Chromatographia*, 41, 43–50.

CHEN Q Z, MOU S F and SONG Q (1997a), 'Determination of sodium saccharin in drinks by high-performance ion-exclusion chromatography', *Fenxi Ceshi Xuebao*, 16, 55–57.

CHEN Q C, MOU S F, LIU K N, YANG Z Y and NI Z M (1997b), 'Separation and determination of four artificial sweeteners and citric acid by high performance anion-exchange chromatography', *J Chromatogr A*, 771(1/2), 135–43.
CHOI M M F, HSU M Y and WONG S L (2000), 'Determination of cyclamate in low-calorie foods by high-performance liquid chromatography with indirect visible photometry', *Analyst*, 125, 217–20.
COFFREY J S, NURSTEN H E, AMES J M and CASTLÉ L (1997), 'A liquid chromatographic method for the estimation of Class III caramel added to foods', *Food Chem*, 58(3), 259–67.
CORRADINI C, CANALI G, COGLIANDO E and NICOLETTI I (1997), 'Separation of alditols of interest in food products by high-performance anion exchange chromatography with pulsed amperometric detection', *J Chromatogr A*, 791(1–2), 343–9.
CSCERHÁTI T and FORGÁCS E (1999), *Chromatography in Food Science and Technology*. Lancaster, Technomic Publishing Co.
CUNNINGHAM A J (1998), *Introduction to Bioanalytical Sensors*. New York, Wiley Press.
CUNNINGHAM W C and WARNER C R (2000), 'Bromine concentration as an indication of pre-baking bromination of bread products', *Food Addit Contam*, 17(2), 143–8.
DA COSTA C T, HORTON D and MARGOLIS S A (2000), 'Review: Analysis of anthocyanins in foods by liquid chromatography, liquid chromatography-mass spectrometry and capillary electrophoresis', *J Chromatogr A*, 881, 403–10.
DALZEL J M (1998), *Ingredients Handbook: Fat Substitutes*. Leatherhead, Leatherhead Food Research Association.
DANIELSON N D, SHERMAN J H and SCHOMAKER D D (2000), 'Determination of cations and anions by HPLC', in Nollet L M L, *Food Analysis*, 2nd edn. New York, Marcel Dekker, pp. 987–1012.
DAVIES A M C and GIANGIACOMO R (2000), 'Near infrared spectroscopy'. Proceedings of the 9th international conference, NIR Publications.
DE LEENHEER A P, LAMBERT W E and VAN BOCXLAER J F (2000), *Modern Chromatographic Analysis of Vitamins*, 3rd edn. Chromatographic Science Series Vol. 84, New York, Marcel Dekker.
DIAZ G and CABANILLAS G (1996), 'Analysis of synthetic food antioxidants', in Nollet L M L, *Handbook of Food Analysis, Vol. 2*, New York, Marcel Dekker, pp. 1769–834.
DIEFFENBACHER A (1998), 'Determination of phenolic antioxidants in foods', *Dtsche Lebensm Rundsch*, 94, 381–5.
DIXON D E (1998), 'Immunoassays', in Nielsen S S, *Food Analysis*, 2nd edn. Gaithersburg, Aspen Publishers Inc., 331–48.
EDER R (2000), 'Pigments', in Noller L M L, *Food Analysis by HPLC*, New York, Marcel Dekker, pp. 825–80.
FATIBELLO F, MARCOLINO J and PEREIRA A V (1999), 'Solid-phase reactor with

copper (II) phosphate for flow-injection spectrophotometric determination of aspartame in tabletop sweeteners', *Anal Chim Acta,* 384, 167–74.

FERNANDES C F and FLICK G J (1998), 'Capillary electrophoresis for food analysis', in Wetzel D L B and Charalambous G, *Instrumental Methods in Food and Beverage Analysis,* Amsterdam, Elsevier, pp. 575–12.

FERNANDES J C B, KUBOTA L T and DE OLIVEIRA N (1999), 'Potentiometric biosensor for L-ascorbic acid based on ascorbate oxidase of natural source immobilized on ethylene-vinylacetate membrane', *Anal Chim Acta,* 385, 3–12.

FERREIRA I M P L V O, LIMA J L F C, MONTENEGRO M C B S M, OLMOS R P and RIOS A (1996), 'Simultaneous assay of nitrite, nitrate and chloride in meat products by flow injection', *Analyst,* 121(10), 1393–6.

FOOD CHEMICALS CODEX (1996), 4th edn., Committee on Food Chemicals Codex, Institute of Medicine, Washington DC, National Academy Press.

FRAZIER R A, AMES J M and NURSTEN H E (2000a), *Capillary Electrophoresis for Food Analysis – Method Development.* Cambridge, Royal Society of Chemistry.

FRAZIER R A, INNS E L, DOSSI N, AMES J M and NURSTEN H E (2000b), 'Development of a capillary electrophoresis method for the simultaneous analysis of artificial sweeteners, preservatives and colours in soft drinks', *J Chromatogr A,* 876, 213–20.

GALENSA R (1998), 'Biosensor-coupling with FIA and HPLC systems', *Lebensmittelchemie,* 52(6), 15.

GALLETTI G C, BOCCHINI P, GIOACCHINI A M and RODA A (1996), 'Analysis of the artificial sweetener aspartame by means of liquid chromatography-electrospray mass spectrometry', *Rapid Commun Mass Spectrom,* 10, 1153–5.

GLÓRIA M B A (2000), 'Intense sweeteners and synthetic colours', in Nollet L M L, *Food Analysis by HPLC,* New York, Marcel Dekker, pp. 523–73.

GONZÁLES M, GALLEGO M and VALCÁRCEL M (1999), 'Gas chromatographic flow method for the preconcentration and simultaneous determination of antioxidant and preservative additives in fatty foods', *J Chromatogr A,* 848, 529–36.

GORDON I and KRISHNAKUMAR V (1999), 'European demand', *Int Food Ingred,* 1, 14–15.

GOUVEIA S T, FATIBELLO O and DE ARAUJO N (1995), 'Flow-injection-spectrophotometric determination of cyclamate in low calorie soft drinks and sweeteners', *Analyst,* 120, 2009–12.

GROMES R, SCHNELLBACHER B, SIEGL T and VATTER T (1995), 'Determination of aspartame levels in foodstuffs with a new enzymatic assay and HPLC', *Dtsch Lebensm Rundsch,* 91, 171–4.

HALL C A, ZHU A and ZEECE M G (1994), 'Comparison between capillary electrophoresis and HPLC separation of food grade antioxidants', *J Agric Food Chem,* 42, 919–21.

HASEBE Y, AKIYAMA T, YAGISAWA T and UCHIYAMA S (1998), 'Enzyme-less amperometric biosensor for L-ascorbate using poly-L-histidine-copper complex as an alternative biocatalyst', *Talanta,* 47, 1139–47.

HASENHUETTL G L and HARTEL R W (1997), *Food Emulsifiers and their Applications.* London, Chapman and Hall.

HAYASHI T, UENO E, ITO Y, OKA H, OZEKI N, ITAKURA Y, YAMADA S, KAGAMI T, KAJITA A, FUJITA H and ONO M (1999), 'Analysis of beta-carotene and paprika colour in foods using reversed-phase thin-layer chromatography-scanning densitometry', *Shokuhin Eiseigaku Zasshi,* 40, 356–62.

HENSHALL A (1997), 'Use of ion chromatography in food and beverage analysis', *Cereal Foods World,* 42, 414–19.

HERMANN H (2000), 'Rapid automated analysis of antioxidants', *Labor Praxis,* 24, 24–7.

HIDIROGLU N, MADERE R and BEHRENS W (1998), 'Electrochemical determination of ascorbic acid and isoascorbic acid in ground meat and processed foods by HPLC', *J Food Comp Anal,* 11, 89–96.

HIROKADO M, KIMURA K, SUZUKI K, SADAMASU Y, KATSUKI Y, YASUDA K and NISHIJIMA M (1999), 'Detection method of madder colour, cochineal extract, lac colour, carthamus yellow and carthamus red in processed foods by TLC', *Shokuhin Eiseigaku Zasshi,* 40, 488–93.

HUANG Y M, ZHANG C, ZHANG X R and ZHANG Z J (1999), 'A sensitive chemiluminescence flow system for the determination of sulfite', *Anal Lett,* 32, 1211–24.

ISMAIL A A, VAN DE VOORT F R and SEDMAN J (1997), 'Fourier Transform Infrared Spectroscopy', in Paré J R J and Bélanger J M R (1997), *Instrumental Methods in Food Analysis.* Series: Techniques and instrumentation in analytical chemistry – Vol. 18, Amsterdam, Elsevier.

INDYK H E et al. (2000), 'Determination of biotin and folate in infant formula and milk by optical biosensor-based immunoassay', *JAOCS,* 83(5), 1141–8.

ISO (1997), 'Processed cheese and processed cheese products – calculation of the content of added citrate emulsifying agents and acidifiers/pH controlling agents, expressed as citric acid. International Organisation for Standardisation, ISO 12082:1997.

ISO (1998a), 'Starch derivatives: Determination of composition of glucose syrups, fructose syrups, and hydrogenated glucose syrups. Method using HPLC', International Organisation for Standardisation, ISO 10505:1998.

ISO (1998b), 'Meat and meat products: Determination of starch and glucose contents. Enzymatic method', International Organisation for Standardisation, ISO 13965:1998.

ITAKURA Y, UENO E and ITO Y (1999), 'Analysis of lac and cochineal colours in foods using reversed phase thin-layer chromatography-scanning densitometry', *Shokuhin Eiseigaku Zasshi,* 40, 183–8.

ITO Y (1999), 'Assay method for sucralose contained in a variety of foods', *Foods Food Ingred J Japan,* 182, 35–41.

JOL C N, NEIESS T G, PENNINKHOF B, RUDOLPH B and DE RUITER G A (1999), 'A novel high-performance anion-exchange chromatographic method for the analysis of carrageenans and agars containing 3,6-anhydrogalactose', *Anal Biochem*, 268, 213–22.

KAROVIČOVÁ J and ŠIMKO P (1996), 'Preservatives in foods' in Nollet L M L, *Handbook of Food Analysis, Vol. 2*. New York, Marcel Dekker, pp. 1745–66.

KAROVIČOVÁ J and ŠIMKO P (2000a), 'Preservatives and antioxidants' in Nollet L M L, *Food Analysis by HPLC*, 2nd edn. New York, Marcel Dekker, pp. 575–620.

KAROVIČOVÁ J and ŠIMKO P (2000b), 'Determination of synthetic phenolic antioxidants in foods by HPLC', *J Chromatogr A*, 882, 271–81.

KARYAKIN A A, KARYAKINA E E and GORTON L (2000), 'Amperometric biosensor for glutamate using Prussian blue-based 'artificial peroxidase' as a transducer for hydrogen peroxide', *Analyt Chem*, 72(7), 1720–3.

KAUFMANN A, PACCIARELLI S V, ROTH S and RYSER B (2000), 'Determination of some food additives in meat products by ion-chromatography', *Mitt Lebensm Hyg*, 91, 581–96.

KITADA Y, OKAYAMA A, IMAI S, NAKAZAWA H and BOKI K (1997), 'Analysis of soluble dietary fibre by electrophoresis and high-performance liquid chromatography in soft drinks', *Jpn J Toxicol Env Health*, 43, 274–9.

KOK W (2000), *Capillary Electrophoresis: Instrumentation and Operation*. Chromatographia CE Series, Volume 4, Vieweg Germany.

KOSWIG S, FUCHS G, HOTSOMMER H J and GRAEFE U (1997), 'The use of HPAE-PAD for the determination of composite sugars of gelling carrageenans and agarose by HPLC', *Seminars in Food Analysis*, 2, 71–83.

KOTSIRA V P and KLONIS Y D (1998), 'Colorimetric assay for lecithin using two co-immobilised enzymes and an indicator dye conjugate', *J Agric Food Chem*, 46, 3389–94.

KRESS-ROGERS E (1998), 'Chemosensors, biosensors and immunosensors', in Kress-Rogers E, *Instrumentation and Sensors for the Food Industry*, Cambridge, Woodhead, pp. 581–669.

KUCHTA T, HAPALA A, MARIASSYOVA M, CANTAGALLI D and GIROTTI S (1999), 'Luminometric determination of the antioxidative activity of natural colour concentrates', *Bull Food Res*, 38(2), 103–8.

KUO K L, HUANG H Y and HSIEH Y Z (1998), 'High performance capillary electrophoretic analysis of synthetic food colours', *Chromatographia*, 47(5/6), 249–56.

LANCASTER F E and LAWRENCE J F (1996), 'High-performance liquid chromatographic separation of carminic acid, alpha- and beta-bixin, and alpha- and beta-norbixin, and the determination of carminic acid in foods', *J Chromatogr A*, 732, 394–8.

LEE H S (1995), 'Liquid chromatographic determination of benzoic acid in orange juice: interlaboratory study', *J AOAC Int*, 78, 80–82.

LIN Y H, CHOU S S and SHYU Y T (2000), 'Simultaneous determination of sweeteners and preservatives in preserved fruits by micellar electrokinetic capillary chromatograpy', *J Chromat Sci*, 38, 345–52.

LIU J, LEE T, BOBIK E, GUZMA-HARTY M and HASTILOW C (1993), 'Quantitative determination of monoglycerides and diglycerides by HPLC and evaporative light scattering detection', *J AOCS,* 70(4), 343–7.

LOMER M C E, THOMPSON R P H, COMMISSO J, KEEN C L and POWELL J J (2000), 'Determination of titanium dioxide in foods using inductively coupled plasma optical emission spectrometry', *Analyst,* 125, 2339–43.

LOZANO A, ALEGRIA A, BARBERA R, FARRE R and LAGARDA M J (1998), 'Evaluation of method for determining ascorbic acid, citric acid, lactose and maltodextrins in infant juices and formulas', *Alimentaria,* 298, 95–9.

MASLOWSKA J (1996), 'Colorants', in Nollet L M L, *Handbook of Food Analysis, Vol. 2.* New York, Marcel Dekker, pp. 1723–43.

MCCABE D R and ACWORTH I N (1998), 'Determination of synthetic phenolic antioxidants in food using gradient HPLC with electrochemical array detection', *Amer Lab News,* 30(13), 18B, 18D, 16.

MOH M H, TANG T S and TAN G H (2000), 'Improved separation of sucrose ester isomers using gradient HPLC with evaporative light scattering detection', *Food Chem,* 69, 105–10.

MOPPER B (1996), 'Determination of cations and anions by capillary electrophoresis', in Nollet L M L, *Handbook of Food Analysis, Vol. 2.* New York, Marcel Dekker, pp. 1867–87.

MYERS P (2000), 'Hype and chips', *LC GC Europe,* 13, 744–50.

NEIER S and MATISSEK R (1998), 'Trennung und Bestimmung von Farbstoffen in Süßwaren mittels HPLC' *Dtsche Lebensm Rundsch,* 94(11), 374–80.

NIELSEN S S (1998), *Food Analysis,* 2nd edn. Gaithersburg, Aspen Publishers Inc.

NMKL (1989), 'Colours, synthetic, water-soluble. Liquid chromatographic determination in foods', Nordic Committee on Food Analysis, Method No. 130.

NMKL (1997a), 'Saccharin. Liquid chromatographic determination in beverages and sweets', Nordic Committee on Food Analysis, Method No. 122.

NMKL (1997b), 'Benzoic acid, sorbic acid and p-hydroxybenzoic acid esters. Liquid chromatographic determination in foods', Nordic Committee on Food Analysis, Method No. 124.

NMKL (1997c), 'Starch and glucose: enzymatic determination in foods', Nordic Committee on Food Analysis, Method No. 145.

NMKL (2000), *Methods of Analysis.* Nordic Committee on Food Analysis.

NOLLET L M L (1996a), *Handbook of Food Analysis.* Vol. 1, New York, Marcel Dekker.

NOLLET L M L (1996b), *Handbook of Food Analysis.* Vol. 2, New York, Marcel Dekker.

NOLLET L M L (2000), *Food Analysis by HPLC,* 2nd edn, New York, Marcel Dekker.

NYYSSÖNEN K, SALONEN J T, PARVIAINEN M T (2000), 'Ascorbic acid', in De Leenheer A P, Lambert W E and van Bocxlaer J F, *Modern Chromatographic Analysis of Vitamins,* 3rd edn. New York, Marcel Dekker.

OKAWA Y, KOBAYASHI H and OHNO T (1998), 'Direct and simultaneous determination of ascorbic acid and glucose in soft drinks with electrochemical filter/biosensor FIA system', *Bunseki Kagaku,* 47, 443–5.

OLIVER J and PALOU A (2000), 'Review: chromatographic determination of carotenenoids in foods', *J Chromatogr A,* 881, 543–55.

PALAROAN W S, BERGANTIN J J and SEVILLA F, III (2000), 'Optical fibre chemiluminescence biosensor for antioxidants based on an immobilized luminol/haematin reagent phase', *Anal Lett,* 33, 1797–810.

PANFILI G, MANZI P, COMPAGNONE D, SCARCIGLIA L and PALLESCHI G (2000), 'Rapid assay of choline in foods using microwave hydrolysis and a choline biosensor', *J Agric Food Chem,* 48, 3403–7.

PANT I and TRENNERY V C (1995), 'The determination of sorbic acid and benzoic acid in a variety of beverages and foods by micellar electrokinetic capillary chromatography', *Food Chem,* 53(2), 219–26.

PARÉ J R J and BÉLANGER J M R (1997), *Instrumental Methods in Food Analysis.* Series: Techniques and instrumentation in analytical chemistry – Vol. 18, Amsterdam, Elsevier.

PATZ C D, GALENSA R and DIETRICH H (1997), 'Determination of sulfite in fruit juices by HPLC-biosensor coupling', *Dtsch Lebensm Rundsch,* 93, 347–51.

PÉREZ-RUIZ T, MARTINÉZ-LOZANO C, SANZ A and BRAVO E (2000), 'Analysis of glutamate in beverages and foodstuffs by capillary electrophoresis with laser-induced fluorescence detection', *Chromatographia,* 52, 599–602.

PÉREZ-URQUIZA M and BELTRÁN J L (2000), 'Determination of dyes in foodstuffs by capillary zone electrophoresis', *J Chromatogr A,* 898, 271–5.

PERIS-TORTAJADA M (2000), 'HPLC determination of carbohydrates in foods', in Nollet L M L, *Food Analysis by HPLC,* New York, Marcel Dekker, pp. 287–320.

PIZZOFERRATO L, DI LULLO G and QUATTRUCCI E (1998), 'Determination of free, bound and total sulphites in food by indirect photometry-HPLC', *Food Chem,* 63, 275–9.

POMERANZ Y and MELOAN C E (1994), 'X-ray methods', in Pomeranz Y and Meloan C E, *Food Analysis: Theory and Practice.* London, Chapman and Hall, pp. 158–71.

PREUß A and THEIR H P (1996), 'Isolation of natural thickeners and gums from food for determination by capillary column gas chromatography', *Z Lebensm Unters Forsch,* 176, 5–11.

PRICE B J and MAJOR H W (1990), 'X-ray fluorescence proves useful for quality control', *Food Technol,* 44(9), 66–70.

PRODOLLIET J (1996), 'Intense sweeteners', in Nollet L M L, *Handbook of Food Analysis, Vol. 2.* New York, Marcel Dekker, pp. 1835–66.

PRODROMIDIS M I, TZOUWARA-KARAYANNI S M, KARAYANNIS M I and VADGAMA P M (1997), 'Bioelectrochemical determination of citric acid in real samples using a fully automated flow injection manifold', *Analyst,* 122(10), 1101–6.

QU F, QI Z H, LIU K N and MOU S F (1999), 'Determination of aspartame by ion chromatography with electrochemical integrated amperometric detection', *J Chromatogr A*, 850, 277–81.

QUEMENER B, MAROT C, MOUILLET L, DA RIZ V and DIRIS D (2000a), 'Quantitative analysis of hydrocolloids in food systems by methanolysis coupled to reverse HPLC. Part 1. Gelling hydrocolloids', *Food Hydrocolloids*, 14, 9–17.

QUEMENER B, MAROT C, MOUILLET L, DA RIZ V and DIRIS D (2000b), 'Quantitative analysis of hydrocolloids in food systems by methanolysis coupled to reverse HPLC. Part 2. Pectins, alginates and xanthans', *Food Hydrocolloids*, 14, 19–28.

RAMSAY G (1998), *Commercial Biosensors*. New York, J. Wiley and Son.

ROBERTS M A and QUEMENER B (1999), 'Measurement of carrageenans in food: challenges, progress, and trends in analysis', *Trends in Food Sci and Technol*, 10, 169–81.

ROYLE L, AMES J M, CASTLE L, NURSTEN H E and RADCLIFFE C M (1998), 'Identification and quantification of class IV caramels using capillary electrophoresis and its application to soft drinks', *J Sci Food Agric*, 76(4), 579–87.

RUSSEL L F (2000) 'Quantitative determination of water soluble vitamins', in Nollet L M L, *Food Analysis by HPLC*, 2nd edn. New York, Marcel Dekker, pp. 403–76.

SÁDECKA J and POLONSKÝ J (1999), 'Determination of inorganic ions in foods and beverages by capillary electrophoresis', *J Chromatogr A*, 834, 401–17.

SÁDECKA J and POLONSKÝ J (2000), 'Review: electrophoretic methods in the analysis of beverages', *J Chromatogr A*, 880, 243–79.

SANDER L C, SHARPLESS K E and PURSCH M (2000), 'Review: C30 stationary phases for the analysis of food by liquid chromatography', *J Chromatogr A*, 880, 189–202.

SCHREIER C J and GREENE R J (1997), 'Determination of ethoxyquin in feeds by liquid chromatography: collaborative study', *J AOAC Int*, 80, 725–31.

SCHÜEP W and SCHIERLE J (1997), 'Determination of beta carotene in commercial foods: interlaboratory study', *J AOAC Int*, 80, 1057–64.

SCHULTE E (1995), 'Vereinfachte Bestimmung von Sorbinsäure, Benzoesäure und PHB-Estern in Lebensmitteln durch HPLC', *Dtsche Lebensm Rundsch*, 91(9), 286–9.

SCOTT A O (1998), *Biosensors for Food Analysis*, Cambridge, The Royal Society of Chemistry.

SEKIGUCHI Y, MATSUNAGA A, YAMAMOTO A and INOUE Y (2000), 'Analysis of condensed phosphates in food products by ion chromatography with an on-line hydroxide eluent generator', *J Chromatogr A*, 881, 639–44.

SHERMA J (2000), 'Review: thin-layer chromatography in food and agricultural analysis', *J Chromatogr A*, 880, 129–47.

SHI R, STEIN K and SCHWEDT G (1996), 'Flow-injection analysis of carbonate, sulfite and acetate in food', *Dtsch Lebensm Rundsch*, 92, 323–8.

SIU D C and HENSHALL A (1998), 'Ion chromatographic determination of nitrite and nitrate in meat products', *J Chromatogr A*, 804, 157–60.
SONG W O, BEECHER G R and EITENMILLER R R (2000), *Modern Analytical Methodologies In Fat- and Water-soluble Vitamins*. Chichester, Wiley.
SPRENGER C, GALENSA R and JENSEN D (1999), 'Simultaneous determination of cellobiose, maltose and maltotriose in fruit juices by high-performance liquid chromatography with biosensor detection', *Dtsch Lebensm Rundsch*, 95, 499–504.
SU X L, WEI W Z, NIE L and YAO S Z (1998), 'Flow-injection determination of sulfite in wines and fruit juices by using a bulk acoustic wave impedance sensor coupled to a membrane separation technique', *Analyst*, 123, 221–4.
SUN D H, WATERS J K and MAWHINNEY T P (2000), 'Determination of thirteen common elements in food samples by inductively coupled plasma atomic emission spectrometry: Comparison of five digestion methods', *J AOAC Int*, 83(5), 1218–24.
SWEDBERG S (1997), 'Capillary electrophoresis: Principles and applications', in Paré J R J and Bélanger J M R, *Instrumental Methods in Food Analysis*. Series: Techniques and instrumentation in analytical chemistry – Vol. 18, Amsterdam, Elsevier.
THYSSEN K and ANDERSEN K S (1998), 'Determination of sorbitan tristearate in vegetable oils and fats', *J AOCS*, 75(12), 1855–60.
TRENERRY V T (1996), 'The determination of the sulfite content of some foods and beverages by capillary electrophoresis', *Food Chem*, 55, 299–303.
TSAFACK V C, MARQUETTE C A, LECA B and BLUM L J (2000), 'An electrochemiluminescence-based fibre optic biosensor for choline flow-injection analysis', *Analyst*, 125, 151–5.
TUNICK M H, PALUMBO S A and FRATAMICO P M (1998), *New Techniques in the Analysis of Foods*. New York, Kluwer Academic/Plenum.
VAN DER MEEREN P and VANDERLEENEN J (2000), 'Phospholipid analysis by HPLC', in Nollet L M L, *Food Analysis by HPLC*, 2nd edn. New York, Marcel Dekker, pp. 251–85.
VRBOVA E, PECKOVA J and MAREK M (1993), 'Biosensor for determination of starch', *Starke*, 45(10), 341–4.
WATANABE T and TERABE S (2000), 'Review: analysis of natural pigments by capillary electrophoresis', *J Chromatogr A*, 880, 311–22.
WEISS J (1995), *Ion Chromatography*, 2nd edn. Weinheim, Germany, VCH.
WETZEL D L B (1998), 'Analytical near infrared spectroscopy', in Wetzel D L B and Charalambous G, *Instrumental Methods in Food and Beverage Analysis*. Amsterdam, Elsevier, pp. 141–94.
WETZEL D L B and CHARALAMBOUS G (1998), *Instrumental Methods in Food and Beverage Analysis*. Amsterdam, Elsevier.
WOJCIECHOWSKI C, DUPUY N, TA C D, HUVENNE J P and LEGRAND P (1998), 'Quantitative analysis of water-soluble vitamins by ATR-FTIR spectroscopy', *Food Chem*, 63, 133–40.
WOODWARD J R, BRUNSMAN A, GIBSON T D and PARKER S (1998), 'Practical

construction and function of biosensor systems for quality control in the food and beverage industry', in Scott A O, *Biosensors for Food Analysis.* Cambridge, Royal Society of Chemistry.

WYGANT M B, STATLER J A and HENSHALL A (1997), 'Improvements in amperometric detection of sulfite in food matrices', *J AOAC Int*, 80, 1374–80.

YEBRA-BIURRUN M C (2000), 'Flow injection determination of artificial sweeteners: a review', *Food Addit Contam*, 17(9), 733–8.

YOSHIDA A (1995), 'Enzyme immunoassay for carminic acid in foods', *J AOAC Int*, 78(3), 807–11.

ZHAO R R and JOHNSON B P (2000), 'Capillary electrochromatography: analysis of sucralose and related carbohydrate compounds', *J Liq Chromat Rel Technol*, 23(12), 1851–7.

7

Adverse reactions to food additives

M. A. Kantor, University of Maryland

7.1 Introduction

Consumers have a paradoxical view of food additives. The term itself often has a negative connotation among consumers, who frequently state a preference for foods that are 'natural' and made without chemical additives and preservatives. However, these same individuals also claim to prefer foods that are nutritious, convenient and maintain freshness, precisely the foods which are likely to contain nutritional additives and preservatives such as antioxidants (Sloan, 1998).

Misconceptions about food additives are perpetuated in the media and popular press, and recently have been disseminated via the Internet. There is confusion about the sources and functions of these compounds. Consumers are confused, for example, about the relative safety of 'natural' as opposed to 'artificial' food ingredients. A number of studies have shown consumers' suspicion of synthetic chemicals in foods which are seen as posing a higher health risk than natural ingredients (Sloan *et al.*, 1986; McNutt *et al.*, 1986; Crowe *et al.*, 1992). Unrealistic fears about food additives may be attributed in part to the public's fundamental lack of understanding of toxicology, including the failure to appreciate the concept of 'dose' or the body's capacity to metabolize and detoxify the myriad of food constituents people are exposed to daily (Jones, 1992).

Although food additives have been blamed for a variety of ills, including epileptic seizures, brain tumors, and even criminal behavior, only a few of the approximately 2,800 food additives used in the United States have ever been directly linked to adverse events. New food additives undergo extensive safety testing prior to approval, and existing additives suspected of causing health problems are periodically re-evaluated by regulatory agencies responsible for food safety, such as the US Food and Drug Administration (FDA). Consequently, there exists a vast body of research indicating that exposure to most dietary food additives, at levels typically found in the food supply, poses no hazard for the general population. However, individuals in certain groups clearly are sensitive to certain food additives such as sulfiting agents. It would not be surprising if some individuals were susceptible to certain food additives considering the population's biological diversity, which includes immuno-compromised individuals, people with inherited metabolic disorders, and people with different capacities to metabolize xenobiotics. Moreover, a nearly infinite number of synergistic or antagonistic interactions are possible between food additives, other dietary components, and pharmaceuticals (Borzelleca, 1995).

Since most of the adverse reactions attributed to food additives involve non-protein substances, affected individuals are usually considered to have a 'sensitivity' or an 'intolerance' rather than an allergy. An adverse reaction to a food additive is also known as a 'food idiosyncrasy' because symptoms such as asthma and headache occur through unknown mechanisms, and may be of psychosomatic origin (Taylor and Hefle, 2001). In contrast, adverse reactions to foods such as eggs, wheat, peanuts and other protein sources are likely to be a true allergy (or 'hypersensitivity') because they involve an antigen (food glycoprotein)/antibody reaction mediated though an IgE immune response. Food allergy symptoms include gastrointestinal disturbances (nausea, vomiting), respiratory symptoms (breathing difficulty, rhinitis), cutaneous symptoms (urticaria or hives, rash), headache, or dizziness. These symptoms range in severity and may be life-threatening if anaphylactic shock occurs (Taylor and Hefle, 2001).

This chapter begins by discussing how consumer perceptions of risks associated with food additives compare to other food-related health risks. It then addresses the US government's system for monitoring adverse reactions from food additives. The bulk of the chapter focuses on additives that have been blamed for causing health problems and which have generated controversy in the US and other countries. Finally, there is a brief discussion of future trends and additional sources of information are provided.

7.2 Consumer attitudes about food additives

Two of the best known surveys that monitor consumer attitudes, concerns, and behaviors on various nutrition and food safety issues are the 'Trends' survey developed by the Food Marketing Institute (FMI) and the 'Prevention Index' published by Rodale Press (1996).

FMI, a trade association for food retailers and wholesalers, began tracking consumer trends in 1973. In its most recent survey (FMI, 2000), 7% of food shoppers with an interest in nutrition (n=870) indicated a concern about food or 'chemical' additives. This percentage has remained fairly constant during the past four years and ranked seventh among all the nutrition/health concerns voluntarily mentioned (fat content was the leading concern mentioned by 46% of respondents). The percentage of shoppers expressing concerns about 'preservatives' steadily declined from 8% in 1996 to 4% in 2000. However, 59% of shoppers in 2000 said they had looked for and purchased products labeled as 'natural,' which may reflect a desire to avoid food additives (FMI, 2000).

Results of this type of survey need careful interpretation, however, because responses that people voluntarily give to open-ended questions often differ from their responses to specific items. For example, when a sample of 2000 consumers were queried about their perceptions of food safety threats, 5% voluntarily mentioned 'additives,' 7% mentioned 'preservatives,' and 2% mentioned 'artificial coloring' (FMI, 1994). However, when specifically asked about their perceptions of 'additives and preservatives,' 'nitrites in food,' and 'artificial coloring,' 25%, 34%, and 22% of consumers, respectively, said they considered these substances to be serious hazards. In 1997, these percentages declined to 21% for 'additives and preservatives' and 28% for 'nitrites in food' ('artificial coloring' was not included in the 1997 survey, presumably because it was no longer considered to be an important consumer issue). These data suggest that consumer concerns about food additives may be waning in comparison to other safety issues, such as bacterial contamination (FMI, 1997).

The Prevention Index, which annually tracked a variety of health-related behaviors in a national sample between 1983 and 1996, also indicated a decline in consumer concerns about food additives. The percentage of consumers indicating they 'try a lot to avoid foods that contain additives, such as colorings and artificial flavorings' steadily declined from a high of 36% in 1987 to 27% in 1995 (Rodale, 1996). In contrast, according to the recent 'Shopping for Health' national telephone survey, a joint FMI/Prevention survey, 52% of shoppers said they purchased foods without artificial additives or preservatives for health reasons (FMI and *Prevention Magazine*, 2000). However, 73% of respondents also said they purchased fortified foods (which typically contain substantial amounts of food additives) to maintain health. This apparent contradiction suggests that consumers distinguish between what they see as beneficial as opposed to harmful additives, accepting some foods with additives but not others.

7.3 Reporting adverse reactions

The Food Additives Amendment to the Federal Food, Drug, and Cosmetic Act of 1938, which was adopted in 1958, requires that proof of safety of a new food additive be furnished by the manufacturer based on extensive scientific research. Since it is impossible to conclusively prove the safety of a new food additive through animal and clinical trials, manufacturers routinely conduct post-marketing surveillance and long-term follow-up studies to monitor adverse events.

The Adverse Reaction Monitoring System (ARMS) is a passive, post-marketing surveillance system designed to collect and analyze reports of adverse reactions allegedly due to food additives and other ingredients. The system was set up by the FDA in early 1985, partly as a response to the large number of complaints the agency received after aspartame was approved for soft drinks in 1983. Originally established to monitor adverse reactions associated with food additives and color additives (which are regulated separately by the FDA), the system was later expanded to include all food products. Today, the FDA has similar programs to monitor adverse events for pharmaceuticals, medical products and devices, vaccines, blood components, cosmetics, veterinary products, and dietary supplements.

The FDA collects complaints not only from consumers but also from physicians, food industry personnel, and others. Most reports describe symptoms of acute reactions presumably due to a food intolerance or sensitivity. The FDA attempts to determine if there actually is an association between the onset of specific symptoms and the ingestion of a food additive, as well as the strength and consistency of the relationship. If warranted, the agency may recommend that the implicated additive be investigated through challenge tests or other clinical studies (Tollefson, 1988). A key function of ARMS is to serve as a sentinel or early-warning system for identifying potential public health threats so that appropriate actions, such as implementing a warning label or product recall, may be taken. In deciding whether to initiate regulatory actions, the FDA is assisted by a group of independent scientists serving on a Health Hazard Evaluation Board. Board members assess the potential short- and long-term risks associated with an additive, particularly to vulnerable groups such as infants, pregnant women, or the elderly. The FDA is not authorized to ban an approved food additive automatically if a large number of adverse reactions are reported. However, if a convincing body of scientific evidence emerges that indicates a safety problem, the FDA can restrict or prohibit the use of an additive or institute a warning label, as it has done for nitrite and sulfiting agents.

If a complaint reported to ARMS contains sufficient detail, it is classified according to severity of the reaction as well as to frequency and consistency of the symptoms associated with the suspected ingredient (Tollefson and Barnard, 1992). Type 1 or severe reactions include anaphylactic shock, hypotensive episodes, severe gastrointestinal distress (such as protracted vomiting and diarrhea leading to dehydration), cardiac arrhythmia, severe respiratory distress, fainting, seizures, and any reaction requiring emergency medical treatment. These cases are investigated by FDA field personnel, who attempt to interview the complainant or a close family member. The FDA also may contact the attending physician, examine medical records, and analyze portions of the food suspected of causing the reaction. Type 2 reactions are more moderate in nature. These include such conditions as mild gastrointestinal or respiratory distress, localized rash or edema, insomnia, fatigue, and mild neurological symptoms such as mild anxiety, dizziness, and headache.

Cases are classified into one of four categories (A = strongest association; D = weakest association) according to the strength of the association between the ingested food additive or ingredient and the resulting symptoms. Individuals are assigned to Group A if they experienced identical symptoms each time they consumed different food products containing the additive of interest. They are assigned to Group B if the same symptoms recur each time a single product is consumed containing the additive in question. Individuals are placed in Group C if symptoms occurred once when a food product containing the additive of interest was ingested, but Group C individuals did not re-challenge themselves to see if these symptoms would recur each time they ate the same food. Complainants assigned to Group D either did not have recurring symptoms each time they ingested the ingredient, or they consulted a physician who told them it was unlikely their symptoms were being caused by the suspected ingredient.

Because ARMS is a form of passive surveillance that is based on spontaneous reports, it has a number of limitations. Most importantly, it cannot establish a definitive, causal relationship between the ingestion of an incriminated substance and the occurrence of symptoms. Since people are exposed daily to a myriad of food ingredients, it is inherently difficult to attribute an adverse event to a specific food substance. Moreover, symptoms reported are often vague or general in nature. Other confounders, which were discussed by Bradstock *et al.* (1986), include:

- Symptoms may be caused by an underlying disease or co-morbidity, or by a drug, rather than by an ingested substance.
- The segment of the population reporting adverse reactions cannot be considered representative of the entire population ingesting the product due to selection bias.
- The incidence of adverse symptoms cannot be determined, since the number of people ingesting the substance (population at risk) is not known.
- Events are under-reported in a passive surveillance system because of the failure to recognize an adverse event as being related to a food additive, the amount of effort involved, or other reasons.
- Serious adverse reactions are more likely to be reported than mild reactions, even if the former occur less frequently.
- Symptoms occurring shortly after a product is used are more likely to be reported, whereas delayed reactions may not be associated with a product and not reported.
- Data that are voluntarily reported are often incomplete and not standardized.
- Reporting may be influenced by publicity from advocacy or opposition groups.

Furthermore, there is a recognized association between the number of complaints received by the FDA and the length of time a product has been on the market. There is likely to be a large number of complaints immediately after a new food additive is introduced into the marketplace because of extensive media coverage and heightened awareness of potential risks. The number of complaints tends to taper off dramatically over time, a trend observed repeatedly by the FDA.

Although the US is one of the few countries to have a formal system for monitoring adverse reactions from food additives as part of its regulatory structure, other countries also periodically review additives. In addition, the Food and Agriculture Organization (FAO) of the United Nations and the World Health Organization (WHO) also monitors food additives. Since 1956, the Joint FAO/WHO Expert Committee on Food Additives has developed specifications for the purity of additives, evaluated toxicological data, and recommended safe levels of use.

7.4 Controversial food additives

From 1980 through June 2000, the FDA received a total of 34,011 reports describing adverse reactions attributed to food additives, food products, infant formula, medical foods, dietary supplements, and other substances in food. Adverse reaction reports were submitted by consumers, food companies, consumer advocacy groups, and by individuals commenting on various FDA initiatives and proposals. The food additives receiving the most complaints are shown in Table 7.1. The fat substitute olestra accounted for more than half of all complaints received by ARMS. The following six sections pertain to food additives that have generated controversy because of safety concerns. For each additive, pertinent data from the ARMS database are presented.

7.4.1 Food dyes

The notion that colors, flavors, and other common food additives could adversely affect children's behavior was widely publicized in the 1970s. According to Feingold (1974), 40–70% of children who exhibited impulsive behavior, learning disabilities, short attention spans, and other symptoms characteristic of 'attention deficit hyperactivity disorder' or ADHD (formerly called hyperkinesis, hyperactivity, or minimal brain dysfunction) showed dramatic improvement when placed on diets without food dyes and other common additives.

Because this hypothesis generated considerable anxiety and had broad public health implications, the Nutrition Foundation (formerly affiliated with the International Life Sciences Institute but now defunct) convened a committee to evaluate the scientific evidence relative to the Feingold hypothesis, and if warranted, provide guidelines for further research. The committee, known as the National Advisory Committee on Hyperkinesis and Food Additives (1975), found no conclusive link between diet and behavior, but could not rule out a possible association since no studies appropriate for testing the hypothesis had been carried out. Based on the committee's recommendations, several double-blind crossover studies were subsequently conducted, but these produced

Table 7.1 ARMS data. Food additives receiving adverse reaction reports, 1980–June 2000. (Data include adverse reactions reported to FDA before ARMS was officially established in 1985)

Substance monitored	Total number of complaints	Percent of total records in ARMS
Olestra	18,309	53.8
Aspartame	7,335	21.6
	Searle/NutraSweet: 4,428	*13.0*
	Consumers and other sources: 2,256	*6.7*
	Received prior to ARMS: 649	*1.9*
Sulfiting agents	1,141	3.4
Monosodium glutamate	905	2.7
Simplesse	74	0.2
FD&C Yellow No. 5 (tartrazine)	34	0.10
Psyllium	27	0.07
Saccharin	11	0.03
Nitrite	10	0.02

inconsistent results and failed to conclusively prove or disprove the Feingold hypothesis (Conners *et al.*, 1976; Goyette *et al.*, 1978; Harley *et al.*, 1978a). Furthermore, some of the data were difficult to interpret because of small sample sizes and methodological problems involved in objectively evaluating behavior.

Additional studies were carried out in which children were 'challenged' with cookies formulated either with or without a mixture of color additives (Harley *et al.*, 1978b). Although these more rigorously designed studies also provided little evidence that food additives adversely affected behavior, a small subgroup of children appeared to be sensitive to the additives (Lipton *et al.*, 1979). However, a Consensus Development Panel of the National Institutes of Health concluded that there was no convincing scientific evidence that food dyes or other food additives were related to hyperactivity (NIH, 1982). This panel recommended that children exhibiting hyperactive behavior should not be given special diets devoid of food additives, because such diets may be nutritionally restrictive. Further, it was not possible to predict which, if any, children might benefit. A more recent NIH Consensus Development Statement (1998) also failed to find a clear association between diet and ADHD, and focused mainly on treating the disorder with psychostimulant medications rather than through diet manipulation. Panel members recognized that some dietary strategies had produced intriguing results, but stopped short of specifically endorsing the need for additional research on food additives. It is noteworthy that the American Academy of Pediatrics (2000) also does not mention dietary factors in its clinical practice guidelines for diagnosing and evaluating children with ADHD.

The color additive most often associated with adverse effects is FD&C Yellow No. 5, or tartrazine (Jones, 1992; Fuglsang *et al.*, 1993). From 1980

152 Food chemical safety

through June 2000, a total of 34 complaints about tartrazine were reported to ARMS. FDA's Advisory Committee on Hypersensitivity to Food Constituents concluded in 1986 that tartrazine may cause hives in fewer than 0.01% of the population, but the committee found no evidence that it provoked asthma attacks, nor that aspirin-intolerant individuals were more sensitive to it. FD&C Yellow No. 5 was the first certified color required to be listed on food ingredient labels, but the FDA extended this labeling requirement to all certified colors in 1991.

7.4.2 Aspartame

The sweetener aspartame (L-aspartyl-L-phenylalanine-methyl ester) has been mired in controversy since its initial approval in 1974. In recent years, opponents of aspartame have accused aspartame of causing a wide range of conditions including systemic lupus, multiple sclerosis, vision problems, headaches, fatigue, disruptive or aggressive behavior in children, adverse effects on the developing fetus, and Alzheimer's disease. One report claimed that aspartame-sweetened soft drinks provided to American military personnel during the Persian Gulf War may have precipitated Gulf War syndrome.

In fact, the results of nearly 200 toxicological and clinical studies have demonstrated the safety of aspartame. Its use has been endorsed by the Joint FAO/WHO Expert Committee on Food Additives, American Medical Association, American Heart Association, and numerous other health agencies. It is consumed in more than 90 countries worldwide and is an ingredient in over 1,000 products. Nevertheless, perhaps because of its widespread use (half the US population regularly consumes products sweetened with aspartame) or negative publicity, it is second only to olestra in the number of adverse reaction complaints it has generated through ARMS (Table 7.1).

Approved in 1981 as a table top sweetener and for dry foods, aspartame was permitted in carbonated soft drinks in 1983, and in 1996 its approval was extended to all foods and beverages. There has been controversy over the role a Public Board of Inquiry played in the approval process, but this has been discounted in reports by the American Medical Association (Council of Scientific Affairs, 1985) and Stegink (1987). Reports of adverse reactions began almost immediately after approval in the 1980s, and by mid-1984 more than 600 complaints had been received by the FDA. Reports of adverse reactions peaked in 1985, when over 1,500 complaints were received by ARMS, and have been declining since then. As of June 2000, ARMS had received a total of 7,335 complaints about aspartame, with 47% of complaints linked to diet soft drinks, followed by 27% of complaints attributed to table top sweeteners. All other product categories were mentioned in fewer than 10% of complaints.

A total of 92 different symptoms have been reported by individuals claiming to have experienced an adverse reaction to aspartame, with the most frequent one being headache, reported by 28.8% of individuals. This was followed by dizziness or poor equilibrium (11.4% of reports), change in mood (10.6%), vomiting or nausea (10.1%), abdominal pain and cramps (7.0%), change in vision (5.9%), diarrhea (5.2%), seizures and convulsions (4.5%), memory loss (4.3%), fatigue (3.9%), rash (3.4%), sleep problems (3.2%), and hives (3.0%). Menstrual irregularities, breathing difficulties, and grand mal seizures each accounted for less than 2.0% of complaints. The variety of these alleged reactions makes it unlikely that a single ingredient could be responsible for them. Furthermore, 76% of the complainants who provided information on gender were female, although there is no apparent reason that women should be more susceptible than men.

A recent report by Olney *et al.* (1996) linking aspartame to glioblastoma brain tumors in the US and several industrialized European countries generated considerable media attention but was criticized by mainstream scientists as being seriously flawed. As an example, Olney provided no data on the aspartame intake of brain cancer patients. As a proposed mechanism, he suggested that aspartame undergoes nitrosation in the stomach to form a carcinogenic compound, but this reaction has not been shown to occur in humans. There is little biological plausibility for aspartame to produce adverse reactions or increase disease risks. Upon ingestion, aspartame is rapidly metabolized in the intestine by hydrolytic and proteolytic enzymes to aspartic acid, phenylalanine, methanol, and to its dipeptide base, alpha-Asp-Phe. The latter compound is hydrolyzed in intestinal mucosal cells by brush border enzymes. Thus, neither aspartame nor its dipeptide base enters the portal blood or the general circulation (Stegink, 1987). Methanol derived from aspartame does not accumulate in the body, as it is converted to formaldehyde and then to formate, which is further metabolized to carbon dioxide and water and excreted. The concentration of methanol (10% of aspartame by weight) in aspartame-containing beverages is negligible, and is substantially lower than that found in many fruit and vegetable juices (Butchko and Kotsonis, 1989).

Stegink (1987) reviewed some of the major studies investigating the highest potential ingestion levels of aspartame by the general population as well as by sensitive segments of the population. He concluded that minimal risk was associated with aspartame consumption at the 99th percentile of projected intake levels (34 mg/kg body weight), since postprandial plasma levels of aspartate, phenylalanine, and formate all were consistently within their normal range. Even at 'abuse' levels of intake (200 mg/kg body weight) these plasma concentrations were far below values associated with adverse effects. Furthermore, among individuals with phenylketonuria, the rare hereditary disorder which affects approximately 1 out of 16,000 people, a single large dose of aspartame transiently raised serum phenylalanine levels only slightly, and levels remained substantially below the toxic threshold. Stegink also found no link in test animals fed the sweetener at very high doses. Fernstrom (1994) reviewed evidence indicating that aspartame does not adversely affect brain chemistry. In particular, it does not raise the ratio of phenylalanine to other large neutral amino acids in plasma to values beyond the normal range. There is no evidence that the combination of aspartame and glutamate, which may have additive effects, contribute to brain damage, mental retardation, or neuroendocrine dysfunction. Aspartame was also found to have no effect on headaches in a double-blind placebo study (Schiffman *et al.*, 1987).

According to the National Cancer Institute's Surveillance, Epidemiology, and End Results (SEER) database on cancer incidence, cases of brain and nervous system cancers began increasing in 1973, well before aspartame was approved. The incidence continued to rise through 1985, leveled off and remained constant until about 1990, and has been steadily declining since then for all races and genders combined (Ries *et al.*, 2000). However, aspartame consumption steadily increased during the late 1980s, as it was introduced into a wider array of products. If a link between aspartame and brain tumors existed, a continued rise in brain tumor rates would have been expected during the late 1980s and 1990s.

7.4.3 Olestra

The fat substitute olestra contains a backbone molecule of sucrose, rather than glycerol, that is esterified to either six, seven, or eight fatty acids. Olestra is impervious to digestive enzymes and is unabsorbed, contributing no calories to the diet. However, the lack of absorption is responsible for gastrointestinal side effects in some people. The FDA's approval of olestra on January 24, 1996, 25 years after it was patented, culminated a laborious process that, as in the case of aspartame, was steeped in controversy. Assessing olestra's safety, including the results of more than 150 studies submitted by the Procter & Gamble Company (P&G), the manufacturer of olestra, was challenging because olestra is a 'macro-ingredient' that is incorporated into foods at levels far greater than traditional food additives. Therefore, it could not be evaluated using standard toxicology procedures that involve feeding large doses (e.g., 100 times the likely human intake) to laboratory animals. This and other problems with testing macro-ingredients was reviewed by Gershoff (1995). The FDA's Food Advisory Committee, a group of 25 experts which recommended approval of olestra, felt that the gastrointestinal symptoms did not represent a serious health risk, even in potentially sensitive persons. However, as a condition of approval, the committee recommended that products containing olestra should bear a label statement warning about the possibility of intestinal discomfort.

Snack foods containing olestra were first introduced into test markets in five states during 1996, and went on sale nationally in February 1998. From April 1996 through June 2000, the FDA received a total of 18,309 adverse reports associated with olestra, the most of any product or ingredient in the ARMS database (Table 7.1). The most commonly reported symptoms were diarrhea, abdominal pain, flatulence, and nausea. Only 72 of these complaints were reported directly by consumers. The majority (15,665) were submitted by P&G, which conducted a post-marketing surveillance program to monitor adverse reactions. The Center for Science in the Public Interest (CSPI), a consumer advocacy group that mounted an extensive opposition campaign against olestra's approval, also collected 2,572 consumer complaints via its toll-free telephone hotline and Internet web site. Anecdotal reports received by CSPI indicated that the onset time of stomach cramps ranged from 30 minutes to 16 hours after ingestion of olestra, with some consumers experiencing symptoms after eating only half an ounce of olestra-containing snack chips. The symptoms, which ranged from mild to severe, occasionally persisted for a week and resulted in time lost from school or work (Jacobson, 1997).

The large number of adverse reports associated with olestra may have been related to the FDA's own warnings about intestinal discomfort and widespread media attention which emphasized the novelty of the ingredient and potential side effects. During test marketing in Columbus, Ohio, CSPI flew airplane banners over the city to advertise its toll-free hotline. Frito-Lay, Inc, which was the first major company to use olestra in a snack product, took out a full-page ad in the *Washington Post* (Jan. 15, 1998) in support of the additive. Some scientists expressed reservations about the approval process. For example, Blackburn (1996) stated that during its hearings, the Food Advisory Committee lacked objectivity, focused too much on short-term toxicity and metabolic studies, and failed to adequately consider epidemiological evidence. He felt the approval process was inherently flawed because the FDA had been supportive and collaborative of P&G's petition throughout the approval process, although this is in accordance with the FDA's statutory authority. Moreover, he intimated that the FDA was under pressure to approve olestra because P&G's patent, which had already been extended, was set to expire on the day following the committee hearings.

Clydesdale (1997), however, stressed that the review process was conducted with complete integrity and fairness. He pointed out that the FDA approval process did not need to consider whether the benefits of an additive outweighed the risks, but only that all relevant scientific issues were identified. In addition, there needed to be sufficient safety data to indicate a 'reasonable certainty of no harm' from the proposed use of the additive. Most members of the Food Advisory Committee agreed that these conditions had been satisfied. In addition to requiring a warning label, however, the FDA approval stipulated that olestra-containing foods be fortified with the four fat-soluble vitamins (A, D, E, and K) to compensate for potential metabolic losses, and that P&G conduct a comprehensive, post-marketing surveillance program to monitor adverse health effects.

Despite the large number of complaints reported to ARMS, recent randomized, double-blind trials found no association between ad libitum consumption of olestra-containing snack chips and clinically significant symptoms (Cheskin *et al.*, 1998; Sandler *et al.*, 1999). In their review, Thomson *et al.* (1998) reported that diarrhea and abdominal cramping were no more likely to result from olestra-containing snacks than from consuming similar snacks made with conventional triglycerides. One of the remaining controversies over the approval of olestra was whether it would inhibit the absorption of carotenoids and thus increase the risk of age-related macular degeneration or other chronic diseases resulting from free radical oxidative stress. Most members of the Food Advisory Committee felt that olestra would not significantly affect carotenoid metabolism. Recent reports from the Olestra Post-marketing Surveillance Study found no evidence that olestra consumption resulted in decreased serum concentrations of carotenoids (Thornquist *et al.*, 2000; Rock *et al.*, 1999). Furthermore, there was no association between olestra intake and serum levels of lutein and zeaxanthin, carotenoids found in the eye's macula, or with macular pigment optical density (Cooper *et al.*, 2000). However, Kelly *et al.* (1998) reported that a high dietary intake of olestra (20–40 g/day) was associated with a reduction in plasma concentrations of vitamin E and several carotenoids.

7.4.4 Monosodium glutamate

The flavor enhancer monosodium glutamate (MSG) has been controversial since Kwok (1968) described adverse reactions attributed to MSG-containing foods eaten in Chinese restaurants. Symptoms of numbness at the back of the neck, weakness, and palpitations typically began about 15–20 minutes after the first dish was consumed, and continued for about two hours. Originally called 'Chinese restaurant syndrome' because symptoms were associated with dining at Asian restaurants, the syndrome is now called 'MSG symptom complex.' Since this first report appeared, many other articles, including anecdotal ones, have been published in the technical and popular literature (Filer and Steginck, 1994). Some of the controversy about MSG stems from early animal studies which are not relevant to humans. For example, injection of glutamate directly into the brains of laboratory animals resulted in neuronal damage (Olney *et al.*, 1972), but oral consumption of glutamate has not been found to produce this effect.

From 1980 through June 2000, the FDA compiled 905 reports describing adverse reactions attributed to MSG in its ARMS database. The peak year for MSG complaints was 1990, when ARMS received 200 reports. Only six MSG complaints were reported during the period of July 1999–June 2000. A total of 75 different symptoms were attributed to MSG. Headache was the most frequently reported symptom, occurring in 48.1% of the reports. This was followed by vomiting and nausea (in 22.0% of reports), diarrhea (16.1%), changes in mood (15.8%), changes in heart rate (13.7%), abdominal pain and cramps (13.4%), dizziness or problems with balance (9.9%), difficulty breathing (9.0%), and fatigue or weakness (8.0%). Other symptoms linked to MSG were reported by fewer than 8% of the complainants.

There is little biological plausibility for MSG to produce adverse effects at normal levels of consumption. Free glutamate is found in many common foods, such as mushrooms, tomatoes, and parmesan cheese, and l-glutamic acid and its monosodium salt (MSG) are chemically and biologically equivalent. There is no scientific evidence that the glutamate in manufactured MSG or in hydrolyzed protein ingredients produce effects different from glutamate normally found in foods (Taliaferro, 1995). As a result, MSG remains on the FDA's original list of GRAS ('generally recognized as safe') ingredients, but the agency has sponsored several reviews on the safety of MSG, other glutamates, and hydrolyzed protein as part of its routine investigation of pre-1958 GRAS additives. One such review in 1980 concluded that MSG was safe but recommended further studies to evaluate safety at high intake levels. In 1986, the FDA's Advisory Committee on Hypersensitivity to Food Constituents concluded that MSG posed no threat to the general public, but that reactions of brief duration might occur in some people. Other professional bodies have also stated a belief in the general safety of MSG. In 1987, the Joint FAO/WHO Expert Committee on Food Additives placed MSG in its most favorable category for a food ingredient, assigning an acceptable daily intake of 'not specified.' A similar safety assessment was reported in 1991 by the European Communities' (EC) Scientific Committee for Food, which is comprised of independent scientists from EC member states. The EC report also indicated that MSG was safe for infants.

Because of the longstanding public interest in MSG and adverse reports appearing on several widely-viewed television shows, in the mid-1990s the FDA contracted with the Federation of American Societies for Experimental Biology (FASEB) to conduct a comprehensive review of MSG and other free glutamates (i.e., glutamate not bound to a protein molecule). The resulting report (FASEB/LSRO, 1995), which reviewed about 600 studies, reaffirmed the safety of MSG for the general population at levels normally consumed. There was no evidence that MSG acts as a neurotoxin or that adverse reactions from MSG are immunologically mediated and involve IgE. However, the report stated that an unknown percentage of the population may develop MSG symptom complex after consuming large quantities (about 3 g) of MSG or other free glutamates. The symptoms were more likely to occur when MSG was ingested on an empty stomach, such as in soup served at the beginning of a meal. The report listed the following symptoms as being characteristic of this syndrome: burning sensation in the back of the neck, forearms and chest; numbness in the back of the neck, radiating to the arms and back; tingling, warmth and weakness in the face, temples, upper back, neck and arms; facial pressure or tightness; chest pain; headache; nausea; rapid heartbeat; drowsiness; and weakness.

In addition, the FASEB report stated that a small percentage of individuals with severe, poorly controlled asthma may be prone to MSG symptom complex and suffer temporary bronchospasm or other symptoms for 6–12 hours after consuming MSG in doses of 0.5–2.5 grams. However, Woessner *et al.* (1999) found no evidence that MSG could induce asthma attacks in asthmatic subjects and advised maintaining a 'healthy skepticism' about the existence of MSG sensitivity in asthma patients. In a similar vein, Stevenson (2000) criticized previous studies which suggested that MSG could induce asthma, pointing out that asthmatic patients who were challenged with MSG failed to experience symptoms.

Yang *et al.* (1997) found that oral challenges of MSG resulted in headache, numbness, tingling, and flushing. However, more recent studies generally have not supported the existence of MSG symptom complex. Simon (2000)

158   Food chemical safety

found little evidence that MSG could induce or exacerbate urticaria. He emphasized the importance of conducting well-designed, placebo-controlled double-blind challenges when investigating adverse reactions attributed to MSG and food additives in general, and criticized previous studies for failing to do so. A recent multicenter, double-blind, placebo-controlled study involved 130 subjects who believed they were sensitive to MSG (Geha *et al.*, 2000). These authors administered multiple MSG challenges to investigate if the additive could induce reproducible symptoms. The results indicated that large amounts (5 g) of MSG given without food produced symptoms more often than placebo. However, the symptoms reported were not persistent or serious, and the responses were not consistent upon re-testing. These authors concluded that there was no evidence that MSG in food was associated with adverse reactions.

7.4.5   Sodium nitrite

Sodium nitrite ($NaNO_2$) and sodium nitrate ($NaNO_3$) have a long history of use as curing agents in meats, fish, and poultry because they impart desirable effects on color, flavor, texture, and preservation. Nitrite is particularly important as a preservative because it inhibits *Clostridium botulinum*. Nitrite became controversial when it was recognized it could be transformed into carcinogenic and mutagenic N-nitroso compounds such as nitrosamines. Formation of nitrosamines from nitrite occurs both in food and in the body, particularly in the stomach. However, most of the human exposure to pre-formed nitrosamines comes from non-dietary sources, as nitrosamines also exist in cigarette smoke and certain rubber products (IFT, 1998).

Assessing the potential adverse effect of nitrite as a food additive is difficult because people are exposed to nitrite from a variety of sources. For example, nitrite is derived from nitrate, which occurs naturally in vegetables such as beets, lettuce, and spinach. Nitrates also may be found in drinking water, especially well water, as a result of fertilizer run-off. Nitrates present in saliva are chemically reduced to nitrite by oral microbial flora. Nitrite also is formed through a variety of endogenous metabolic pathways involving other nitrogen compounds. Thus, ingestion of nitrite from cured foods accounts for only a small percentage of the body's total exposure to nitrite (Cassens, 1995).

During the 1970s, concerns about nitrite and nitrosamines led to a report by the National Academy of Sciences which recommended lowering the public's exposure to nitrate, nitrite, and N-nitroso compounds (NAS/NRC, 1981). In a follow-up report (NAS/NRC, 1982), alternatives to nitrite were explored, but an acceptable substitute for use in food processing was not identified. However, food manufacturers began to reduce the use of nitrate in food. As a result of this and other developments, the public's exposure to nitrite and nitrosamines has been declining. Nitrate was largely banned from cured meat in 1978, except for use in a few specialty products, thus removing a source of nitrite. Cassens (1997) reported that retail cured meats have a residual nitrite content of about 10 ppm, which is an 80% reduction since the mid-1970s. Further, antioxidants such as ascorbic acid, erythorbate, and tocopherol, which inhibit nitrosation, are routinely used in nitrite-cured foods. An antioxidant inhibitor is a mandatory ingredient in bacon, since the high temperature to which bacon is typically exposed during cooking augments the formation of nitrosamines from nitrite (IFT, 1998).

The ARMS database contains only ten adverse reaction complaints attributed to nitrite. Nitrate derived from vegetables, such as spinach and beets, and from well water have been linked to methemoglobinemia in infants, a condition in which fetal hemoglobin is oxidized and incapable of transporting oxygen, but nitrate and nitrite additives have not been implicated in this condition. The fatal dose of sodium nitrite is about 22 mg/kg of body weight. Lower doses of sodium nitrate or sodium nitrite have caused methemoglobinemia. However, it is virtually impossible for adults or children to consume enough nitrite from cured meats for this additive to have an acute toxic effect. The question of whether nitrite increases the risk of cancer has been investigated extensively. Carcinogenicity studies in animals have been mostly negative, suggesting that in-vivo formation of N-nitroso compounds does not occur to a significant extent. In a recent review, Eichholzer and Gutzwiller (1998) found little epidemiologic evidence linking dietary intake of nitrate, nitrite, or N-nitroso compounds to various cancers. The presence of pre-formed N-nitroso compounds in cured meats and fish is probably a greater health risk than the formation of N-nitroso compounds in the digestive system.

The World Cancer Research Fund and American Institute for Cancer Research (1997) issued a report suggesting that N-nitroso compounds formed in cured meats through reactions with nitrite 'possibly' increase the risk of colon and rectal cancer, but stated that epidemiologic evidence is weak. The report found insufficient evidence that cured foods increase the risk of stomach or pancreatic cancer. The report further pointed out that it is difficult to draw conclusions on the association between cured foods and cancer, because cured meats and fish contain salt and other constituents in addition to nitrite, and when cooked, heterocyclic amines are formed, which are carcinogenic. The issue of nitrite and cancer recently emerged when several published studies suggesting an association between cured meats and childhood leukemia (Peters *et al.*, 1994) and brain tumors (Sarasua and Savitz, 1994; Bunin *et al.*, 1994) received media attention. However, none of these studies provided strong or convincing evidence, and the American Cancer Society concluded in its 1996 dietary guidelines that 'nitrites in foods are not a significant cause of cancer among Americans.'

7.4.6 Sulfites
Sulfites, a group of six compounds that have been on the GRAS list since 1959, include sulfur dioxide (SO_2), sodium sulfite (Na_2SO_3), sodium metabisulfite (NaS_2O_5), potassium metabisulfite (KS_2O_5), sodium bisulfite ($NaHSO_3$), and

potassium bisulfite ($KHSO_3$). They are also known as sulfiting agents. Sulfites have many applications in food processing, serving as:

- antimicrobials in wine making and wet milling of corn
- inhibitors of oxidizing enzymes in enzymatic browning (fresh fruits, guacamole)
- inhibitors of non-enzymatic (Maillard) browning (dried fruits, dehydrated potatoes)
- antioxidants
- dough conditioners for some baked goods (frozen pizza pie doughs)
- bleaching agents (food starches, maraschino cherries and hominy)
- fungicides during grape storage to prevent mold growth.

Reports of severe adverse reactions among some asthmatics exposed to sulfites began appearing in the mid-1970s. Although symptoms such as wheezing and hives were often treated successfully with epinephrine, some victims lost consciousness or suffered fatal anaphylactic shock. From the mid-1970s to 2000, 27 deaths attributed to sulfites have been reported to ARMS. In 1982, CSPI began petitioning the FDA to ban sulfites. As reports of adverse reactions mounted, the FDA contracted with FASEB to investigate sulfite-related health problems. In 1985, FASEB concluded that although sulfites are safe for most people, the additives pose a hazard of unpredictable severity to asthmatics and other sensitive people.

To add to this information, a total of 1,141 complaints attributed to sulfiting agents were reported to ARMS from 1980 through June 2000. The number of complaints declined from the peak year of 1985, when more than 500 reports were received, to approximately 19 reports or less per year since 1990. Of the 809 reports with adequate information on the intensity of the reaction, 393 (48.6%) were classified as severe. Of the 781 reports with information on the consistency of the reaction following the consumption of sulfite-containing foods, 314 (40.2%) were classified as Group A reactions because several episodes of the reaction were observed and more than one product containing sulfiting agents was implicated. A total of 133 (17%) reports were classified as Group B reactions, as multiple adverse episodes were attributed to one product. The most frequently reported symptom attributed to sulfites was breathing difficulty, which occurred in 34.5% of the complainants. This was followed by vomiting and nausea (13.1%), diarrhea (13.0%), abdominal pain and cramps (10.5%), difficulty swallowing (8.4%), hives (8.1%), dizziness (7.9%), local swelling (7.3%), itching (7.1%), headache (6.7%) and change in heart rate (6.6%).

Adverse reactions to sulfites appear to occur mainly among a small percentage of asthmatics, but it is possible for individuals without asthma to be sulfite sensitive. It is typically more of a problem in individuals with severe asthma who are also taking corticosteroid drugs to control their disease. Among these individuals, the prevalence of sulfite sensitivity is about 8%, while it is about 1% in asthmatics who are not dependent on steroids (Taylor and Bush, 1986).

The mechanism of sulfite-induced asthma is not well-understood. Reactions to sulfited foods probably depend on the sulfite residue level in the food, the sensitivity threshold of the individual, the type of food consumed, and whether sulfite exists in the free (more toxic) form or combined (less toxic) form. The toxicology of sulfites has been reviewed by Madhavi and Salunkhe (1995). Sulfite sensitivity is not a true allergic reaction (Taylor *et al.*, 1988). The FDA initially estimated that more than 1 million Americans are sensitive to sulfites, but more recent estimates lowered the number of asthmatics who may be sulfite sensitive to 80,000–100,000 (Bush *et al.*, 1986).

About a quarter of the adverse reactions from sulfites were associated with salad bars, mainly from those in restaurants rather than in supermarkets. The leading foods implicated in adverse reactions resulting from sulfiting agents are listed in Table 7.2. Based on the FASEB report of 1985, in 1986 the FDA prohibited the use of sulfites for maintaining color and crispness of salad bars and other fresh produce (pre-cut or peeled potatoes, used to make french fries or hash browns, were exempt). The agency also required that sulfites used specifically as preservatives must be listed on the ingredient label, regardless of the amount in the finished product. Sulfites used for other purposes must be listed on the product ingredient label if present at levels of at least 10 parts per million (the limit of detection). In 1988, labels stating 'contains sulfites' were required to appear on all wines containing more than 10 ppm of sulfites that were bottled after mid-1987 (Papazian, 1996). Unpackaged bulk foods, such as dried fruit or loose, raw shrimp, require a point of purchase label disclosing the presence of sulfites, but current FDA regulations do not require that restaurants and other food service establishments disclose whether sulfites are used in food preparation. Sulfites continue to be a problem, especially when they inadvertently contaminate a product or are omitted from the ingredient list of products containing them. In

Table 7.2 ARMS data. Products responsible for adverse reaction complaints attributed to sulfiting agents, 1980–June 2000. (Data include adverse reactions reported to FDA before ARMS was officially established in 1985)

Product type	Number of complaints	Percent of records ($n = 1141$)
Salad bar	295	25.9
Non-salad bar fresh fruits and vegetables (excluding potatoes)	222	19.5
Wine	145	12.7
Seafood	144	12.6
Potatoes (fresh)	72	6.3
Potatoes (other than fresh)	61	5.3
Dried fruit	59	5.2
Baked goods	51	4.5
Fruit and vegetable juices	26	2.3
Beer	21	1.8
Alcoholic beverages (other)	18	1.6

recent years, products that have been recalled because they were found to contain undeclared sulfites include canned tuna (March 1997), a fruit and nut mixture (May 1998), an apple pastry product (January 2001), and an imported product containing ginger and vinegar (March 2001).

7.5 Summary and conclusions

In Europe the commonest food additives thought to cause adverse reactions are tartrazine (E102), sunset yellow (E110), annatto, aspartame, benzoic acid and sulfites (Fuglsang et al., 1993). Key epidemiological studies are shown in Table 7.3. Adverse reactions to food additives can occur at any age. A UK study showed a higher reporting of adverse reactions to food additives in the first ten years of life, and more often occurring in females (Young et al., 1987). The mechanism of these reactions is often unknown, and IgE-mediated reactions are rare. Questionnaire-based studies give a high (6.6–7.4%) prevalence of self-reported adverse reactions to food additives in the general population. However, when food challenges are used to make the diagnosis, the prevalence falls to around 0.23%. One study shows the risk to be greatest in the atopic population (those with a genetic tendency to respond with IgE to exogenous proteins), with no reactions observed in non-atopic individuals (Fuglsang et al., 1994). Virtually all reactions are minor and limited to the skin (worsening of eczema/urticaria) with serious systemic reactions rarely reported.

The United States is one of the only countries that has established a formal surveillance system for post-approval monitoring of adverse reactions to food additives. Despite the obvious limitations of a passive surveillance system, ARMS provides an important venue for the public and others to report health problems perceived to be related to food additives. It provides a means of spotting potential longer-term problems with additives which can then be investigated further.

As the evidence from ARMS shows, there is no doubt that some additives do cause adverse health effects in some people. This fact is not surprising. Although countries such as the US have robust approval systems for additives, there is always the possibility that certain individuals within particularly vulnerable groups, for instance asthmatics or other immuno-compromised groups, may be at risk, usually from high doses of an additive. In the great majority of such cases adverse reactions are mild and transient, for example the gastrointestinal discomfort associated with consumption of olestra. In a very few cases, reactions are more severe, as in the case of sulfites. In the latter case, it has been estimated that 1% of asthmatics are at risk of some reaction and, within that group, 8% of asthmatics taking steroids. Within these groups under 50% are at risk of a severe reaction. In general, it has been estimated that less than 0.3% of the general population is at risk of some adverse effect from food additives.

Whilst the levels of risk for the general population are low, consumer concerns about additive safety remain high. Although there has been a decline

Table 7.3 Epidemiology of adverse food reactions to food additives – key studies

Author, date	Type of study	Country	No. of subjects	Point prevalence	Definition
Young et al. 1987	Population based, cross-sectional	UK	18 582	7.4% at > 6 months old	History
Fuglsang et al. 1994	Population based, cross-sectional	Denmark	4274	6.6% at 5–16y	History
Fuglsang et al. 1994	Non-atopic population, cross-sectional	Denmark	4274	0% at 5–16y	Open challenge
Fuglsang et al. 1994	Atopic population, cross-sectional	Denmark	4274	9.8% at 5–16y	Open challenge
Young et al. 1987	Population based, cross-sectional	UK	649	0.23% at > 4y	History and DBPCFC

since the first expressions of concerns about additives in the 1970s and 1980s, some surveys still suggest that up to one quarter of consumers remain concerned about the safety of some food additives. This concern has fueled an often ill-founded readiness to attribute a range of adverse health effects to them. This phenomenon can be seen most clearly in the wide discrepancy between numbers of reported reactions and those validated by clinical tests. It is also reflected in the wide range of symptoms sometimes reported and the tendency for reported symptoms to rise and fall according to media coverage of an additive.

Confusion about the levels of risk posed by food additives can also be complicated by methodological problems. As has been seen, the data collected by ARMS inevitably provides a distorted picture of the adverse health effects caused by additives. It is sometimes difficult to investigate a potential link, for example because consumers might be exposed to a substance like nitrite from various sources, or because it is difficult to measure effects such as hyperactivity amongst children and relate it clearly to a single factor such as exposure to food dyes. However, in most cases it is possible to test the strength of such links. Epidemiological evidence is often very helpful in establishing the strength or weakness of a link between an additive and a particular condition, as in the case of aspartame. It is possible to investigate the suggested biological mechanisms at work, for example through animal or, in some cases, human trials. The use of randomized double-blind placebo studies is perhaps the most powerful single method for testing the strength of a suggested link. The weight of such evidence is usually sufficient to show whether an additive is safe or not for the great majority of the population, though it can open up further debate about what constitutes an acceptable level of risk for that minority of consumers who may be vulnerable to adverse health effects.

Given the discrepancy between consumer concerns about the safety of synthetic additives in particular and the actual levels of risk they pose, it is clear that consumers need a better understanding of the nature, use and approval of additives. Food manufacturers, retailers and government all have a role to play here. Consumers also need a clearer understanding of risk. The fact that some additives may, in some circumstances, cause adverse reactions amongst specific groups can create an impression that additives present a significant risk to consumers in general. This impression has been strengthened by some media reporting and the activities of pressure groups. The Internet has now become a powerful tool for disseminating information, some of which can be inaccurate and misleading. Consumer fears can only be countered by government and the food industry demonstrating their effectiveness in monitoring and investigating complaints as they occur, and doing so in a transparent and accountable way. Systems such as ARMS provide a model to follow in this respect. The relationship between government (FDA), an independent scientific body (FASEB) and a consumer organization (CSPI) in investigating adverse reactions to additives such as sulfites also provides lessons on the ways that different parties can work together in improving food safety.

7.6 Future trends and directions

New additives are emerging all the time. In particular, it is likely that functional foods will grow in popularity as manufacturers develop products for promoting health or preventing diseases. Such foods will probably contain a variety of nutritional additives and new phytochemical ingredients, as well as a new generation of preservatives and other additives developed from natural ingredients, required to maintain product quality. The emergence of antioxidants developed from plants, which both help to preserve foods and may have a protective role against cardiovascular disease and certain types of cancer, is a good example.

Although the benefits of many functional ingredients have yet to be proven, there is a possibility for new health problems to arise if the market for fortified functional foods continues to expand. Some consumers may ingest excessive amounts of certain nutritional food additives such as iron, which could lead to an increased incidence of hemachromatosis in genetically predisposed people. Fortification with specific carotenoids may competitively inhibit the bioavailability of other carotenoids, perhaps leading to adverse physiological consequences.

An expanding global economy and the increasing cultural diversity of many nations may result in the availability of foods containing a variety of additives not previously consumed. The use of preservatives in processed foods may increase if consumers continue to prefer foods that are convenient, fresh, free of pathogens, and have an extended shelf life. Also, as the population of industrialized countries ages, the proportion of people who take medications and who are considered vulnerable to adverse reactions is likely to increase. All these factors make it more likely that some additives may have unforeseen and adverse health effects on some segments of the population. Likely future developments may include refining analytical techniques used in the approval process to anticipate and assess potential impacts of a new additive on particularly vulnerable groups, including the improvement of risk assessment and modeling techniques.

Another future area of research for monitoring adverse reactions to food additives might involve the development of clinical tests to screen for genetic biomarkers indicating food sensitivities. Perhaps vaccines will be developed and administered to sensitive individuals, preventing the occurrence of adverse reactions from food additives.

7.7 Sources of further information and advice

US government agencies providing information on adverse reactions to food additives:

- Food and Drug Administration <http://vm.cfsan.fda.gov/list.html>. Contains information on food additives and premarket approval, including guidance in designing and interpreting studies to assess safety ('FDA Redbook').
- Centers for Disease Control and Prevention, Morbidity and Mortality Weekly Reports <www.cdc.gov/mmwr>.

Research bodies:

- National Institutes of Health Consensus Development Program <http://odp.od.nih.gov/consensus/default.html>.
- National Academy of Sciences <www.nationalacademies.org>.

Key journals:

- *FDA Consumer*
- *Journal of Nutrition* (which published a supplement in August 1997 (vol 127. No. 8S) devoted entirely to studies assessing the nutritional effects of olestra).

Public interest group:

- Center for Science in the Public Interest <www.cspinet.org>.

Major professional organizations:

- Institute of Food Technologists <www.ift.org>.
- American Dietetic Association <www.eatright.org>.
- International Food Information Council <www.ificinfo.org>.
- Institute of Food Science & Technology <www.ifst.org>.

7.8 References

AMERICAN ACADEMY OF PEDIATRICS (2000), 'Clinical practice guideline: diagnosis and evaluation of the child with attention-deficit/hyperactive disorder', *Pediatrics*, 105(5), 1158–70.

BLACKBURN H (1996), 'Olestra and the FDA', *N Engl J Med*, 334(15), 984–6.

BORZELLECA J F (1995), 'Post-marketing surveillance of macronutrient substitutes', *Food Technol*, 49(9), 107–13.

BRADSTOCK M K, SERDULA M K, MARKS J S, BARNARD R J, CRANE N T, REMINGTON P L and TROWBRIDGE F L (1986), 'Evaluation of reactions to food additives: the aspartame experience', *Amer J Clin Nutr*, 43, 464–9.

BUNIN G R, KUIJTEN R R, BOESEL C P, BUCKLEY J D and MEADOWS A T (1994), 'Maternal diet and risk of astrocytic glioma in children: A report from the Children's Cancer Group (United States and Canada)', *Cancer Causes and Control*, 5, 177–87.

BUSH R K, TAYLOR S L, HOLDEN K, NORDLEE J A and BUSSE W W (1986), 'Prevalence of sensitivity to sulfiting agents in asthmatic patients', *Am J Med*, 81(5), 816–20.

BUTCHKO H H and KOTSONIS F N (1989), 'Aspartame: review of recent research', *Comments Toxicology*, 3(4), 253–78.

CASSENS R G (1995), 'Use of sodium nitrite in cured meats today', *Food Technol*, 49(7), 72–80.

CASSENS R G (1997), 'Residual nitrite in cured meat', *Food Technol*, 51(2), 53–5.

CHESKIN L J, MIDAY R, ZORICH N and FILLOON T (1998), 'Gastrointestinal

symptoms following consumption of olestra or regular triglyceride potato chips: a controlled comparison', *JAMA*, 279(2), 150–2.
CLYDESDALE F M (1997), 'Olestra: the approval process in letter and spirit', *Food Technol*, 51(2), 104, 85.
CONNERS C K, GOYETTE C H, SOUTHWICK D A, LEES J M and ANDRULONIS P A (1976), 'Food additives and hyperkinesis: a controlled double blind experiment', *Pediatrics*, 58, 154–66.
COOPER D A, CURRAN-CELENTANO J, CIULLA T A, HAMMOND B R, DANIS R B, PRATT L M, RICCARDI K A and FILLOON T G (2000), 'Olestra consumption is not associated with macular pigment optical density in a cross-sectional volunteer sample in Indianapolis', *J Nutr*, 130(3), 642–7.
COUNCIL ON SCIENTIFIC AFFAIRS, AMERICAN MEDICAL ASSOCIATION (1985), 'Aspartame – review of safety issues', *JAMA*, 254(3), 400–2.
CROWE M, HARRIS S, MAGGIORE P and BINNS C (1992), 'Consumer understanding of food additive labels', *Australian Journal of Nutrition and Dietetics*, 49, 19–22.
EICHHOLZER M and GUTZWILLER F (1998), 'Dietary nitrates, nitrites, and N-nitroso compounds and cancer risk: A review of the epidemiologic evidence', *Nutr Rev*, 56(4), 95–105.
FASEB/LSRO (1995), 'Analysis of adverse reactions to monosodium glutamate (MSG).' Federation of American Societies for Experimental Biology, Life Sciences Research Office, Bethesda, MD. American Institute of Nutrition, Bethesda, MD.
FEINGOLD B F (1974), *Why your child is hyperactive*. New York, Random House.
FERNSTROM J D (1994), 'Dietary amino acids and brain function', *J Am Diet Assoc*, 94, 71–7.
FILER L J and STEGINK L D (eds) (1994), 'A report of the proceedings of an MSG workshop held August 1991', *Critical Rev Food Sci and Nutr*, 34(2), 159–74.
FMI (1994, 1997, 2000), 'Trends in the United States – consumer attitudes and the supermarket' (published annually). Food Marketing Institute, Washington, DC.
FMI AND PREVENTION MAGAZINE (2000), 'Shopping for health 2000. Self-care needs and whole health solutions.' Food Marketing Institute and Prevention Magazine, Emmaus, Pennsylvania.
FUGLSANG G, MADSEN C, SAVAL P and OSTERBALLE O. (1993), 'Prevalence of intolerance to food additives among Danish children', *Pediatr Allergy Immunol*, 4, 123–9.
FUGLSANG G, MADSEN C and HALKEN S (1994), 'Adverse reactions to food additives in children with atopic systems', *Allergy*, 49, 31–7.
GEHA R S, BEISER A, REN C, PATTERSON R, GREENBERGER P A, GRAMMER L C, DITTO A M, HARRIS K E, SHAUGHNESSY M A, YARNOLD P R, CORREN J and SAXON A (2000), 'Multicenter, double-blind, placebo-controlled, multiple-challenge evaluation of reported reactions to monosodium glutamate', *Allergy Clin Immunol*, 106(5 Pt 1), 973–80.

GERSHOFF S N (1995), 'Nutrition evaluation of dietary fat substitutes', *Nutr Rev*, 53(11), 305–13.
GOYETTE C H, CONNERS C K, PETTI T A and CURTIS L E (1978), 'Effects of artificial colors on hyperactive children: a double blind challenge study', *Psychopharmacol Bull*, 14, 39–40.
HARLEY J P, RAY R S, TOMASI L, EICHMAN P L, MATTHEWS C G, CHUN R, CLEELAND C S and TRAISMAN E (1978a), 'Hyperkinesis and food additives: testing the Feingold hypothesis', *Pediatrics*, 61, 818–28.
HARLEY J P, MATTHEWS C G and EICHMAN P (1978b), 'Synthetic food colors and hyperactivity in children: a double-blind challenge experiment', *Pediatrics*, 62, 975–83.
IFT (1998), 'Nitrite and Nitrate: An Update. A backgrounder of the Institute of Food Technologists', February 1998.
JACOBSON M (1997), 'Olestra (Letter)', *Nutr Today*, 32(3), 135.
JONES J M (1992), *Food Safety*. St. Paul, Eagan Press, p. 268.
KELLY S M, SHORTHOUSE M, COTTERELL J C, RIORDAN A M, LEE A J, THURNHAM D I, HANKA R and HUNTER J O (1998), 'A 3-month, double-blind, controlled trial of feeding with sucrose polyester in human volunteers', *Br J Nutr*, 80(1), 41–9.
KWOK R H M (1968), 'Chinese restaurant syndrome', *N Engl J Med*, 278, 796.
LIPTON M A, NEMEROFF C B and MAILMAN R B (1979), 'Hyperkinesis and food additives', in Wurtman R J and Wurtman J J (eds), *Nutrition and the Brain, Vol. 4*. New York, Raven Press, pp. 1–27.
MADHAVI D L and SALUNKHE D K (1995), 'Antioxidants', in Maga J A and Tu A (eds), *Food Additive Toxicology*. New York, Marcel Dekker, pp. 127–9.
MCNUTT K, POWERS M and SLOAN A E (1986), 'Food colors, flavors and safety: a consumer viewpoint', *Food Technology*, 40(1), 72–8.
NAS/NRC (1981), 'The health effects of nitrate, nitrite, and N-nitroso compounds', National Research Council, National Academy Press, Washington, DC.
NAS/NRC (1982), 'Alternatives to the current use of nitrite in foods', National Research Council, National Academy Press, Washington, DC.
NATIONAL ADVISORY COMMITTEE ON HYPERKINESIS AND FOOD ADDITIVES (1975), 'Report to the Nutrition Foundation', June 1.
NIH CONSENSUS STATEMENT (1982), 'Defined diets and childhood hyperactivity', National Institutes of Health, Jan. 13–15, 4(3), 1–11.
NIH CONSENSUS STATEMENT (1998), 'Diagnosis and treatment of attention deficit hyperactivity disorder (ADHD)', National Institutes of Health, Nov. 16–18, 16(2).
OLNEY J W, SHARPE L G and FEIGIN R D (1972), 'Glutamate-induced brain damage in infant primates', *J Neuropathol Exp Neurol*, 31(3), 464–88.
OLNEY J W, FARBER N B, SPITZNAGEL E and ROBINS L N (1996), 'Increasing brain tumor rates: is there a link to aspartame?', *J. Neuropathol Exp Neurol*, 55(11), 1115–23.
PAPAZIAN R (1996), 'Sulfites safe for most, dangerous for some', *FDA Consumer*, 30(10), 11–14.

PETERS J M, PRESTON-MARTIN S, LONDON S J, BOWMAN J D, BUCKLEY J D and THOMAS D C (1994), 'Processed meat and risk of childhood leukemia (California USA)', *Cancer Causes and Control*, 5, 195.
RIES L A G, EISNER M P, KOSARY C L, HANKEY B F, MILLER B A, CLEGG L and EDWARDS B K (eds) (2000), 'SEER Cancer Statistics Review, 1973–1997', National Cancer Institute, Bethesda, MD.
ROCK C L, THORNQUIST M D, KRISTAL A R, PATTERSON R E, COOPER D A, NEUHOUSER M L, NEUMARK-SZTAINER D and CHESKIN L J (1999), 'Demographic, dietary and lifestyle factors differentially explain variability in serum carotenoids and fat-soluble vitamins: baseline results from the sentinel site of the Olestra Post-Marketing Surveillance Study', *J Nutr*, 129(4), 855–64.
RODALE (1996), 'The Prevention Index – a report card on the nation's health, 1996 summary report.' Rodale Press, Inc, Emmaus, Pennsylvania.
SANDLER R S, ZORICH N L, FILLOON T G, WISEMAN H B, LIETZ D J, BROCK M H, ROYER M G and MIDAY R K (1999), 'Gastrointestinal symptoms in 3181 volunteers ingesting snack foods containing olestra or triglycerides', *Annals Internal Med*, 130, 253–61.
SARASUA S and SAVITZ D A (1994), 'Cured and broiled meat consumption in relation to childhood cancer: Denver, Colorado (USA)', *Cancer Causes and Control*, 5, 141.
SCHIFFMAN S S, BUCKLEY E, SAMPSON H A, MASSEY E W, BARANIUK J N, FOLLETT J V and WARWICK Z S (1987), 'Aspartame and susceptibility to headache', *N Engl J Med*, 317(19), 1181–5.
SIMON R A (2000), 'Additive-induced urticaria: experience with monosodium glutamate (MSG)', *J Nutr*, 130, 1063S–6S.
SLOAN A E (1998), 'Food industry forecast: Consumer trends to 2020 and beyond', *Food Technology*, 52(1), 37–44.
SLOAN A E, POWERS M and HOM B (1986), 'Consumer attitudes towards additives', *Cereal Foods World*, 31(8), 523–32.
STEGINK L D (1987), 'The aspartame story: a model for the clinical testing of a food additive', *Am J Clin Nutr*, 46, 204–15.
STEVENSON D D (2000), 'Monosodium glutamate and asthma', *J Nutr*, 130, 1067S–73S.
TALIAFERRO P J (1995), 'Monosodium glutamate and the Chinese restaurant syndrome: a review of food additive safety', *J Environ Health*, 57(10), 8–12.
TAYLOR S L and BUSH R K (1986), 'Sulfites as food ingredients, a Scientific Status Summary by the Institute of Food Technologists Expert Panel on Food Safety & Nutrition', *Food Technol*, 40(6), 47–52.
TAYLOR S L and HEFLE S L (2001), 'Food allergies and other sensitivities, a Scientific Status Summary by the Institute of Food Technologists Expert Panel on Food Safety & Nutrition', *Food Technol*, 55(9), 68–83.
TAYLOR S L, BUSH R K, SELNER J C, NORDLEE J A, WIENER M B, HOLDEN K, KOEPKE J W and BUSSE W W (1988), 'Sensitivity to sulfited foods among sulfite-sensitive subjects with asthma', *J Allergy Clin Immunol*, 81(6), 1159–67.

THOMSON A B, HUNT R H and ZORICH N L (1998), 'Olestra and its gastrointestinal safety', *Aliment Pharmacol Ther,* 12(12), 1185–200.
THORNQUIST M D, KRISTAL A R, PATTERSON R E, NEUHOUSER M L, ROCK C L, NEUMARK-SZTAINER D and CHESKIN L J (2000), 'Olestra consumption does not predict serum concentrations of carotenoids and fat-soluble vitamins in free-living humans: early results from the sentinel site of the olestra post-marketing surveillance study', *J Nutr,* 130(7), 1711–18.
TOLLEFSON L (1988), 'Monitoring adverse reactions to food additives in the US Food and Drug Administration', *Regulatory Toxicology and Pharmacology,* 8, 438–66.
TOLLEFSON L and BARNARD R J (1992), 'An analysis of FDA passive surveillance reports of seizures associated with consumption of aspartame', *J Am Diet Assoc,* 92, 598–601.
WOESSNER K M, SIMON R A and STEVENSON D D (1999), 'Monosodium glutamate sensitivity in asthma', *J Allergy and Clin Immunol,* 104, 305–10.
WORLD CANCER RESEARCH FUND, AMERICAN INSTITUTE FOR CANCER RESEARCH (1997), *Food, Nutrition and the Prevention of Cancer: a Global Perspective,* Washington DC, American Institute for Cancer Research.
YANG W H, DROUIN M A, HERBERT M, MAO Y and KARSH J (1997), 'The monosodium glutamate symptom complex: assessment in a double-blind, placebo-controlled, randomized study', *J Allergy Clin Immunol,* June (6 Pt 1), 757–62.
YOUNG E, PATEL S, STONEHAM M, RONA R and WILKINSON J D (1987), 'The prevalence of reaction to food additives in a survey population', *J Royal College Phys,* 21(4), 241–9.

Part III
Specific additives

8
Colorants

F. J. Francis, University of Massachusetts, Amherst

8.1 Introduction

The appreciation of color and the use of colorants dates back to antiquity. The art of making colored candy is shown in paintings in Egyptian tombs as far back as 1500 BC. Pliny the Elder described the use of artificial colorants in wine in 1500 BC. Spices and condiments were colored at least 500 years ago. The use of colorants in cosmetics is better documented than colorants in foods. Archaeologists have pointed out that Egyptian women used green copper ores as eye shadow as early as 5000 BC. Henna was used to redden hair and feet, carmine to redden lips, faces were colored yellow with saffron and kohl, an arsenic compound, was used to darken eyebrows. More recently, in Britain, in the twelfth century, sugar was colored red with kermes and madder and purple with Tyrian purple.

Until the middle of the nineteenth century, the colorants used in cosmetics, drugs and foods were of natural origin from animals, plants and minerals. That changed with the discovery of the first synthetic dyestuff, mauve, by Sir William Henry Perkin in 1856. The German dyestuff industry rapidly developed a large number of 'coal tar' colorants and they rapidly found applications in the food and cosmetic industries. At the turn of the century, over 700 synthetic colorants were available for use in foods in the US. The potential for fraud and personal harm was obvious and horror stories abounded. For example, Marmion[1] described a situation where a druggist gave a caterer copper arsenite to make a green pudding for a public dinner and two people died. History is rife with anecdotes about adulteration of food, from the recipe for bogus claret wine[2] in 1805 to the attacks on synthetic colorants today. Two centuries ago, adulteration had become a very sophisticated operation. Accum[3] commented, 'To such perfection of ingenuity has the system of counterfeiting and adulterating various commodities of life arrived in this country, that spurious articles are everywhere to be found in the market, made up so skilfully, as to elude the discrimination of the most experienced judges – the magnitude of an evil, which in many cases, prevails to an extent so alarming, that we may exclaim – There is death in the pot'. Many of the adulterants involved color and flavor. Elderberries and bilberries were added to wine. Copper acetate was used to color artificial tea leaves. Red lead was a colorant for cheese. Obviously government regulation was essential and this led to a long series of publications on the safety of colorants for food, drugs and cosmetics.

This chapter is devoted to a description of the chemistry, applications, and safety of the wide variety of natural and synthetic colorants available today. But another aspect has entered into consideration. Food safety of colorants has usually been considered to be a negative if we ignore the many benefits of making food more attractive in appearance. The recent meteoric rise of the nutraceutical industries has made it possible to claim health benefits for many categories of food including the colorants. Where appropriate, the health claims will be included in this chapter.

8.2 Food, drug and cosmetic colorants

8.2.1 Introduction

The chaotic situation existing in the synthetic colorant industry was evident in the 80 colorants available in 1907 to the paint, plastic, textile, and food industries. Obviously very few of them had been tested for safety. Dr Bernard Hesse, a German dye expert employed by the US Department of Agriculture, was asked to study the situation and he concluded that, of the 80 colorants available, only 16 were more or less harmless and he recommended only seven for use in food. This led to the US Food and Drug Act of 1906 which set up a certification procedure which ensured the identity of the colorant and the levels of impurities specifications for each food, drug and cosmetic (FD&C) color permitted for each colorant.[4]

The Federal Food, Drug, and Cosmetic Act of 1938 set up three categories:

1. FD&C colors for use in foods, drugs and cosmetics
2. D&C colors for use in drugs and cosmetics when in contact with mucous membranes or ingested
3. Ext. D&C colors for use in products applied externally.

The 1938 law required colorants on the permitted list to be 'harmless and suitable for food' but the FDA interpreted 'harmless' to mean harmless at any level and this proved to be unworkable. The Color Additive Amendment of 1960 eliminated the 'harmless per se' interpretation and resulted in a list of nine permitted colorants. The list was reduced to eight with the delisting of FD&C

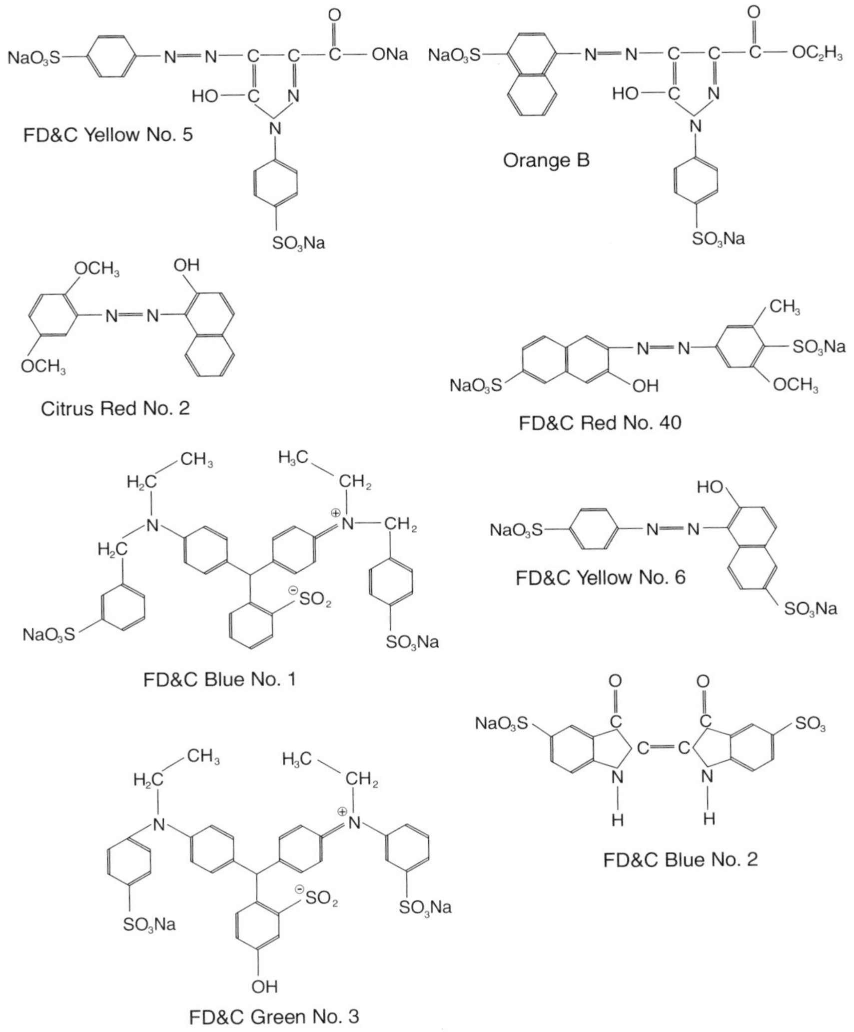

Fig. 8.1 Structures of eight permitted food, drug and cosmetic colorants.

Red No. 3 in 1998 (Fig. 8.1). Colorants in the three categories above were termed 'certified' colorants but the Color Additive Amendment also set up a category of 'exempt' colorants which were not subject to the rigorous requirements of the certified colorants. There are 26 colorants in this category (Table 8.1) and they comprise most of the preparations which would be called 'natural' in other countries. The US does not officially recognize the term 'natural' but it is often used in the popular press.

Table 8.1 Regulatory and safety status of colorants exempt from certification in the US[a]

Color additive	US food use limit	EU status	JECFA ADI (mg/kg/bw)
Algal meal, dried	GMP[b] for chicken feed	NL[c]	NE[d]
Annatto extract	GMP[e]	E160b	0–0.065
Dehydrated beets	GMP	E162	NE
Ultramarine blue	Salt for animal feed up to 0.5//5 by weight	NL	None
Canthaxanthin	Not to exceed 30 mg/lb of solid/semisolid food or pint of liquid food or 4.41 mg/kg of chicken feed	E161g	None
Caramel	GMP	E150	0–200
Beta-apo-8-carotenal	Not to exceed 15 mg/lb of solid or semisolid food or 15 mg/pint of liquid food	E160a	0-5
Beta carotene	GMP	E150	0–5
Carrot oil	GMP	NL	NE
Cochineal extract or carmine	GMP	E120	0–5
Corn endosperm oil	GMP for chicken feed	NL	NE
Cottonseed flour, Toasted partially defatted, cooked	GMP	NL	NE
Ferrous gluconate	GMP for ripe olives only	NL	NE
Fruit juice	GMP	NL	NE
Grape color extract	GMP for non-beverage foods	E163	0–2.5
Grape skin extract (Enocianina)	GMP for beverages	E163	0–2.5
Iron oxide, synthetic	Pet food up to 0.25%	E172	0–2.5
Paprika	GMP	E160c	none
Paprika oleoresin	GMP	E160c	Self-limiting as a spice
Riboflavin	GMP	E101	0–0.5
Saffron	GMP	NL	Food ingredient
Tagetes meal and extract (Aztec Marigold)	GMP	NL	NE
Titanium dioxide	Not to exceed 1% by weight of food	E171	none
Turmeric	GMP	E100	Temporary ADI
Turmeric oleoresin	GMP	E100	Temporary ADI
Vegetable juice	GMP	NL	NE

[a] Adapted from Francis, F. J., 1999. Chap. 4, Regulation of Colorants in *Colorants,* Eagan Press. St. Paul, MN. pp. 223–32.
[b] Good Manufacturing Practice.
[c] Not listed.
[d] Not evaluated.
[e] Calculated as bixin.

8.2.2 Chemistry and usage

The certified and exempt colorants is the most important group of colorants in the US both from a poundage and diversity of applications point of view. The chemical structures of the certified group are shown in Fig. 8.1. This group also includes the lakes which are prepared by reacting the pure FD&C colorants with alumina and washing and drying the resultant precipitate.[5] FD&C colorants are available in many formulations: dry powders, solutions in a variety of solvents, blends, formulations with a variety of carriers, and lakes. The colorant preparations are used in a wide variety of foods available in the food markets. Lakes are used in food formulations where it is desirable to minimize the 'bleeding' of color from one ingredient to another. Stability and solubility of FD&C colorants under a variety of conditions was published by Marmion[4] and Francis.[5]

8.2.3 Safety

The history of the safety of the synthetic colorants has been replete with controversies and contradictions. The decisions by Bernard Hesse at the turn of the century were based primarily on the concept that the colorants should be harmless when consumed in everyday food. The authors of the 1938 Act were primarily concerned with developing a list of colorants which were allowed and would be harmless under the conditions of toxicological testing accepted in the 1930s. These included consideration of the anticipated levels of consumption with the levels of usage likely to be found across the range of expected products (the Acceptable Daily Intake, ADI). It became evident that this concept could be abused when, in 1950, some children became ill after eating popcorn with excessive levels of colorants. The FDA then launched a new round of toxicological investigations. The original list of seven colorants, which had grown to sixteen, was reduced to seven, and it included only two of the original seven. The toxicological protocols were based on weight changes, tumor production, biochemical and physical changes, etc. and teratological, immunological, multi-generation changes, allergenicity, carcinogenicity, and other changes were added. With the possible exception of saccharin, the FD&C colorants became the most tested group of additives in foods. The case of FD&C Red No. 2 (Amaranth) is an example. Amaranth was approved for use in foods in 1907 and survived the batteries of tests until a Russian study in 1972 concluded that it was carcinogenic. It was difficult to repeat the study because the Russian material was of textile grade and contained about 9% impurities, but a flurry of activity resulted. Two studies indicated a problem with carcinogenicity and 16 had negative effects. Regardless, FD&C Red No. 2 was delisted in 1976 in spite of the controversy over the interpretations. For example, the Canadian authorities allowed FD&C Red No. 2 to be used in foods and banned FD&C Red No. 40 (Allura Red) which largely replaced Red No. 2. The American authorities banned FD&C Red No. 2 and allowed FD&C Red No. 40. Similar discrepancies are found in the approved colorants for many countries. A detailed discussion of the safety of each FD&C colorant[6] and the exempt colorants[7] was published but space does not allow a discussion of the safety considerations here. The general protocols for toxicological testing may be found in the FDA publication *Toxicological Principles for the Safety Assessment of Direct Food and Color Additives Used in Food*. Information on specific additives may be found in the *US Code of Federal Regulations – CFR 21*. Listings of the available colorants are also available.[8,9]

8.2.4 Future prospects

Synthetic colorants were severely criticized in the US in the 1960s by consumer groups critical of the 'junk food' concept. Activists had difficulty criticizing the foods themselves but the colorants which were essential to the formulation of convenience foods were vulnerable. The belief that natural adddititives were superior to synthetic compounds from a health point of view is a little naive since 5,000 years is too short a time for humans to develop genetic resistance. Regardless, the popularity of natural formulations increased worldwide and shows no signs of decreasing. The last synthetic food colorant to be approved by the FDA was FD&C Red No. 40 (Allura Red) in 1971 and there are not likely to be any more in the near future. A case can be made that there is no need for any more since, from a tristimulus approach with three primary colors such as red, blue and green, any color can be matched. This is not quite true since the current primary colors available do not cover the desired spectrum. Industry would like a wider choice and several new colorants are being considered.[5] Conventional wisdom would suggest that approvals for new colorants are likely to be in the natural group.

8.3 Carotenoid extracts

8.3.1 Introduction

Carotenoids are probably the best known of the colorants and certainly the largest group of pigments produced in nature with an annual production estimated at 100,000,000 tons. Most of this is fucoxanthin produced by algae in the ocean and the three main pigments, lutein, violaxanthin and neoxanthin in green leaves.[10] Over 600 carotenoid compounds have been reported.

8.3.2 Chemistry and usage

The chemical structure of some typical carotenoids is shown in Fig. 8.2. Beta-carotene occurs in nature usually associated with a number of chemically closely related pigments and extracts have been used as food colorants for many years. For example, palm oil has a high concentration of carotenoid pigments, primarily beta-carotene and about 20 others. Crude palm oil has been used extensively as a cooking oil because of its desirable flavor and as a general

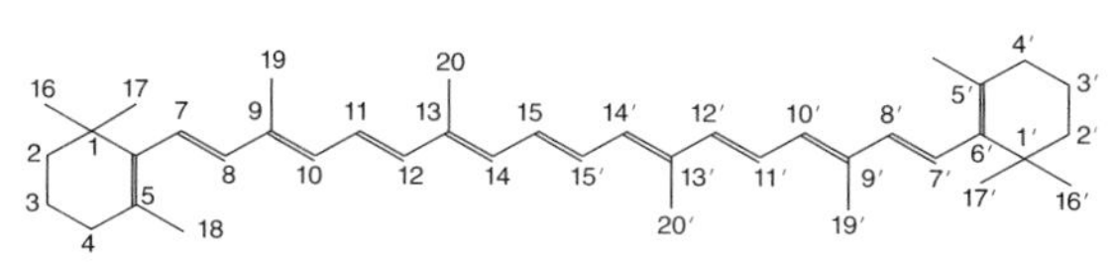

Beta-carotene

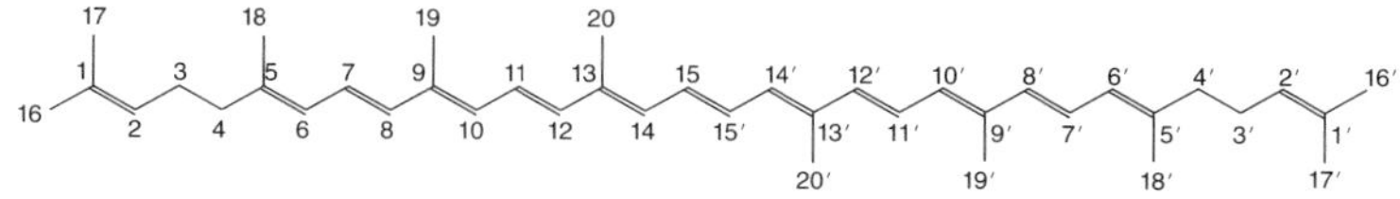

Lycopene

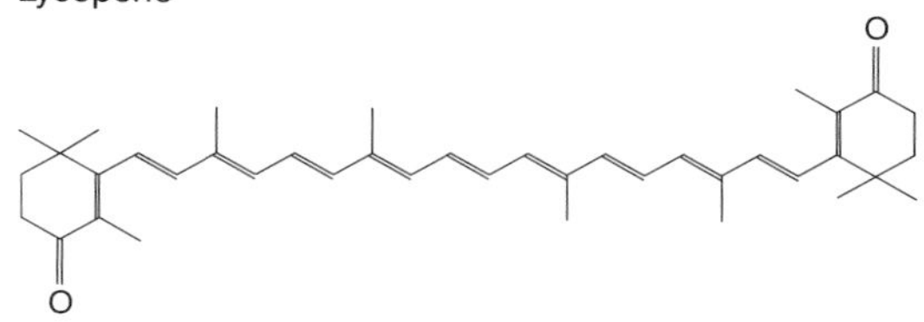

Canthaxanthin

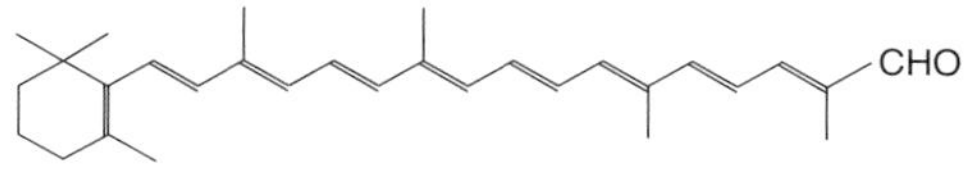

Beta-apo-8'-carotenol

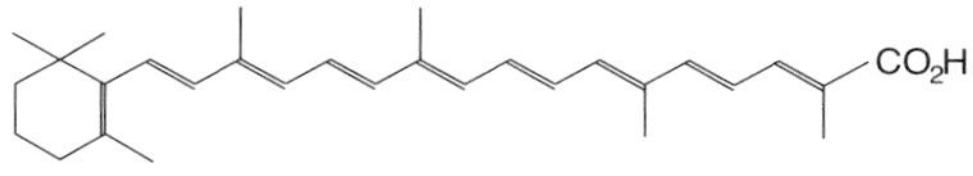

Beta-apo-8'-carotenoic acid

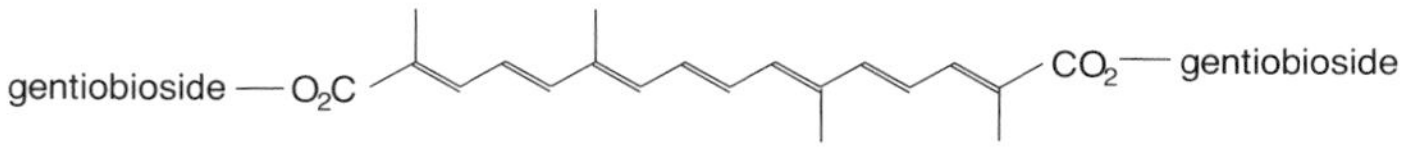

Crocin

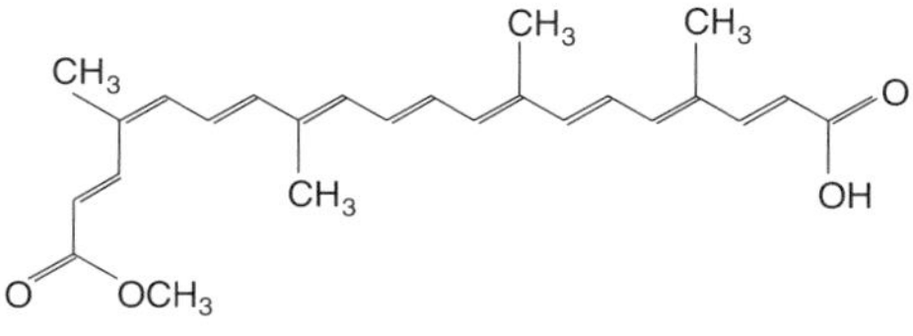

Bixin

Fig. 8.2 Structures of some typical carotenoid compounds.

180   Food chemical safety

edible oil after purification. Both the crude oil and the semi-purified oil are effective colorants. Xanthophyll pastes, well known in Europe, consist of extracts of alfalfa (lucerne), nettles, broccoli and other plants. Unless saponified, they are green because of the chlorophyll content. Many xanthophyll pastes contain as much as 30% carotenes with the major pigments being those of green leaves, lutein, beta-carotene, neoxanthin, and violaxanthin. Extracts of carrots contain about 80% beta-carotene and up to 20% alpha carotene with traces of other compounds including lycopene. Extracts of citrus peels have been suggested for coloring orange juice since the more highly colored juices command a higher price. Astaxanthin is a desirable colorant for trout and salmon in aquaculture and the usual source is byproducts of the shrimp and lobster industries, but demand has led to the cultivation of the red yeast *Phaffia rhodozyma* to produce astaxanthin.[11] Extracts of marigold, *Tagetes erecta*, are well known commodities used primarily as colorants in poultry feed. Extracts are available in three main forms, dried ground petals, crude oleoresins and purified oleoresins in a wide variety of formulations. Preparations from tomatoes have been used to provide flavorful and colorful food ingredients for many years. Recent increases in the demand for natural beta-carotene has led to the production of beta-carotene extracts from microalgae, and some species of *Dunaliella* can accumulate up to 10% d.w. of beta-carotene.[11]

All carotenoid extracts are effective yellow to orange colorants and can be used in a variety of foods depending on government regulations. These include vegetable oils, margarine, salad dressings, pastas, baked goods, dairy products, yoghurt, ice cream, confectionery, juices, and mustard.

8.3.3 Toxicology

There is little toxicological data available for extracts of carrots, alfalfa, corn oil, palm oil, tomatoes, etc. The JECFA had no objections to their use as food colorants provided that the levels of use did not exceed that normally present in vegetables. A number of toxicity experiments were conducted on *Dunaliella* algae in view of its increasing importance in the health food area. Twelve studies on *D. salina* indicated no problems. *Cis* beta-carotene was absorbed to a lesser extent than *trans* beta-carotene. Furahashi suggested a no-observed-effect level (NOEL) of 2.5 g kg/day for extracts from *D. Hardawil*.[12] The Joint Expert Committee on Food Additives of the World Health Organization/United Nations (JECFA) did not establish an NOEL or an ADI because of the variation in the composition of the products.

8.3.4 Health aspects

Carotenoids are of physiological interest because some of them are precursors of vitamin A. They have been in the news recently because many exhibit radical or single oxygen trapping ability and as such have potential antioxidant activity *in vivo*. They may reduce the risk of cardiovascular disease, lung cancer, cervical dysplasia, age-related macular degeneration, and cortical cataract.[13] The beneficial effects of beta-carotene are thought to occur through one of several modes: singlet oxygen quenching (photoprotection), antioxidant protection, and enhancement of the immune response. Evidence suggests that a diet rich in carotenoids reduces the risk of coronary heart disease but supplementing the diet with synthetic beta-carotene did not produce the same benefit. Possibly other carotenoids are important in the diet and this has led to increased interest in carotenoids such as lutein. Interest in lycopene has increased dramatically in recent years due to epidemiological studies implicating lycopene in the prevention of cardiovascular disease, and cancers of the prostate and gastrointestinal tract.

8.3.5 Future prospects
The dramatic increase in the health aspects of the carotenoids has spurred a great deal of interest in these compounds as colorants. The prospect of having both a health and a colorant aspect is very appealing to merchandisers so we can expect an increase in the number of carotenoid extracts available. But with over 600 carotenoids existing in nature, it will be difficult to determine which compounds exhibit health effects.

8.4 Lycopene

8.4.1 Introduction
Lycopene is the major pigment in tomatoes and is one of the major carotenoids in the human diet. It also accounts for 50% of the carotenoids in human serum. Tomato products are widespread in diets around the world and are highly prized for their flavor and color contributions.

8.4.2 Chemistry and usage
The major source of lycopene is tomato products but it also occurs in water melons, guavas, pink grapefruit, and in small quantities in at least 40 plants.[14] The structure of lycopene is shown in Fig. 8.2. It is a long chain conjugated hydrocarbon and its structure suggests that it would be easily oxidized in the presence of oxygen and isomerized to *cis* compounds by heat. Both of these reactions occur in purified solutions of lycopene but in the presence of other compounds normally present in tomatoes, lycopene is more stable. Actually the absorption of lycopene in the human gut is increased by heat treatment probably because the breakdown of the plant cells makes the pigment more accessible. Preparations from tomatoes are widely used in pizza, pasta, soups, drinks and any product compatible with the flavor and color of tomatoes.

182 Food chemical safety

8.4.3 Toxicology
There is little safety data available for tomato products probably because they have been a major food for so long.

8.4.4 Health aspects
A recent review of 72 independent epidemiology studies revealed that intake of tomatoes and tomato products was inversely related to the risk of developing cancers at several sites including the prostate gland, stomach and lung. The data were also suggestive of a reduced risk for breast, cervical, colorectal, esophageal, oral cavity and pancreas.[13] Obviously, the role of lycopene is going to get more research attention in the future.

8.5 Lutein

8.5.1 Introduction
Lutein is a major component of many plants. It is a component of most of the carotenoid extracts suggested as food colorants.

8.5.2 Chemistry and usage
Lutein has a stucture similar to beta-carotene with a hydroxyl group on the ionone ring at each end of the molecule. It is somewhat less sensitive to oxidation and heat degradation than beta-carotene. It contributes a yellow color.

8.5.3 Toxicology
Little data is available but it would be expected to be non-toxic by comparison to similar carotenoids.

8.5.4 Health effects
Several studies have linked lutein to a lower risk for eye, skin and other health disorders, probably through its antioxidant activity. Lutein is apparently metabolized to zeaxanthin, an isomer, and several other compounds which protect the macula from ultraviolet radiation. The suggestion is that lutein may play a positive role in reducing macular degeneration. Other reports have linked lutein to a reduction of risk of cancer.[13] Regardless, lutein is currently being promoted as an important dietary supplement.

8.6 Annatto and saffron

8.6.1 Introduction
Annatto is one of the oldest colorants, dating back to antiquity for coloring food, cosmetics and textiles. Annatto is produced from the seeds of the tropical shrub *Bixa orellana*. Saffron is also a very old colorant dating back to the 23rd century BC. It is produced from the dried stigmas of the flowers of the crocus bulb, *Crocus sativa*. Saffron is known as the gourmet spice because it produces a desirable flavor and color. Its high price is assured because it takes about 150,000 flowers to produce one kilogram of saffron.

8.6.2 Chemistry and usage
The main pigments in annatto are bixin (Fig. 8.2) and norbixin. Bixin is the monomethyl ester of a dicarboxyl carotenoid. Norbixin is the saponified form, a dicarboxyl acid of the same carotenoid. The carboxylic acid portion of the molecule contributes to water solubility and the ester form contributes to oil solubility. Annatto is available in both water soluble and oil soluble liquids and powders. Annatto is somewhat unstable to light and oxygen but, technically, it is a good colorant. The principal use of annatto is as colorant for dairy products due to its water solubility but it is also used to impart a yellow to red color in a wide variety of products. The main pigments in saffron are crocin (Fig. 8.2) and crocetin. Crocin is the digentiobioside of the dicarboxylic carotenoid crocetin. The carboxylic and the sugar portion of the molecule contribute to water solubility. It is more stable to light and oxygen than annatto but, technically, it is a good colorant and is used in a variety of gourmet foods.[15]

8.6.3 Toxicology
A series of studies have shown annatto to be non-genotoxic[11,15] but others have suggested some mitotic aberrations[16] and some genotoxicity.[17] The acute oral toxicity is very low. The oral LD_{50} for rats is greater than 50 g/kg for the oil soluble form and 35 g/kg for the water soluble form. Lifetime toxicity studies in rats at the level of 26 mg/kg/day showed no toxic effects. Rats showed no reproductive problems when fed at 500 mg/kg/day for three generations. It was concluded that annatto was not carcinogenic. The JECFA established an ADI of 0–0.065 mg/kg/day for annatto based on studies nearly 40 years ago. Current research is under way to increase the ADI. Little data is available for saffron but the chemical similarity to the pigments in annatto, and other carotenoids, would suggest that saffron would pose no problems in the food supply.

8.6.4 Health effects
Annatto seeds have long been used by the South American Indians as a traditional medicine for healing of wounds, skin eruptions, healing of burns, and

184 Food chemical safety

given internally for diarrhea, asthma and as an antipyretic.[15] Annatto is claimed to have strong antioxidative potency, as shown by inhibition of lipid peroxidation and lipoxidase activity.

8.6.5 Future propects
Annatto is well established in the market and its use is increasing in poundage probably due to its superior technological properties. If some of the health claims prove to be true, annatto will enjoy increased interest. Saffron is well established in the gourmet markets but its use will be restrained because of its high price.

8.7 Paprika

8.7.1 Introduction
Paprika is a very old colorant and spice. It is a deep red, pungent powder prepared from the dried pods of the sweet pepper, *Capsicum annum*.

8.7.2 Chemistry and usage
Paprika contains capsorubin and capsanthin (Fig. 8.3) which occur mainly as the lauric acid esters, and about 20 other carotenoid pigments. Paprika is produced in many countries which have developed their own specialties. Cayenne or cayenne pepper, produced from a different cultivar of C. *annum*, is usually more pungent. C. *frutescens* is the source of the very pungent Tabasco sauce. Paprika oleoresin is produced by solvent extraction of the ground powder. Obviously paprika supplies both flavor and color and its use is limited to those products compatible with the flavor. The recent rise in demand for tomato products in the form of pizza, salsa, etc., has increased the demand for paprika. Paprika is used in meat products, soups, sauces, salad dressings, processed cheese, snacks, confectionery and baked goods.[10,18]

8.7.3 Toxicology
The acute oral toxicity of paprika is very low with an LD_{50} for mice of 11 g/kg. Several studies have indicated that paprika is not genotoxic. The JECFA did not establish an ADI because they considered that the levels of paprika and its oleoresins in foods would be self-limiting.[11]

8.7.4 Future prospects
Paprika is well established worldwide and will probably increase in volume due to the popularity of tomato products and possibly by analogy to the health effects being attributed to the carotenoids.

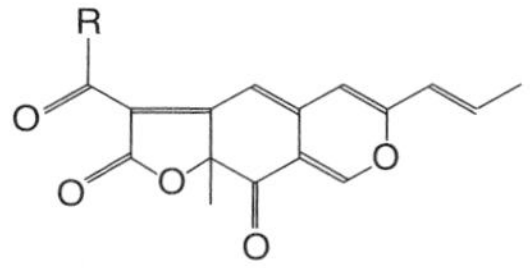

Monascin $R = C_5H_{11}$
Ankaflavin $R = C_7H_{15}$

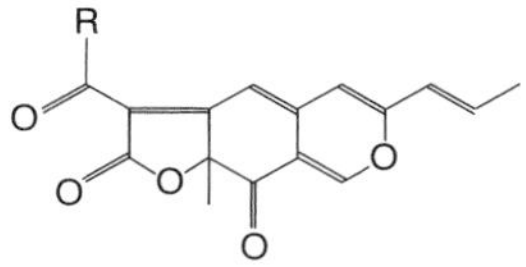

Rubropunctatin $R = C_5H_{11}$
Monascorubin $R = C_7H_{15}$

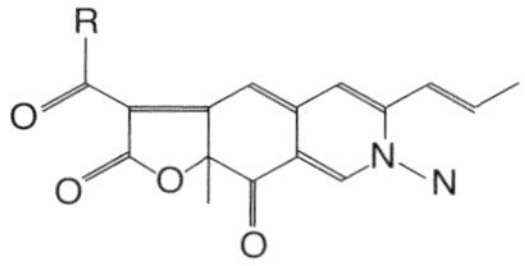

Rubropunctamine $R = C_5H_{11}$
Monascorubramine $R = C_7H_{15}$

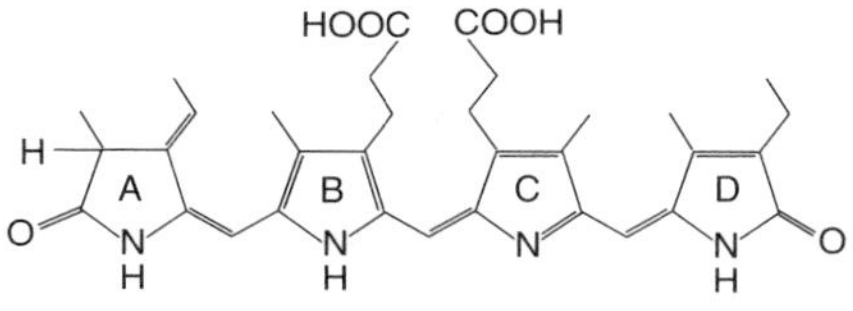

Phycocyanobilin

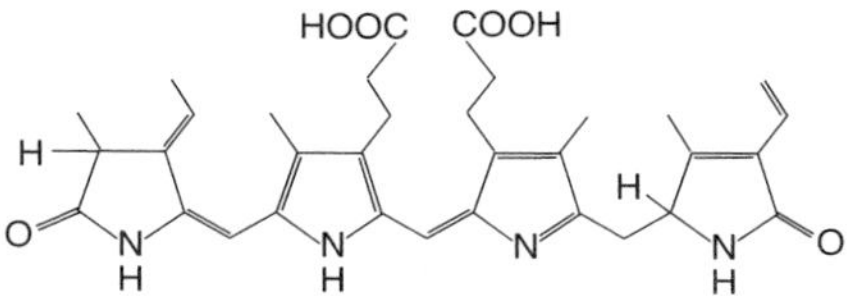

Phycoerythrobilin

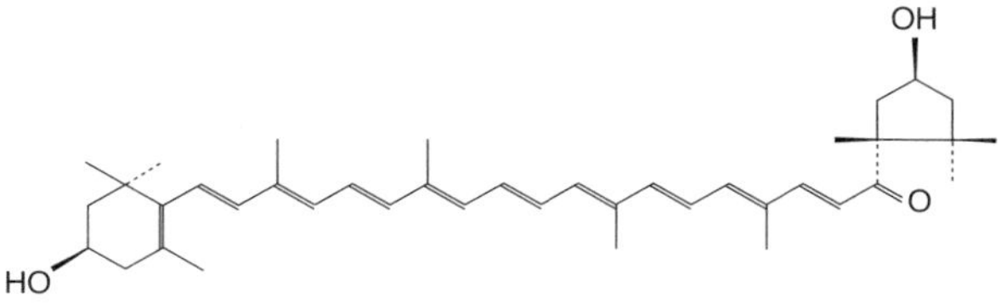

Capsanthin

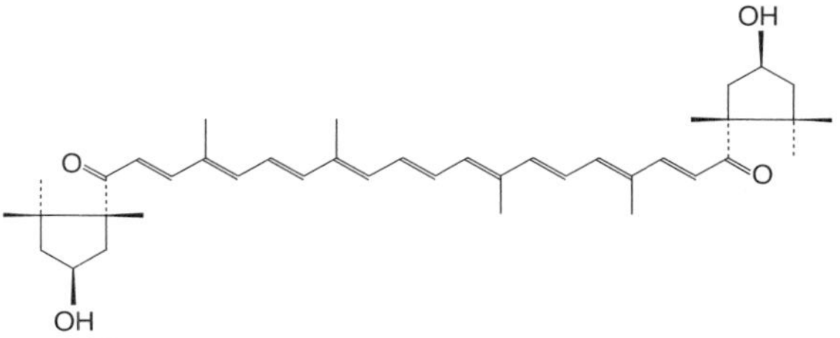

Capsorubin

Fig. 8.3 Top. Pigments in Monascus. Middle. Phycobilins in algae. Bottom. Pigments in paprika.

8.8 Synthetic carotenoids

8.8.1 Introduction
The success of the carotenoid extracts led to the commercialization of synthetic carotenoids, some with the same chemical structure as those in the plant extracts and others with modifications to improve their technological properties. The yellow beta-carotene was synthesized in 1950, followed by the orange beta-8-carotenal in 1962 and the red canthaxanthin in 1964. A number of others soon followed, methyl and ethyl esters of carotenoic acid, citraxanthin, zeaxanthin, astaxanthin, and recently lutein.

8.8.2 Chemistry and usage
The structures of four of the synthetic carotenoids (beta-carotene, canthaxanthin, beta-apo-8'-carotenol, beta-apo-8'-carotenoic acid) are shown in Fig. 8.2. By virtue of their conjugated double bond structure, they are susceptible to oxidation but formulations with antioxidants were developed to minimize oxidation. Carotenoids are classified as oil soluble but most foods require water soluble colorants; thus three approaches were used to provide water dispersible preparations. These included formulation of colloidal suspensions, emulsification of oily solutions, and dispersion in suitable colloids. The Hoffman-LaRoche firm pioneered the development of synthetic carotenoid colorants and they obviously chose candidates with better technological properties. For example, the red canthaxanthin is similar in color to lycopene but much more stable. Carotenoid colorants are appropriate for a wide variety of foods.[10] Regulations differ in other countries but the only synthetic carotenoids allowed in foods in the US are beta-carotene, canthaxanthin, and beta-8-carotenol.

8.8.3 Toxicology
When the first synthetic carotenoid colorant, beta-carotene, was suggested as a food colorant, it was subjected to an extensive series of tests[11] despite the belief that carotenoid extracts containing beta carotene were non-toxic. It was thought that synthetic preparations might contain a different profile of minor contaminants than those in plant extracts. Beta-carotene has a very low acute oral toxicity. The LD_{50} for dogs is greater than 1000 mg/kg and a single intramuscular injection of 1000 mg/kg in rats had no significant effect. Lifetime dietary administration showed no carcinogenicity with a NOEL of 100 mg/kg/day for rats and 1000 mg/kg/day for mice. No teratogenic or reproductive toxicity was shown when four generations of rats were fed up to 100 mg/kg/day. No cytogenic or teratogenic effects were seen in the offspring of rabbits given 400 mg/kg/day by stomach tube. Similar experiments with rats given up to 1000 mg/kg/day by intubation showed no embryo toxicity, teratogenicity or reproductive effects. Apparently rodents can tolerate large amounts of beta-carotene but extrapolation to humans is difficult because rats and humans

metabolize beta-carotene in different ways. With humans, the absorbed beta-carotene is largely converted to vitamin A, esterified, and transported in the lymph. With rats, the beta-carotene is largely converted to non-saponifiable compounds. Regardless, there is ample evidence that beta-carotene in reasonable quantities is harmless to humans. There is, however, evidence that humans with a high intake of beta-carotene may develop hypercarotenemia leading to an orange skin coloration. The JECFA established an ADI of 0–5 mg/kg/day for the sum of all carotenoids used as colorants.

Canthaxanthin was also subjected to an extensive series of toxicological tests[11] which indicated that it was essentially non-toxic. The acute oral toxicity was very low with an oral LD_{50} for mice greater than 10 g/kg. No effect was seen in dogs fed 500 mg/kg/day for 15 weeks. Canthaxanthin was not carcinogenic at feeding levels of 1000 mg/kg/day for 104 weeks for rats and 98 weeks for mice and it may have been anti-carcinogenic. No reproductive or teratogenic effects were seen when rats were fed up to 1000 mg/kg/day for three generations. Ingestion of large amounts of canthaxanthin caused deposition of canthaxanthin crystals, producing retinal impairment, in the eyes of humans, cats and rabbits, but not in rats, mice and dogs. In humans the visual impairment was reversible in a few months. Effects of high doses are available because canthaxanthin is used as a tanning aid. Other adverse effects were hepatitis and urticaria. The JECFA was unable to assign an ADI because of the problem with retinal crystal deposition in humans. Canthaxanthin, in the amounts required for appropriate coloration, is believed to be completely safe.

Beta-8-carotenal and the methyl and ethyl esters of carotenoic acid were also tested for toxic effects with results similar to beta-carotene.

8.8.4 Health effects
There is considerable current research under way to determine if the synthetic carotenoids have the same physiological effects, and consequent health benefits, as the naturally occurring compounds. If this proves to be true, we can expect increased interest in this group of colorants.

8.9 Anthocyanins

8.9.1 Introduction
Anthocyanins are ubiquitous in the plant kingdom. They are responsible for many of the orange, red, blue, violet, and magenta colors. They have been the object of intensive research from a taxonomic point of view and this has resulted in about 275 known structures and about 5,000 of the chemically closely related flavonoid compounds. Their use as colorants dates back to antiquity since the Romans used highly colored berries to augment the color of wine.

8.9.2 Chemistry and usage

Anthocyanins consist of an aglycone combined with one or more sugars. Twenty-two aglycones are known but only six are important as colorants (Fig. 8.4). In view of the ubiquity and high tinctorial power it is not surprising that many sources have been suggested as colorants. Francis listed pigment profiles and methods of extraction for over 40 plants[19,20] and also 49 patents on anthocyanin sources as potential colorants.[21] However, despite the large number of sources, only one dominated the supply for many years. Colorants from grape skins as a by-product of the wine industry is the major source. Grapes are the largest fruit crop for processing and since 80% of the estimated 60,000,000

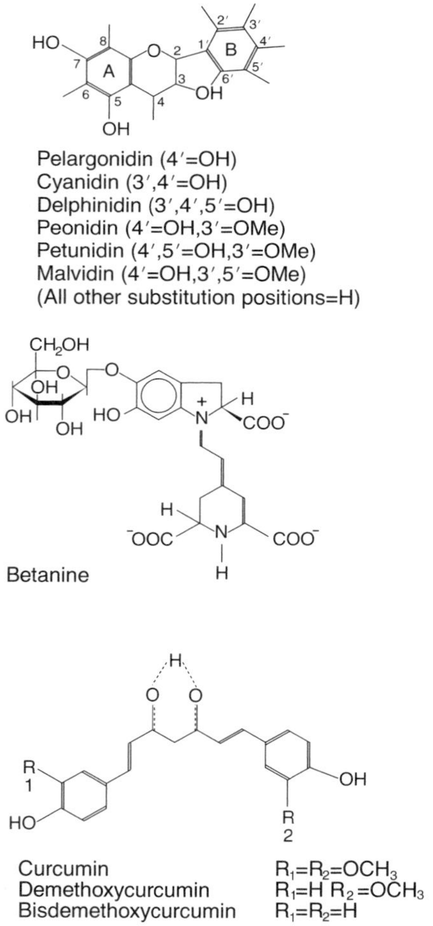

Pelargonidin (4'=OH)
Cyanidin (3',4'=OH)
Delphinidin (3',4',5'=OH)
Peonidin (4'=OH,3'=OMe)
Petunidin (4',5'=OH,3'=OMe)
Malvidin (4'=OH,3',5'=OMe)
(All other substitution positions=H)

Betanine

Curcumin $R_1=R_2=OCH_3$
Demethoxycurcumin $R_1=H\ R_2=OCH_3$
Bisdemethoxycurcumin $R_1=R_2=H$

Fig. 8.4 Top. Structures of the major anthocyanidins in foods. Middle. Structure of betanine. Bottom. Pigments in turmeric.

metric tons is used annually for wine production, this situation is not likely to change. Recently anthocyanins from red cabbage have enjoyed some success.

The sugars attached to the anthocyanin molecule are in order of relative abundance glucose, rhamnose, galactose, xylose, arabinose, and glucuronic acid. The molecule may also contain one or more of the acyl acids p-coumaric, caffeic, and ferulic or the aliphatic acids malonic and acetic esterified to the sugar molecules. Extracts of anthocyanins invariably contain flavonoids, phenolic acids, catechins and polyphenols. The net result is that it is impossible to express the chemical composition accurately. Specifications usually present tinctorial power, acidity, per cent solids, per cent ash and other physical properties.

The major market for the colorants from grapes (generic term enocyanin) is in fruit drinks. Anthocyanins are pH sensitive and show the greatest tinctorial power around pH 3–3.5 and most fruit drinks are in this range. Anthocyanin colorants have been used in a wide variety of food products such as beverages, jams, jellies, ice cream, yoghurt, gelatin desserts, canned fruits, toppings, confections, and many others.

8.9.3 Toxicology
Anthocyanins are not genotoxic as shown by a number of studies.[19] The oral toxicity of mixed anthocyanins was greater than 20 mg/kg/day for rats. Dogs fed a diet containing 15% anthocyanins from Concord grapes for 90 days showed no significant toxic effects. Another multigeneration study on rats fed 15% powder from Concord grapes showed no effects on reproduction. A study on Roselle, a popular drink made from the calyces of *Hibiscus sabdariffa*, showed no toxic effects.[22] The lack of toxic effects is not surprising in view of the long history of wine consumption.

8.9.4 Health aspects
Anthocyanins and the related flavonoids have been very much in the news lately for a wide variety of health claims. These include anticarcinogenic, anti-inflammatory, antihepatoxic, antibacterial, antiviral, antiallergenic, antithrombotic, and antioxidant effects.[13] Antioxidation is believed to be one of the most important mechanisms for preventing or delaying the onset of major diseases of aging including cancer, heart disease, cataracts and cognitive disfunction. The antioxidants are believed to block oxidative processes and free radicals that contribute to the causation of these chronic diseases. Anthocyanins are non-competitive inhibitors of lipid peroxidation comparable in potency to the classical antioxidants such as butylated hydroxy anisole (BHA), butylated hydroxytoluene (BHT), and alpha tocopherol. Anthocyanins were reported to have anti-inflammatory properties comparable to commercial pharmaceuticals. The property of anthocyanins to decrease the fragility and permeability of blood capillaries is common to many flavonoids and was the basis for the original

definition of Vitamin P by Szent Gyorgi in 1936. This claim has been commercialized by the marketing of extracts of bilberry (*Vaccinium myrtillis*).

Interest in the health effects of anthocyanins was piqued by the 'French paradox' in which the mortality from cardiovascular disease was lower than that predicted from the intake of dietary saturated fatty acids. The beneficial effects were greater in association with alcohol taken in the form of wine suggesting that there may be a protective effect of other components of wine. Needless to say the wine industry was pleased with this research.

8.9.5 Future prospects

The potential health effects of anthocyanins and flavonols has stimulated much research in this area but, in view of the chemical complexity of the plant extracts, we are a long way from determining the chemical compounds responsible for the wide variety of claims. Regardless, a colorant with associated health benefits is a very desirable situation from an industry point of view. This is a very active research area.

8.10 Betalains

8.10.1 Introduction

The betalains are confined to ten families of the order *Caryophyllales*.[20] The only foods containing betalains are red beet (*Beta vulgaris*), chard (*B. vulgaris*), cactus fruit (*Opuntia ficus-indica*) and pokeberries (*Phytolacca americana*). They also occur in the poisonous mushroom *Amanita muscaria* but this is not a normal food source. The importance of the betalains as colorants is confined to preparations from red beet.

8.10.2 Chemistry and usage

The betalains have two main groups; the red betacyanins and the yellow betaxanthins. The main pigment of red beets is betanine and its structure is shown in Fig. 8.4. Beets usually contain both betacyanins and betaxanthins and the ratio depends on the cultivar. Some cultivars contain only the yellow betaxanthins and this makes it possible to formulate a range of colorants from yellow to red. Betanine is relatively stable to changes in pH as contrasted with the anthocyanins and this makes it preferable for foods in the pH range 5–6. Both the red and yellow betalains are susceptible to degradation by heat, light and the presence of metal ions. Within these limitations, betalains are ideally used to color products that have a short shelf life, are packaged to reduce exposure to light, oxygen and high humidity, do not receive extended or high heat treatment, and are marketed in the dry state. Despite these limitations, betalains have been suggested for coloring ice cream, yoghurt, cake mixes, gelatin desserts, meat substitutes, gravies, frostings and many others.[20,23]

8.10.3 Toxicology
Betalain pigments have been tested on rodents by feeding 50 mg/kg pure betanin, 2000 ppm betanine in the diet and several other conditions.[11] No carcinogenic or other toxic effects were observed and the authors concluded that red beet extracts were safe as food colorants.

8.10.4 Future prospects
Betalain colorants are well established in the food chain and will probably continue in a limited capacity.

8.11 Chlorophylls

8.11.1 Introduction
The chlorophylls are a group of naturally occurring pigments produced in all photosynthetic plants including algae and some bacteria. Hendry[24] estimated annual production at about 1,100,000,000 tons with about 75% being produced in aquatic, primarily marine, environments. Obviously as a source of raw material for food colorants, chlorophylls present no problem with supply.

8.11.2 Chemistry and usage
Five chlorophylls and five bacteriochlorophylls are known but only two, chlorophylls a and b are important as food colorants. Chlorophyll a has a complex structure with a magnesium ion in the molecule which is easily removed in acid media to form pheophytin a. Removal of the phytyl portion of the molecule produces chlorophyllide a whereas removal of both magnesium and phytyl produces pheophorbide a. Chloropyll b reacts in the same manner. Chlorophyll and chlorophyllide are both bright green in color but pheophytin is olive green and pheophorbide is brown. Attempts to produce a food colorant from chlorophyll are centered around trying to stabilize the molecule by retaining or replacing the magnesium ion. Treatment with copper and zinc salts substitutes copper or zinc for the magnesium and the derivatives are bright green in color. Commercial colorants are usually made from lucerne (alfalfa), *Medicago sativa*, or nettles, *Urtica dioica*, and a series of pasture grasses. The plants are dried, extracted with a solvent and dried resulting in a mixture of chlorophyll, pheophytin and other degraded compounds. The dry residue can be purified to obtain an oil-soluble preparation or treated with an acidified copper solution to prepare a more stable water-soluble copper chlorophyllin. It is not commercially feasible to prepare a colorant containing pure chlorophyll because of the instability of the molecule. The major portion of the chlorophyll colorants are in the water soluble forms and are used in dairy products, soups, oils, sugar confections, drinks and cosmetics.[25]

8.11.3 Toxicology

The JECFA classified chlorophyll under List A which means that the colorant has been fully cleared and its use is not limited toxicologically since when used with 'good manufacturing practice' does not represent a hazard to health. A subchronic oral toxicity study showed no adverse effects.[26]

8.11.4 Health aspects

Chlorophyllin is an effective antimutagenic agent and has been used as a dietary supplement to diminish the intensity of the uncomfortable side effects of cyclophosphamide treatments.[13] Cyclophosphamide is a potent antitumor agent and is used against many forms of cancer and other diseases. Chlorophyllin protects against radiation induced DNA single strand breaks possibly by its ability to scavenge OH^+ and ROO^+ groups.

8.11.5 Future prospects

Chlorophyllins are approved for use in Europe and Asia but only for dentifrices in the US. Since there are no other commercially available natural green colors, these colorants are attracting interest.

8.12 Turmeric

8.12.1 Introduction

Turmeric is a very old colorant produced from the rhizomes of several species of *Cucuma longa*, a perennial shrub grown in many tropical areas around the world.

8.12.2 Chemistry and usage

Turmeric is the ground dried rhizomes and contains three main pigments, curcumin, demethoxycurcumin, and bisdemethoxycurcumin (Fig. 8.4) together with about four potent flavor compounds.[27,28] Turmeric oleoresins are prepared by extracting the dried ground rhizomes with a variety of chemical solvents and concentrating the resins perhaps with the addition of oils and other carriers. Turmeric and turmeric oleoresins are unstable to light and alkaline conditions and a number of substances are sometimes added to stabilize the molecule. Curcumin is insoluble in water but a water-soluble form can be made by complexing the compound with tin or zinc to form an intensely orange colorant but it is not allowed in most countries. The major applications of turmeric are to color cauliflower, pickles and mustard but it is also used in combination with annatto in ice cream, yoghurt, baked goods, oils, salad dressings and confectionery.

8.12.3 Toxicology

Turmeric has been subjected to a number of safety studies because its consumption in Europe, India and the Middle East is very high. Consumption in India was estimated at up to 3.8 g/day.[11] The acute oral toxicity is very low. The oral LD_{50} for the oleoresin in rats and mice is greater than 10 g/kg and for curcumin in mice is greater than 2 g/kg. Rats fed turmeric for 52 weeks at 500 mg/kg/day showed no significant toxicity. Similar experiments with dogs and monkeys showed similar results. There was no evidence of carcinogenicity, reproductive toxicity or teratology. The JECFA did not allocate an ADI for turmeric because they considered it to be a food. A temporary ADI of 0.01 mg/kg for turmeric oleoresin and 0–0.1 mg/kg for curcumin was established in 1990.[12] In 1995, the temporary ADI for curcumin was increased to 1 mg/kg.

8.12.4 Health effects

It has been known since the early 1950s that turmeric had strong antioxidant effects with curcumin being the major compound responsible followed by demethoxycurcumin and bisdemethoxycurcumin. All three inhibit lipid peroxidation and have a positive anti-oxidant effect for hemolysis and lipid peroxidation in mouse erythrocytes.[11] Curry pills containing turmeric are being marketed as a prevention for colon cancer.[29]

8.12.5 Future prospects

Turmeric is well established in the food supply and if it is proven to have a health effect as well as a colorant and flavor component, its future would seem assured.

8.13 Cochineal and carmine

8.13.1 Introduction

Cochineal is a very old colorant. References go back as far as 5000 BC when Egyptian women used it to color their lips. It was introduced to Europe by Cortez who found it in Mexico. Production peaked around 1870 and then declined due to the introduction of synthetic colorants, but it is still a major commodity for Peru, Mexico, and the Canary Islands.[25,30]

8.13.2 Chemistry and usage

Cochineal extract is obtained from the bodies of the female cochineal insects, particularly *Dactylopius coccus* Costa, by treating the dried bodies with ethanol. After removal of the solvent, the dried residue contains about 2–4% carminic acid, the main colored component. The cochineal insects grow on cactus and, since it takes about 50,000–70,000 insects to produce one pound of the colorant, production will always be labor intensive.

Solutions of carminic acid, at pH4 show a range of colors from yellow to orange depending on the concentration. When complexed with aluminum, a series of stable brilliant red hues ranging from 'strawberry' to 'blackcurrant' can be produced depending on the ratio of aluminum to carminic acid. Purified extracts of cochineal have been termed 'carmine' but the term usually refers to a lake of carminic acid with aluminum, calcium or magnesium. Carmine usually contains about 50% carminic acid. Carmine is considered to be technologically a very good food colorant. It is ideally suited to foods with a pH above 3.5 such as comminuted meat and poultry products, surimi, and red marinades. It is also used in a wide variety of other products such as jams, gelatin desserts, baked goods, confections, toppings, dairy products, non-carbonated drinks, and many others.

8.13.3 Toxicology

A number of studies have shown that cochineal extract and carmine are not carcinogenic.[11] Rats fed carmine up to 100 mg/kg/day showed only reduced growth at the higher levels. Other studies show no reproductive problems in a single generation study or reproductive/teratology problems in a multigeneration study. The JECFA assigned a combined ADI of 0–5 mg/kg/day for cochineal and carmine.

8.13.4 Future prospects

Carmine is well entrenched in the food industry and probably will remain there because of its superior technological properties. It is, however, a very labor intensive industry because of the hand harvesting of the insects with a consequent high price.

8.13.5 Related compounds

Carmine belongs to the anthraquinone class of compounds and several other chemically closely related compounds are also used as colorants.[25] Kermes is a well known colorant in Europe. It is obtained from the insects, *Kermes ilicis* or *Kermococcus vermilis*, which grow on oak trees. It contains kermisic acid, the aglycone of carminic acid, and its isomer ceroalbolinic acid. Its properties are very similar to carmine. Lac is a red colorant obtained from the insect *Laccifera lacca* which is found on several families of trees in India and Malaysia. The lac insects are better known for their production of shellac. They contain a complex mixture of anthraquinones. Alkanet is a red pigment from the roots of *Alkanna tinctoria* Taush and *Alchusa tinctoria* Lom. All three have been cleared for food use in Europe but not in the US.

8.14 Monascus

8.14.1 Introduction
Monascus colorants are well known in Asia and in Chinese medicine date as far back as 1590. The colorants are produced by several fungal species of the genus *Monascus* which grow readily on a number of carbohydrates, especially rice, but also on wheat, soybeans, corn and other grains. The Koji process involves inoculating the solid grain mass with the fungus, primarily *Monascus purpureus*, and drying the substrate to produce 'red rice' which is used as a colorant for many foods.[31,32]

8.14.2 Chemistry and usage
Monascus species produce six pigments (Fig. 8.3). Monascin and ankaflavin are yellow, rubropunctatin and monascorubin are red, and rubropunctamine and monascorubramine are purple. The pigments are very reactive and have been reacted with a variety of compounds such as polyamino acids, amino alcohols, chitin amines or hexamines, proteins, sugar amines, aminobenzoic acid and many others[21,28] to produce compounds with greater water solubility, thermostability and photostability than the parent compounds. The ability to grow on a solid substrate has led to a large body of data on optimization of pigment production in solid state fermentation and later in submerged fermentation.[31] The conditions of growth can be modified to optimize the ratio of the different pigments and also for several other compounds with health implications. Monascus pigments are soluble in ethanol and the derivatives are soluble in water. A range of colors from yellow to purple can be produced by manipulating the ratio of pigments and the pH. Their stability in neutral media is a real advantage. The colorants are suitable for processed meats, marine products, ice cream, jam, toppings, and tomato ketchup, and traditional oriental products such as koji, soy sauce, and kambuki. Their solubility in alcohol makes them appropriate for alcoholic beverages such as saki.

8.14.3 Toxicology
Monascus colorants have been consumed for hundreds of years and are believed to be safe for human consumption. Tests with a series of microorganisms have demonstrated no mutagenicity. No toxicity was observed in rodents or in fertile chicken eggs. The yellow pigments have an LD_{50} for mice of 132 mg/20g. No ADI is available.[11] There has been some concern that some of the strains produced antibiotics which is obviously undesirable for a food colorant, but it is possible to choose strains of Monascus which do not produce antibiotics.

8.14.4 Health effects
A large number of reports describe the use of Monascus preparations in herbal medicines and food supplements. These include the suppression of tumor

196 Food chemical safety

production, regulation of immunoglobulin production, lowering of lipids in hyperlipidemia, and reduction of aminoacetaphen-induced liver toxicity by antioxidase action. One example would be Cholestin distributed by the Pharmanex company. It is obtained from red rice imported from China as a neutraceutical and contains a natural inhibitor of the rate-limiting synthesis of cholesterol. It is claimed to reduce total cholesterol and low-density lipoprotein and triglyceride levels while increasing high-density lipoprotein levels. A neutraceutical with a beneficial effect for heart disease is a very desirable product so Cholestin was soon followed by several other products which may have been produced to maximize the content of the active ingredient. One problem was that the active ingredient was identical to that in Mevacor, a prescription drug patented by Merck. The FDA ruled that Cholestin was a drug, not a food supplement, and banned it. The Pharmanex company sued the FDA and the courts made a temporary decision in favor of the company and reversed the ban.[31] This situation seems to permit an end run around the strict rules governing approval of prescription drugs by simply finding a natural source, manipulating the growing conditions to maximize the content of the drug, and calling the formulation a food supplement.

8.14.5 Future prospects
Monascin colorants are well entrenched in Asia, particularly China, Japan, and Taiwan and probably will continue to be an important product in view of their long history. They are not allowed in the US and there seems to be little interest in them. Certainly, the situation illustrated by the Cholestin debate will have to be settled, probably by new legislation, before a commercial firm petitions to have Monascin colorants pemitted in the US.

8.15 Iridoids

8.15.1 Introduction
The colorants from saffron have enjoyed good technological success as colorants and spices but their high price has led to searches for other sources of the same pigments. The pigments, but not the flavor, can be obtained in much larger quantities from the fruits of the gardenia or Cape jasmine plant.[33]

8.15.2 Chemistry and usage
The fruits of gardenia, *Gardenia jasminoides*, contain three groups of pigments, crocins, iridoids and flavonoids. Structures of six of the nine iridoid pigments are shown in Fig. 8.5. The formulas for five flavonoids from G. *fosbergii* are also shown. This is a different species but botanically closely-related species tend to have similar pigment profiles. The crocins are orange and the flavonoids are pale yellow. The iridoids are interesting because they can be reacted with amino acids or proteins to produce a range of colors from green to yellow, red,

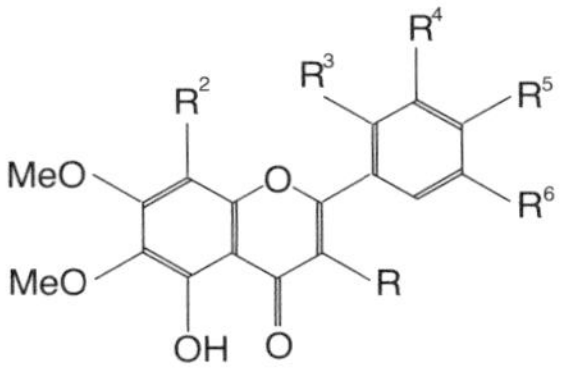

Flavonoid compounds

Compound	R¹	R²	R³	R⁴	R⁵	R⁶
1	H	H	H	OMe	OMe	OMe
2	OMe	H	H	OH	OMe	OH
3	H	H	OMe	OMe	H	OH
4	OMe	H	H	OMe	OMe	OH
5	OMe	OMe	H	H	OH	H

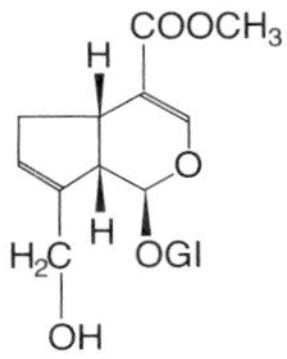

Geniposide 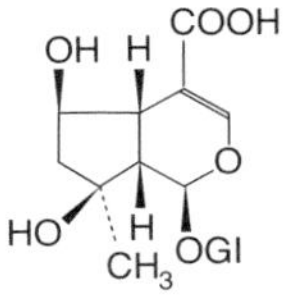   Shanzhiside 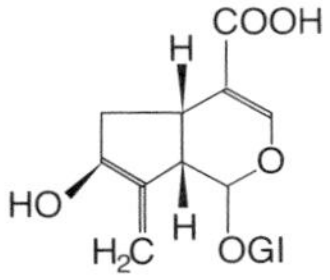   Gardoside

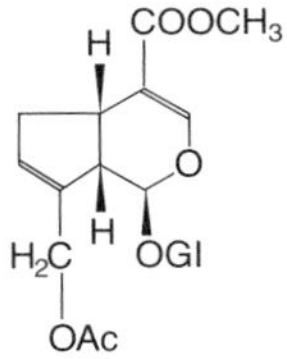

Acetylgeniposide 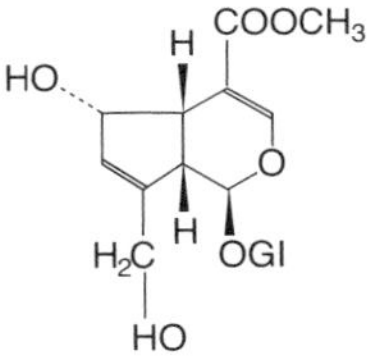   Methyldeacetylasperuloside 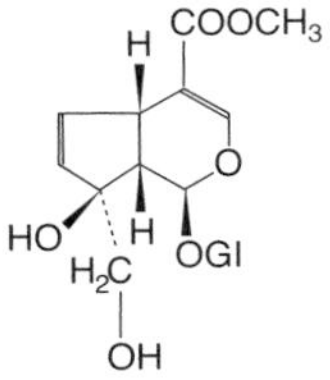   Gardenoside

Fig. 8.5 Top. Some flavonoid pigments in gardenia. Bottom. Six of the nine iridoid pigments in gardenia.

and blue. A number of patents[21] have described the manipulation of the reaction conditions such as time, temperature, pH, oxygen content, degree of polymerization, reaction with selected microorganisms, etc. The compounds can be hydrolyzed to produce genipin which reacts readily with taurine to produce an attractive blue color. Four greens, two blues and one red colorant have been commercialized in Japan. They have been suggested for use with candies, condiments, ices, noodles, imitation crab, fish eggs, chestnuts, beans, dried fish substitutes, liqueurs, baked goods, etc.

8.15.3 Toxicology
The geniposides from gardenia were found to have some hepatoxicity due to the aglycone genipin produced by hydrolysis of the geniposides.[34] The yellow, green, red, and blue colorants were studied extensively and were found to be safe for human consumption as food colorants.[35,36]

8.15.4 Future prospects
The yellow crocins from gardenia have received some success for the same colorant applications as saffron,[37] but the iridoid derivatives have not received the same promotion. The range of colorants available from the same source would seem to make them attractive possibilities.

8.16 Phycobilins

8.16.1 Introduction
The phycobilins belong to the heme group of pigments which include the green chlorophylls in plants and the red hemoglobins in animals. The phycobilins are major biochemical components of the blue-green, red, and cryptomonad algae.[38]

8.16.2 Chemistry and usage
Phycobilins are colored, fluorescent, water-soluble pigment-protein complexes. They can be classified into three groups according to color: phycoerythrins are red with a bright orange fluorescence; phycocyanins and allophycocyanins are both blue and fluoresce red. The structures of two are shown in Fig. 8.3. Phycocyanins and allophycocyanins share the same chromophore but differ in the protein portion. The attachment of the bilin chromophore to the protein is very stable and this makes them desirable from a colorant point of view. Phycobilin preparations can be made by simply freeze drying algal cell suspensions which can be grown in ponds or sophisticated tubular reactors. Suggested applications involve chewing gums, frozen confections, dairy products, soft drinks, and ice cream.

8.16.3 Toxicology
One study on toxicity of phycocyanin from *Spirolina platensis* reported no adverse effects.[39] There is little other data available on toxicology but no toxic effects would be expected in view of the long history of algal consumption.

8.16.4 Future prospects
The future of the phycobilins looks promising for two reasons. First, there are no other blue natural colorants available and, admittedly, blue is not a favorite food

color but niche markets are evolving. Second, spirolina is becoming an attractive product in the health food area, and the same facilities used to produce spirolina products could be used to produce phycobilins.

8.17 Caramel

8.17.1 Introduction

Caramel is a brown colorant obtained by heating sugars.[33] The official FDA definition is as follows: 'The color additive caramel is the dark brown liquid or solid resulting from carefully controlled heat treatment of the following food grade carbohydrates: dextrose, invert sugar, lactose, malt syrup, molasses, starch hydrolysates and fractions thereof, or sucrose.' Heating sugar preparations to produce brown flavorful and pleasant-smelling products has been practised in home cooking for centuries but the first commercial caramel colorants appeared in Europe about 1850.

8.17.2 Chemistry and usage

Commercial caramel is a very complex mixture of heat degraded carbohydrates. In 1980, the JECFA recommended that further information on the chemical properties be obtained in order to establish a suitable classification and specification system. The International Technical Caramel Committee attempted to provide this information and undertook an extensive research program.

The complexity of the mixtures made it impossible to define the chemical composition so the commercial preparations were divided into four groups (Table 8.2) on the basis of a series of sophisticated chemical assay procedures. Caramel colorants must be compatible with the food products in which they are used, which usually means the absence of flocculation and precipitation in the food. These undesirable effects result from charged macromolecular components of caramel which react with the food. Hence the net ionic charge of the caramel macromolecules at the pH of the intended food product is the prime determinant of compatibility. Caramel colorants are used in a variety of foods (Table 8.2) but over 80% of the caramel produced in the US is used to color soft drinks particularly colas and root beers.

8.17.3 Toxicology

One of the major considerations of the research requested by JECFA was the safety aspects which was not surprising in view of the chemical complexity of the caramels. The program resulted in the publication of 11 papers in the same issue of the journal *Food and Chemical Toxicology* 1992 (Vol. 30) and seven of them were on toxicology. Caramel colorants were given a clean bill of health and JECFA assigned an ADI of 0–200 mg/kg/day.

Table 8.2 Caramel formulations

Class	Charge	Reactants	Usage
1	−	No ammonium or sulphite compounds	Distilled spirits, desserts, spice blends
2	−	Sulphite compounds	Liqueurs
3	+	Ammonium compounds	Baked goods, beer, gravies
4	+	Both sulphite and ammonium compounds	Soft drinks, pet foods, soups

8.17.4 Future prospects
Caramel colorants are well established in food formulations and probably will remain that way in the foreseeable future.

8.18 Brown polyphenols

8.18.1 Introduction
There are two important sources of brown polyphenols, cocoa and tea, used as colorants for foods. Both are very old and date back to antiquity.

8.18.2 Chemistry and usage
The cacao plant, *Theobroma cacao,* is the source of chocolate which is well known and highly prized in international commerce. The cacao pods contain beans which are fermented and pressed to provide a brown liquid which is the raw material for chocolate. The press cake is ground and sold as cocoa and it also provides a brown colorant. The pods, beans, shells, husks and stems have also been suggested as colorants. They contain a very complex mixture of acyl acids, leucoanthocyanins, flavonoid polymers, tannins, and catechin-type polymers.[33]

The tea plant, *Thea sinensis,* has provided a desirable beverage for centuries but it is also used as a colorant. Extracts of tea contain a very complex mixture of glycosides of myricetin, quercitin and kaempferol, epicatechin, epigallocatechins, acyl acids and many other polyphenol compounds.[33] In black tea, the above compounds may act as precursors to the poorly defined compounds thearubin and theaflavin.

Both cocoa and tea are used in a variety of food products, including beverages, bakery products, confections, toppings, dry mixes, etc.

8.18.3 Toxicology
Tarka[40] wrote an extensive review of the toxicology of cocoa and the methylxanthenes, theobromine, caffeine and theophyllin. The review involved

biological and behavioral effects, metabolism, species variation, pregnancy, lactation, reproduction, teratology, mutagenicity, fibrocystic breast diseases, and drug and dietary factors. No implications for human use as a beverage or a colorant were discussed. The toxicology of tea is similar to cocoa and no adverse effects for humans have been implied.

8.18.4 Future prospects
Both cocoa and tea are well entrenched and this is not likely to change. Neither group is permitted as a food colorant in the US but they are permitted as food ingredients and this accomplishes the same end.

8.19 Titanium dioxide

8.19.1 Introduction
Titanium dioxide is a large industrial commodity with world production over 4,000,000 tons but only a very small proportion is used as a food colorant. Commercial TiO_2 is produced from the mineral ilmanite, which occurs in three crystalline forms, but the only one approved for food use is synthetic anatase. Anatase occurs in nature but only the synthetic version is approved because it contains fewer impurities.[41]

8.19.2 Chemistry and usage
Titanium dioxide is a very stable compound with excellent stability towards light, oxidation, pH changes, and microbiological attack. It is virtually insoluble in all common solvents. It is available in oil-dispersible and water-dispersible forms with a wide variety of carriers. Titanium dioxide is a very effective whitener for confectionery, baked goods, cheeses, icings, toppings, and numerous pharmaceuticals and cosmetics.

8.19.3 Toxicology
Titanium dioxide has been subjected to a number of safety tests[11] and found to be non-genotoxic, non-carcinogenic and exhibited no adverse effects in rats, mice, dogs, cats, guinea pigs, and rabbits. The LD_{50} values are greater than 25 g/kg/day for rats and 10 g for mice. Apparently titanium dioxide is poorly absorbed and non-toxic. JECFA has not established an ADI since they consider titanium dioxide to be self regulating under GMP. In the US, it is allowed up to 1% by weight in food.

8.19.4 Future prospects
Titanium dioxide is well established and will probably remain that way.

202 Food chemical safety

8.20 Carbon black

8.20.1 Introduction
Carbon black is a large volume industrial commodity but its food use is very small.[40]

8.20.2 Chemistry and usage
Carbon black is derived from vegetable material, usually peat, by complete combustion to residual carbon. The particle size is very small, usually less than 5 μm, and consequently is very difficult to handle. It is usually sold to the food industry in the form of a viscous paste in a glucose syrup. Carbon black is very stable and technologically a very effective colorant. It is widely used in Europe and other countries in confectionery.

8.20.3 Toxicology
In the US in the 1970s when the GRAS list was being reviewed, safety data were requested on carbon black in view of the possibility that it might contain heterocyclic amines. Apparently, the cost of obtaining the data was higher than the entire annual sales of food grade carbon black so the tests were never done. Carbon black is not permitted in the US.

8.21 Miscellaneous colorants

A number of preparations are used in small volume or with minimal effect on color.[41] Ultramarine blue, a synthetic aluminosulphosilicate blue colorant, is widely used in cosmetics and in salt intended for animal consumption. A variety of brown iron oxides are used in cosmetics, drugs and pet foods. Talc is a large industrial commodity with many uses. It is used as a release agent in the pharmaceutical and baking industries as well as a coating for rice grains. Zinc oxide is an effective whitener for food and food wrappers. It is also added as a nutrient. Riboflavin provides an attractive yellow green color to foods as well as a nutrient benefit. Corn endosperm oil is added to chicken feed to enhance the yellow color of the skin and eggs. Dried algal meal is produced from *Spongiococcum spp.* and may be added to chicken feed to enhance the color of skin and eggs. Extracts from other algae are permitted in other countries. Four products from cottonseeds may be added to food but they are usually considered ingredients and impart only a slight yellow color. Shellac, obtained from the insect, *Lassifer lacca*, is added to foods as a surface coating or glaze and it does not affect the appearance. Octopus and squid ink contain mixtures of meloidin polymers and are effective black colorants for pasta for special occasions for some ethnic groups. They are not permitted in the US. All of the above preparations are considered harmless for human consumption by virtue of a long history of use and sometimes because their use is in such small amounts or concentrations as to be considered of no concern even though the toxicological data may be a little hazy.

8.22 Outlook

Colorants present a wide variety of preparations added to foods to increase their visual appeal. Depending on their classification, they may be subjected to extensive or minimal toxicity testing. For example, FD&C Red No. 2 is probably the most tested additive in our food supply, second only to saccharin. Others have been grandfathered in by virtue of a long history of consumption. But our society is moving towards a more formulated food supply with more attention being paid to ingredients available in large quantities but not necessarily in the most appealing form. Colorants and other additives, such as flavorants and texturants, are vital to making our food supply appealing. Interest in colorants, in particular, is increasing as judged by the research activity. Over the last 50 years, there has been a distinct trend towards the use of natural colorants as compared to synthetic 'coal tar dyes'. In one study,[21] the five-year period from 1979–1984 yielded the same number of colorant patents as the previous ten-year span from 1969–1978. In the fifteen-year span, there were 356 patents on natural colorants and 71 on synthetics. This trend has continued and shows no sign of changing.[42–47] The confidence in natural colorants over synthetics may be a little naive but it is realistic.

8.23 References

1. MARMION, D. M. *Handbook of US Colorants For Foods, Drugs, and Cosmetics*, 2nd Edn. John Wiley & Sons, New York, 1984.
2. WALFORD, J. Historical Development of Food Coloration. Chap 1 in *Developments in Food Colors*, Vol. 1, Editor J. Walford. Applied Science Publishers, London, 1980, pp. 1–26.
3. ACCUM, F. *A Treatise on Adulteration of Food.* Ab'm Small Press, 1820.
4. MARMION, D. M. *Handbook of US Colorants For Foods, Drugs and Cosmetics*, 3rd Edn. John Wiley & Sons, New York, 1991.
5. FRANCIS, F. J. FD&C Colorants. Chap 5 in *Colorants*. Eagan Press, St. Paul, MN, 1999, pp. 33–42.
6. BORZELLICA, J. F. and HALLIGAN, J. B. Safety and Regulatory Status of food, drug, and cosmetic color additives. *ACS Synposium Series 484: Food Safety Assessment*, 1992, 377–90.
7. HALLIGAN, J. B., ALLAN, D. C. and BORZELLICA, J. F. The safety and regulatory status of food, drug, and cosmetic color additives exempt from certification. *Food and Chemical Toxicol.* 1995, 515–28.

8. SMOLEY, C. K. *Everything Added to Foods in the United States*. U.S. Food and Drug Administration. CRC Press, Boca Raton, FL, 1993.
9. CLYDESDALE, F. M. *Food Additives; Toxicology, Regulation and Properties*, CD Rom. CRC Press, Boca Raton, FL, 1997.
10. FRANCIS, F. J. Carotenoids. Chap 6 in *Colorants*. Eagan Press, St. Paul, MN, 1999, pp. 43–54.
11. FRANCIS, F. J. Safety of Food Colorants. Chap. 4 in *Natural Food Colorants*, 2nd Edn. Edit. G. A. F. Hendry and J. D. Houghton. Blackie Academic & Professional, Glasgow, Scotland, 1996, pp. 112–30.
12. FURUHASHI, T. World Health Organization. *Food Additive Series*, No. 32, 1993.
13. MAZZA, G. Health Aspects of Natual Colors. Chap 14 in *Natural Food Colorants*. Edit. G. J. Lauro and F. J. Francis. Marcel Dekker, New York, 2000, pp. 289–314.
14. NGUYEN, M. L. and SCHWARTZ, S. J. Lycopene. Chap. 7 in *Natural Food Colorants*, Edit. G. J. Lauro and F. J. Francis. Marcel Dekker, New York, 2000, pp. 153–92.
15. LEVY, L. W. Annatto. Chap. 6 in *Natural Food Colorants*. Edit. G. J. Lauro and F. J. Francis. Marcel Dekker, New York, 2000, pp. 115–52.
16. ORANEZ, A. T. and RUBIO, R. O. Genotoxicity of pigments from seeds of *Bixa orellana* (atsuete) 1. Determined by the *Allium* test. *Phillipine Jour. Science*, 1996, 125, 259–69.
17. ORANEZ, A. T. and BAYOT, E. Genotoxicity of pigments from the seeds of *Bixa orellana* (atsuete) 2. Determined by lethal test. *Phillipine Jour. Science*, 1999, 126, 163–73.
18. LACEY, C. L. and GUZINSKI, J. A. Paprika. Chap 5 in *Natural Food Colorants*. Edit. G. J. Lauro and F. J. Francis. Marcel Dekker, New York, 2000, pp. 97–114.
19. FRANCIS, F. J. Food Colorants: Anthocyanins. *Critical Reviews in Food Science and Nutrition*, 1987, 28, 273–314.
20. FRANCIS, F. J. Anthocyanins and Betalains. Chap 7 in *Colorants*. Eagan Press, St. Paul, MN, 1999, pp. 55–66.
21. FRANCIS, F. J. *Handbook of Food Colorant Patents*. Food and Nutrition Press, Westport, CT, 1986.
22. ASKARI, A., MIRZA, M. and SOLANGI, M. S. P. Toxicological studies on herbal beverages and seed extracts of *Hibiscus sabdariffa, L.* Roselle. *Pakistan Jour. of Scientific and Industrial Research*, 1996, 39, 28–32.
23. VON ELBE, J. H. and GOLDMAN, L. L. The Betalains. Chap 2 in *Natural Food Colorants*, Edit. G. L. Lauro and F. J. Francis. Marcel Dekker, New York, 2000, pp. 11–30.
24. HENDRY, G. A. F. Chlorophylls. Chap 10 in *Natural Food Colorants*. Edit. G. J. Lauro and F. J. Francis. Marcel Dekker, New York, 2000, pp. 227–36.
25. FRANCIS, F. J. Chlorophylls, Haems, Phycobilins, and Anthraquinones. Chap 8 in *Colorants*. Eagan Press, St. Paul, MN, 1999, pp. 67–76.

26. FURUKAWA, F., KASAKAWA, K., NISHIKAWA, A., IMAZAWA, T. and HIROSE, M. A 13-week subchronic oral toxicity study of chlorophyll in F 344 rats. *Bull. Nat. Inst. Health Sciences*, 1998, 116, 107–12.
27. BUESCHER, R. and YANG, L. Turmeric. Chap 9 in *Natural Food Colorants*. Edit. G. J. Lauro and F. J. Francis. Marcel Dekker, New York, 2000, pp. 205–26.
28. FRANCIS, F. J. Turmeric, Carthamin and Monascus. Chap. 9 in *Colorants*. Eagan Press, St. Paul, MN, 1999, pp. 77–81.
29. HIRSCHLER, B. Curry pills may prevent cancer. *Amer. Soc. Hort. Science Newsletter,* 2001, 19, 7.
30. SCHUL, J. Carmine. Chap. 1 in *Natural Food Colorants*, Edit. G. J. Lauro and F. J. Francis. Marcel Dekker, New York, 2000, pp. 1–10.
31. MUDGETT, R. E. Monascus. Chap. 3 in *Natural Food Colorants*. Edit. G. J. Lauro and F. J. Francis. Marcel Dekker, New York, 2000, pp. 31–86.
32. MARTINKOVA, L., PATAKOVA-JUZLOVA, P., KREN, V., KUCEROVA, Z., HAVLICEK, V., OLSOVSKY, P., HOVORKA, O., RIHOVA, B., VISELA, D., ULRICHKOVA, J. and PRIKRYLOVA, V. Biological activities of oligoketide pigments of *Monascus purpurus*. *Food Additives and Contaminants*, 1999, 16, 15–24.
33. FRANCIS, F. J. Caramel, Brown Polyphenols, and Iridoids. Chap. 10 in *Colorants*, Eagen Press, St. Paul, MN, 1999, pp. 83–8.
34. YAMMONO, T., TSUGIMOTO, Y., NODA, T., SHIMIZU, M., OHMORI, M., MORTA, S. and YAMADA, S. Hepatotoxicity of geniposides in rats. *Food and Chem. Toxicol*. 1990, 28, 515–19.
35. YOSHIZUMI, S., OKUYOMA, H. and TOYAMA, R. Natural coloring agents for noodles. *Shoukuhin Kogyo to Kagaku*, 1980, 23, 102–105, 24.
36. IMAZAWA, T., NISHIKAWA, A., FURUKAWA, F., KASAHARA, H., IKEDA, T., TAKAHASHI, M. and HIROSI, M. Lack of carcinogenicity of gardenia blue colour given chronically in the diet of F 344 rats. *Food and Chemical Toxicol*. 2000, 38, 314–18.
37. KATO, Y. Gardenia Yellow. Chap. 4 in *Natural Food Colorants*. Edit. G. J. Lauro and F. J. Francis. Marcel Dekker, New York, 2000, pp. 87–96.
38. HOUGHTON, J. D. Phycobilins. Chap. 8 in *Natural Food Colorants*, Edit. G. J. Lauro and F. J. Francis. Marcel Dekker, New York, 2000, pp. 193–204.
39. AKHILENDER NADU, M., VISWANATHA, S., NARASINKA MURTHY, K., RAVISHANKAR, G. A. and SRINIVAR, L. Toxicity assessment of phycocyanin, a blue colorant from blue green algae, *Spirulina platensis*, *Food Biotechnol.*, 1999, 13, 51–6.
40. TARKA, S. M. The toxicology of cocoa and methylxanthenes: A review of the literature. *CRC Critical Reviews in Toxicology*, 1982, 9, 275–313.
41. FRANCIS, F. J. Miscellaneous Colorants. Chap. 11 in *Colorants*. Eagan Press, St. Paul, MN, 1999, pp. 89–95.
42. ANON. Natural colorants set for growth in Europe. *Confection.* 1998, Jan, 28.

43. GOYLE, A. and GUPTA, R. G. Use of synthetic colors in food. *Science and Culture*, 1998, 64, 241–5.
44. O'CARROLL, P. Naturally exciting colors. *World of Ingredients*, 1999, Mar./Apr. 39–42.
45. MILO ORR, L. Amazing technicolor candy coats. *Prepared Foods*, 1999, 168, 85–86, 90–91.
46. HOLST, S. Natural colors for food and beverages. *Food Marketing and Technology*, 2000, 14, 14–16.
47. WINTERHOLTER, P. and STRAUBINGER, M. Saffron: Renewed interest in an ancient spice. *Food Reviews International*, 2000, 16, 39–59.

9
Safety assessment of flavor ingredients

K. R. Schrankel and P. L. Bolen, International Flavors & Fragrances, Union Beach and R. Peterson, International Flavors & Fragrances, Dayton

9.1 Introduction: definition and use of flavoring substances

When using the term 'flavor', a certain inherent understanding of the term is evident. However, its use in the technical discussion of food requires a more imprecise definition. A common technical definition of the word 'flavor' is the sum total of the sensory responses of taste and aroma combined with the general tactile and temperature responses to substances placed in the mouth. Flavor can also mean any individual substance or combination of substances used for the principal purpose of eliciting the latter responses. This latter usage will be the way in which the term is used in this chapter.

Flavoring substances can vary considerably in complexity from a chemically simple, single component substance such as ethyl butyrate to a chemically complex, multiple component substance such as ginger oleoresin. The latter, for example, is composed of hundreds of different molecular species both volatile and non-volatile. The selection of the particular flavoring substances to be used in the production of a certain flavor profile are, of course, chosen primarily for their sensory properties. However, the chemical nature of these substances must also be considered, and in some cases are the deciding factor of whether or not they can be employed at all. In addition to the sensory and chemical characteristics of flavoring substances, their safety/ legislative status in various countries must always be considered. This latter consideration may involve defined maximum usage levels of flavoring substances in foods and beverages.

9.2 The range and sources of flavoring ingredients

Consideration will be given first to some of the more chemically complex flavoring materials, which are of natural origin. The definitions which follow are, of necessity, general in nature and not necessarily all inclusive.

Absolute
A water soluble flavoring material made by the dewaxing of a concrete with hot alcohol, e.g. Beeswax absolute.

Concrete
A solid to semi-solid substance containing volatile essential oils and a fatty-waxy material obtained by the extraction of plant tissue with lipophilic solvents like hexane, e.g. Orris concrete.

Essential oil
The volatile oil containing the 'essential' flavor/aroma of the named plant, herb, root, bark, flower, etc. The volatile oil is obtained by distillation or expression, e.g. expressed Orange oil.

Folded oil
A concentrated essential oil. Folding is a gravimetric/volumetric measure of the strength of a concentrated essential oil expressed as a multiple of a standard. A folded Citrus oil would be compared to the expressed oil (steam distilled oil in the case of distilled Lime oil), e.g. 5 × Orange oil.

Extract
A solution of flavoring materials extracted from flowers, roots, bark, fruit, etc. It may be alcoholic or non-alcoholic; solid extract or fluid extract. These materials are frequently concentrated and standardized, e.g. Gentian extract.

Balsam
A natural exudate from a tree or plant.

Isolate/fraction
A chemical substance or simple mixture of substances obtained from natural sources by distillation or extraction such as citral from lemongrass oil or eugenol from clove bud.

Oleoresin
A plant extractive consisting of the essential oil and the flavorful, non-volatile principals of the plant. The latter are frequently pungent, biting or heat bearing materials, e.g. ginger oleoresin.

Tincture
A relatively dilute alcoholic extract of natural raw materials in which the solvent is left (in part) as a diluent, e.g. Civet tincture.

Resinoid
Extractives of resinous materials by a hydrocarbon type solvent. The extractives are both volatile and non-volatile materials, and the resinous substance used is non-cellular in nature, e.g. Olibanum resinoid.

9.3 Basic principles of safety evaluation

The safety evaluation of flavoring substances provides a unique challenge to regulators due to their large number and their generally low level of use. In addition, many flavoring substances occur in natural products (i.e., extracts, oleoresins, and essential oils) and have a long history of safe use.

Assessing the safety of flavoring substances, in principle, is similar to that of other food additives and includes the consideration of four basic elements:

- inherent toxicity of the substance
- exposure to the substance
- chemical structure and structure-activity relationships
- natural occurrence.

Over the last 45 years, the Joint FAO/WHO Expert Committee on Food Additives (JECFA) has been in the forefront of developing a procedure to evaluate the safety of flavoring substances and has recognized the unique issues surrounding their evaluation. Several other organizations including the Commission of the European Communities' Scientific Committee on Food (SCF, 1991), the Flavor and Extract Manufacturers' Association (FEMA) Expert Panel (Woods and Doull, 1991) and the Committee of Experts on Flavoring Substances of the Council of Europe (CE, 1974, 1981, 1992, 2000a, 2000b) have recognized the importance of structure-activity relationships, metabolism, and exposure data in the safety evaluation of flavoring substances.

9.3.1 Evaluation of toxicological data

Ideally, the evaluation of flavoring substances relies on a complete range of safety data including short- and long-term toxicity studies, and special studies which examine possible reproductive, carcinogenic, mutagenic and sensitization characteristics of the flavoring substance. Traditionally, from controlled animal safety studies, the doses, duration, and route of exposure are used to provide information relevant to the determination of a no-observed-effect level (NOEL). The basis for a NOEL is important because a safety factor is applied to extrapolate to a safe exposure level for sensitive human populations. For example, a NOEL derived from a long-term animal study typically has a 100-

fold safety factor applied to it, which represents a 10-fold factor to account for intraspecies sensitivity and an additional 10-fold factor to account for interspecies sensitivity. A long-term study in humans sometimes uses a safety factor of only 10-fold, representing protection of sensitive populations. Larger or smaller safety factors might be used, depending on the strength of the underlying data on which the NOEL was based. In some cases, the acceptable daily intake (ADI) for a flavoring substance is derived in this manner, with the safety factor-adjusted NOEL representing a numerical ADI.

Historically flavoring substances have been tested for safety by testing representative members of a chemically similar group. This means that many flavoring substances have not been subjected to detailed and comprehensive toxicity testing programs. This is also due in part to the large number of flavoring substances that are used in food (>2,500). Many flavoring substances are self-limiting and are typically used at very low concentrations to impart their desired effect, and exposure from their use in food is, generally, very low. In fact, the intake of 95% of flavoring substances used in food in the US is 1 mg/day or less (Munro *et al.*, 1998). Moreover, many flavoring substances (~50%) are simple acids, aldehydes, alcohols, and esters which, for the most part, are rapidly metabolized to innocuous end products.

Regardless, it is important to carry out the safety evaluation for all flavoring substances. In many cases, it is possible to use pharmacokinetic or metabolic data on the substance in addition to toxicity data or other information on structurally-related compounds.

9.3.2 Estimation of exposure

Unlike many substances which are added to foods to achieve a desired effect, flavoring substances are added in amounts that are self-limiting and governed by the potency of the substance used to provide the necessary organoleptic characteristics of the food product. As a result, flavoring substances generally are used in low concentrations and add only small amounts to human intake.

Many methods for calculating intake have been suggested; however, estimating the intake of flavoring substances historically has been performed in two ways. One method involves multiplying the level of use of the flavoring substance in a particular food category and multiplying this value by the amount of that particular food category eaten per person and dividing by the total population to estimate the *per capita* intake. This method of calculating intake, however, typically leads to gross overestimates of consumption because intake data for foods are stratified into broad categories of products, such as baked goods or non-alcoholic beverages. As a result, some flavoring substances which are used only in one of a few products within a category would be assumed to be present in all products within that category. Also, within a food category, it may be that the flavoring substance itself characterizes specific brands of products. Thus, the assumption that all brands or products within an individual food category contain a certain flavoring substance leads to further overestimation of intake. Moreover, the amount of the flavoring substance added to a food is not typically the amount remaining in the food at the time of consumption due to processing losses, evaporation, and waste. Overall, the assumptions inherent to this method of estimating intake may lead to intakes that are exaggerated by several orders of magnitude.

A second method of estimating the intake of flavoring substances is to assume that the total amount of a particular flavoring substance produced has been added to broad categories of foods (i.e., poundage data) and is completely consumed by the total population. This method uses poundage data obtained from surveys of substances reported to be intentionally added to foods by ingredient manufacturers and food processors in the US and in Europe. In the US, these surveys have been conducted by the US National Academy of Sciences/National Research Council (NAS/NRC) between 1970 and 1987. The International Organization of the Flavor Industry (IOFI, 1995) has provided similar data for Europe. Independent reviews of these kinds of surveys have shown that the reported poundage values account for only about 80% of total use. As a result, the intake calculations are corrected upwards to account for any underreporting. As an additional conservative factor, it is assumed that only 10% of the population consumes a particular flavoring substance. The formula for this calculation is as follows:

$$\text{Intake}(\mu g/\text{day}) = \frac{[(\text{annual volume in kg}) \times (1 \times 10^9 \mu g/kg)]}{[(\text{population}/10) \times 0.8 \times 365 \text{ days}]}$$

Both of the methods for estimating the intake of flavoring substances described above tend to exaggerate human intake due to the assumptions incorporated into the calculation. Another reason for the inflated intakes is due to the fact that both estimation methods involve disappearance data (i.e., the amount of the flavoring substance presumed to be used in food) and do not consider losses and waste which occur during food manufacture, storage, preparation, and consumption. For the purposes of safety evaluation, it is preferable to overestimate flavoring substance intake in order to remain conservative. That is, by assuming that the population consumes more than they actually do, there is an added level of protection built into the comparison of intake to the level determined to show no adverse effect in the toxicological assessment of the substance.

A more accurate method for determining intake, but one that is very time, labor, and cost intensive, is to use individual consumer dietary surveys. This method, referred to as the detailed dietary analysis, involves recording in detail the eating habits of a group of individual consumers. Each consumer is responsible for reporting the food items consumed during the designated recording period. To ensure accuracy and applicability with this approach, however, it is necessary to obtain data from a relatively large segment of the population, with sufficient representation from various age, ethnic, and social groups. Also, the data on the foods consumed by the surveyed population must still be correlated with amounts of the flavoring substance that the food contains. When other factors, such as compliance in completing the survey forms, accuracy of reporting, or seasonal variations in eating patterns, are considered in this approach, it is fairly easy to understand why it is not considered practical for generating intake data for a large number of flavoring substances. Consequently, it is most often the case that the broader, but less accurate methods described previously, are used for estimating intakes.

To validate the use of the less accurate, but more practical, methodology, Hall *et al.* (1999) conducted a comparison between the use of poundage data on a *per capita* basis and the use of a detailed dietary analysis in calculating the intake of flavoring substances. Ten flavoring substances were selected as representative examples, and intakes for these substances were calculated using both methodologies. The results clearly showed that the detailed dietary analysis provided good estimates of the distribution of intakes across the population, as well as patterns of intake of individuals, but it was both expensive and labor intensive. The poundage method resulted in intake values that were comparable to, if not greater than, the detailed dietary analysis and was considered to be a more practical approach to intake estimations (Smith *et al.*, 2001).

9.3.3 Evaluation of structurally related substances

Chemical structure determines the inherent toxicity of a substance, its metabolic profile and its pharmacokinetics. Knowledge of structure-activity relationships has expanded over the years. In 1978 Cramer *et al.* incorporated chemical structure, pharmacokinetics and knowledge of metabolic fate to produce a decision tree for classification of flavoring substances into toxicity concern levels. The US Food and Drug Administration (FDA) has also used structure-activity relationships to define concern levels for food additives in its Redbook (FDA, 1982). JECFA has repeatedly demonstrated the use of structure-activity relationships and known metabolic pathways in the evaluation of flavoring substances and has recognized that generation of extensive toxicity data is not necessary when toxicity data are available for one or more members of a homologous series of chemical substances (WHO, 1987).

Knowledge of chemical structure, pharmacokinetics, and metabolic pathways provides a method to assess the safety of flavoring substances that lack a full safety testing profile using data from structurally related substances which have been adequately tested for toxicity.

9.3.4 Natural occurrence

Another consideration in the estimation of the intake of flavoring substances intentionally added to foods is the natural occurrence of the substance in food. For many flavoring substances, natural occurrence is the principal source of human exposure as compared to that which is intentionally added to foods. The natural occurrence of a particular ingredient, however, does not necessarily prove the safety of the substance. In the safety assessment, therefore, the ratio of the natural occurrence of a flavoring substance to the intentional addition to food, referred to as the consumption ratio (CR) (Stofberg and Kirschman, 1985), provides an important perspective on the intentional addition of a flavoring substance to food. A CR of greater than one indicates that the substance is nature predominant, and a CR of greater than 10, which according to Stofberg and Grundschober (1987) applies to most flavoring substances, indicates that the added substance contributes an insignificant amount to the diet (i.e., <10%). Therefore, flavoring substances added to food and contributing <10% of natural dietary sources would be of minimal safety concern.

Given that there is a very large number of flavoring substances used in foods and that they are, in general, added in low and self-limiting quantities, it is impractical to subject each individual flavoring substance to a routine but extensive toxicological evaluation within any set time frame. Consequently, it is desirable to have a rational basis for determining the need for the testing of flavoring substances.

An excellent example of the importance of natural occurrence is in the evaluation of process flavors. Process flavors are flavors generated from interactions between protein nitrogen, carbohydrate, fat, or fatty acid sources during thermal processing. They mimic flavors naturally present in cooked foods, particularly cooked meats. As such, process flavors are a potential dietary source of heterocyclic amines (HCAs), which are reported to be potent mutagens and animal carcinogens (Munro *et al.*, 1993). However, due to strong organoleptic properties, process flavors have a low level of use. Moreover, an analytical study has shown that HCAs are essentially not detectable (detection limit = 50 ppb) in process flavors currently on the market (The US Flavor and Extract Manufacturers Association (FEMA), Washington DC, unpublished). Although toxicological data are available on HCAs, the studies are inadequate to characterize dose-response relationships and do not represent the low levels found in foods. Since HCAs do occur naturally, particularly in cooked meats, a comparison of exposure to HCAs from use of process flavors in food can be made with exposure of HCAs found naturally in the diet. This comparison has shown that the intake to HCAs from process flavors is far lower than that of HCAs naturally occurring in food and it was concluded that the use of process flavors would not pose a significant health risk to humans. Richling *et al.* (1998) has demonstrated the ubiquitous occurrence of HCAs in numerous foods.

9.3.5 Allergens and intolerances
Owing to the extreme sensitivity of certain susceptible individuals, an increasingly important concern among consumers is adverse reactions that occur as a consequence of food allergies or intolerances. Food allergies involve an abnormal immune response while food intolerances do not involve the immune system.

Many consumers believe they have food allergies, but true allergic reactions involving an immune response are known to occur in only 1–2% of the adult population (Sloan, 1986; Taylor, 1985). Food allergies are known to occur more frequently in young infants, i.e. 4–8%, attesting to the fact that many allergies are outgrown (Bock and Atkins, 1990). Probably > 90% of food allergies result from exposure to only a small group of foods (FAO, 1995). These include: cereals containing gluten (e.g. wheat, rye, barley, oats, etc.), crustacea (e.g. shrimp, crab, lobster, etc.), eggs, fish, milk, peanuts, soybeans and tree nuts (almonds, pecans, etc.). In virtually all cases, proteins have been found to be the key elicitors of allergic reactions in these and other foods associated with allergic responses. In contrast, only a small fraction of the proteins found in nature are allergenic. Most ingested proteins are extensively degraded in the intestinal tract and are not taken into the bloodstream in sufficient quantity, either in whole or part, to induce an allergic response.

Safety assessment of flavoring materials as potential elicitors of allergic reactions involves a consideration of the chemistry of the material, the source, the method of manufacture and an evaluation of use level and subsequent exposure (Taylor and Dormedy, 1998). Although flavoring substances may contain proteins, and are, therefore, of concern to susceptible individuals, most contain only small, volatile, organic compounds. Flavors are produced from many synthetic and natural sources, yet they seldom contain allergenic components. The source materials used for production of the flavor is an important consideration, especially when some flavors are produced from known allergen-containing food sources, e.g. peanut and soy protein. The method of manufacture can also have a major impact on the potential of a flavoring substance to elicit a sensitive response. Many flavoring substances are produced using distillation or extraction as a method of purification or production which eliminate or significantly reduce the presence of proteins in the final product. Finally, as mentioned earlier, flavors are typically very intense and are seldom major components of the foods we eat. Many flavoring substances are 'self-limiting,' and others range in use from parts-per-million to ~1% of the final product. An enormous variety of flavoring substances is used in foods such that exposure to any one ingredient would likely be quite limited.

Food intolerances are mostly caused by substances other than proteins and affect a limited number of individuals. Examples include sulfite-induced asthma, lactose intolerance, MSG-induced headache and flushing, etc., yet these substances are only 'associated with' these symptoms and remain unproven as causative agents. Significant levels of these non-flavor ingredients are typically required to elicit this type of reaction, and there is no evidence that flavoring substances are involved (Taylor and Dormedy, 1998).

Assessment of the allergenicity of new foods produced by genetic modification has been specifically addressed by the International Food Biotechnology Council (IFBC) in collaboration with the International Life Sciences Institute (ILSI) Allergy and Immunology Institute (AII) (1996). A rational approach to assess allergenicity has been proposed that utilizes a decision tree that takes into account the source of the gene, comparisons of amino acid sequence with those of known allergens, *in vitro* and *in vivo* immunologic testing and an assessment of the characteristics of the gene product. As a total approach, these assessments are believed to provide reasonable assurances that potentially dangerous antigens will not be introduced into the food supply; or if they are, they must be appropriately labeled.

9.3.6 Biotechnology and flavor production

Advances in modern molecular techniques now allow the development of new genetically modified (GM) plant and animal sources from which flavoring substances can be derived. Additionally, they allow for the creation of microorganisms with specifically engineered pathways that can be used efficiently to produce specific flavor enzymes and chemicals. Some view this enabling technology as the beginning of a new 'green' revolution with potential to provide tremendous benefits in health, agriculture and food production. Yet others call for a halt to its use, charging that these newly developed organisms may cause illness and harm the environment. Governing bodies throughout the world struggle to develop consistent policies that both protect the consuming public yet provide avenues for progress.

A basic premise in the safety assessment of new food products is the idea of 'substantial equivalence', whereby the relative safety of the new products can be judged by comparing it to its traditional counterpart which has had a history of safe use. This approach provides a scientific basis for a quantitative determination of similarities and differences between the new product and the frequently wide range of that component found in comparable traditional products. Several international organizations and national regulatory agencies have endorsed this approach. Two United Nations oganizations, the Food and Agriculture Organization (FAO) and the World Health Organization (WHO), have concluded that a comparative approach as embodied in the concept of 'substantial equivalence' is the most appropriate strategy for assessing the safety and nutritional aspects of GM foods (FAO/WHO, 2000). The US Food and Drug Administration (FDA) employs this concept in conducting the safety evaluation of foods derived from new plant varieties and has identified characteristics that would dictate need for more extensive evaluation. These characteristics include: the appearance of substances new to foods, the occurrence of allergens as unusual components, changes in major dietary nutrients and increases in toxic or anti-nutritional factors. Key among the special considerations involved for GM foods is prevention of the appearance of new allergens because the concentration required to elicit a serious response in hypersensitive people can be quite low. Newly engineered proteins may elicit entirely new allergic responses not seen before.

In a similar approach to that taken by the FDA, the International Food Biotechnology Council (IFBC) addressed issues associated with food and flavor ingredient production from GM materials and concluded that a reasonable safety assessment could be made by determining whether the ingredient had a prior history of safe use in approved foods and, if so, whether the GM substance was

216   Food chemical safety

the same as that conventionally produced (IFBC, 1990). Finally, the FEMA Expert Panel reconsidered the criteria employed in making a GRAS assessment for a flavoring substance in light of GM technology (Hallagan and Hall, 1995b). The FEMA panel concluded that once special issues relating to purity, method of production and transmissibility are addressed, safety assessment of a flavoring substance produced by modern biotechnology should be no different than that employed for safety assessment of conventionally produced ingredients.

From a safety perspective there is no evidence today that GM foods threaten human health, and most enacted legislation is intended to provide 'right-to-know' information to the consumer through labeling. Labeling, however, is a double-edged sword in that it can also be misleading by implying that there is something unsafe about the product. There is concern that heightened public awareness over this new technological advance is easily translated into avoidance of an entirely safe, but adversely labeled product. The labeled product also comes at a cost, a cost to ensure that the material is authenticated as free from contamination. In the case of GM foods this is a daunting requirement because special handling and audit procedures must take place along the entire path from seed to plate. The result is that the labeled product and the associated technology used in its development are unfairly stigmatized.

9.4   Regulatory groups

The safety assessment of flavoring substances has been undertaken by several national and international authorities including JECFA, the FEMA Expert Panel, the SCF and the CE.

9.4.1   JECFA

For over 45 years, JECFA has evaluated food additives and flavoring substances using the principles and criteria outlined in this chapter and discussed in detail in *Environmental Health Criteria 70: Principles for the Safety Assessment of Food Additives and Contaminants in Food* (WHO, 1987). When JECFA began evaluating flavoring substances, it tried to establish an ADI for the substance or sometimes established a group ADI for groups of related flavoring substances. Depending on the type of data available, either an ADI 'not specified'[*] or a numerical ADI is typically established. In some cases, JECFA may request additional data to provide further evidence of safety and, in the interim, a temporary ADI may be established or no ADI is allocated. Using this system, by 1991 JECFA had evaluated approximately 70 flavoring substances, providing numerical ADIs for over 30 substances.

---

* ADI not specified is applied to food substances of very low toxicity that, in the opinion of the Committee, do not represent a hazard to health. Consideration is also given to total dietary intake and use levels necessary to achieve the desired effect.

In 1996, JECFA adopted a procedure for the safety evaluation of flavoring substances (JECFA, 1998). This procedure is described in detail in Annex 5 of the toxicological monograph of the 44th meeting of JECFA (JECFA, 1998) and by Munro et al. (1999). The safety evaluation procedure was established to provide a practical and scientifically based method to evaluate flavoring substances. The basis of the JECFA evaluation procedure is the establishment of human exposure thresholds for three structural classes: Class I represents substances with simple chemical structures having minimal anticipated toxicity; Class II represents substances with structures that are less innocuous than Class I, but do not contain features suggestive of toxicity as in Class III; and Class III represents substances having chemical structures with structural features that may indicate toxic potential. Human exposure thresholds were calculated using a probabilistic approach from a database of over 600 NOELs by Munro et al. (1996). Briefly, a database of over 600 substances with corresponding NOELs was compiled and the substances were classified into one of three structural classes using the Cramer et al. (1978) decision tree. For each structural class, the cumulative distribution of the NOELs was plotted along with the lognormal distributions fitted to these data. The human exposure threshold for each structural class was calculated from the 5th percentile NOEL using a 100-fold safety factor. The 5th percentile NOEL provides 95% confidence that any other substance of unknown toxicity from the same structural class would not have a NOEL less than the 5th percentile for that structural class. These human exposure thresholds were incorporated into the safety evaluation procedure.

In addition, for flavoring substances with limited safety data but extremely low exposures, JECFA adopted a more conservative threshold (JECFA, 1999). This threshold of 1.5 μg/day is the same as the threshold of regulation of 1.5 μg/day based on carcinogenicity endpoints estimated by Rulis (1989) and adopted by the FDA as a threshold of regulation for food contact substances (Federal Register, 1993, 1995). It was calculated based on a 10^{-6} risk using carcinogenicity potency data generated by Gold et al. (1989). As noted by the FDA, an exposure level of 1.5 μg/day would result in a negligible risk even if the substance of unknown toxicity was later shown to be a carcinogen. More information on the threshold of regulation is presented in a recent review article by Barlow et al. (2001).

The JECFA safety evaluation procedure for flavoring substances leads the evaluator through a decision tree with two branches, one for substances that are metabolized to innocuous end products and the second for substances that are not metabolized to innocuous end products (Fig. 9.1). The decision tree is applied using a series of questions:

1. What is the structural class of the flavoring substance?
2. Is the substance predicted to be metabolized to innocuous products?
3. Do the intended conditions of use result in an exposure greater than the human exposure threshold for the structural class?
4. Is the substance or its metabolites endogenous?

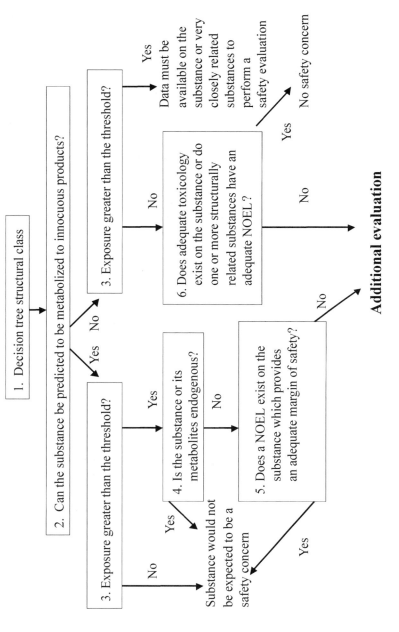

Fig. 9.1 Safety evaluation sequence (adapted from WHO, 1996).

5. Does a NOEL exist on the substance or a structurally related compound?
6. Do the intended conditions of use result in an exposure greater than 1.5 μg/day?

Since the introduction of the JECFA safety evaluation procedure, over 900 flavoring substances have been evaluated by JECFA using this procedure. This procedure provides a practical and efficient scientific means to evaluate flavoring substances by integrating knowledge of toxicity, structure-activity relationships, structural class, metabolic fate and exposure.

9.4.2 FEMA Expert Panel

In 1959, in response to the 1958 Food Additives Amendment to the 1938 Food, Drug, and Cosmetic Act, FEMA initiated a generally recognized as safe (GRAS) assessment program using a panel of independent experts referred to as FEXPAN. These experts meet three to four times a year to carry out GRAS evaluations, and their conclusions are published periodically in the journal *Food Technology*.

As described by Hallagan and Hall (1995a), the GRAS process consists of four key elements:

1. *General recognition of safety*. The US FDA has defined GRAS to mean that there is a reasonable certainty in the views of competent scientists that the substance is not harmful under the intended conditions of use.
2. *Among experts qualified by scientific training and experience to evaluate safety*. Over the years, FEXPAN members have included experts from the field of veterinary medicine, pathology, food chemistry, medicine, chemistry, biochemistry, pharmacology, and toxicology.
3. *Through scientific procedures or through experience based on common use in food if used in food prior to 1958*. FEXPAN uses the same criteria described above to conduct GRAS assessments; namely, exposure, natural occurrence, chemical identity, structure-activity relationships, metabolism, pharmacokinetics, and toxicity data.
4. *Conditions of intended use*. The GRAS status is determined, in part, based on the intended conditions of use of the flavoring substance through the calculation of a possible average daily intake (PADI) and a *per capita* exposure estimate.

The FEMA GRAS assessment program is an integrative approach that incorporates many of the same principles used by JECFA in the safety evaluation of flavoring substances. The FDA recognizes the work of the FEMA Expert Panel lists and incorporates information provided by FEMA on the safety of flavoring substances into its flavor database. In a Federal Register announcement on February 3, 1976, the FDA recognized the FEMA GRAS lists when giving their opinion on bulk labeling of flavors and in other direct ways (Hallagan and Hall, 1995a).

GRAS and the FDA

In 1958, the US Congress enacted the Food Additives Amendment to the 1938 Federal Food, Drug, and Cosmetic Act. The basic intent of the amendment was the requirement for a producer to demonstrate the safety of a new food additive to the FDA prior to its use in food. At the time of the amendment, it was recognized that not all substances intentionally added to food would require a formal pre-market safety review by the FDA. As a result, a two-step definition of food additive was adopted by Congress. The first part includes any substance, the intended use of which results or may reasonably be expected to result, directly or indirectly, in it becoming a component or otherwise affecting the characteristics of food (Federal Register, 1997), whereas, the second part excludes substances that are generally recognized, among experts qualified by scientific training and experience to evaluate their safety (qualified experts), as having been adequately shown through scientific procedures (or, in the case of a substance used in food prior to January 1, 1958, through experience based on common use in food) to be safe under conditions of their intended use (Federal Register, 1997).

Subsequent to the 1958 amendment, the FDA included a list of food substances in its regulations that when used for the purposes indicated and in accordance with current good manufacturing practice, are GRAS (Federal Register, 1997). Further categories, including spices, seasonings and flavorings, were added over the years. As part of this process, producers of new food substances were permitted to submit GRAS petitions for the FDA to file and review, or conduct their own self-determinations of GRAS using scientific principles.

In April 1997, the FDA published a proposal for a notification system intended to replace the petition process for GRAS substances (Federal Register, 1997). Under this system, a detailed summary of the information on the notified substance should be submitted to the FDA rather than all 'supportive' information, itself with the understanding that the information is available upon request. The written notification should include a description of the notified substance, the intended conditions of use, and a description of the basis for the GRAS determination. The FDA is required to acknowledge receipt of the notification within 30 days and provide a response letter within 90 days. A response from the FDA that does not find any issues with the notification is not equivalent to affirmation of GRAS status by the FDA but permits commercial use of the substance. GRAS petitions caught in the transitional period may be converted to GRAS notifications and will be assessed on a case-by-case basis.

The safety requirements for GRAS food substances depend largely on their historical profile. The affirmation of GRAS status is often based on the common use of the substance in foods prior to 1958 coupled with a lack of known or documented adverse effects. Section 170.30 of the Code of Federal Regulations specifies that 'general recognition of safety through experience based on common use in food ... shall be based solely on food use of the substance prior to January 1, 1958, and shall ordinarily be based upon generally available data

Safety assessment of flavor ingredients 221

and information' (21 CFR, Section 170.30(c)). Food substances that have no documented history of use *per se* prior to 1958 may qualify for GRAS status by scientific procedures. The quantity and quality of the scientific evidence required to establish GRAS status for a substance that lacks a documented history of safety in the proposed uses is the same as for a food additive petition. GRAS substances that are to be used in new ways or at different concentrations must also undergo review to reevaluate their status. Affirmation of GRAS substances is generally made by an expert advisory panel convened to review the available scientific data. If there is a lack of convincing evidence that the substance is GRAS, the substance may be considered a food additive and may be subject to a food additive petition.

9.4.3 Europe – Scientific Committee on Food

The Scientific Committee on Food (SCF), established in 1974, provides advice to the European Commission (EC) on matters relating to the protection of public health and safety, with respect to the consumption of food, and, in particular, nutritional, hygienic and toxicological issues. In 1997, the EC reorganized its Scientific Committees and established a Scientific Steering Committee composed of the chairs of the eight new Scientific Committees. This reorganization allowed for the updating and broadening of mandates of the former Committees, including the SCF. As such, the established mandate of the SCF was declared to:

> provide advice on scientific and technical questions concerning consumer health and food safety associated with the consumption of food products, and in particular, questions relating to toxicology and hygiene in the entire food production chain, nutrition, and applications of agrifood technologies, as well as those relating to materials coming in contact with foodstuffs, such as packaging. (Hallagan and Hall, 1995b)

The SCF consists of independent experts, who are considered highly qualified in scientific disciplines associated with medicine, nutrition, toxicology, biology, chemistry, or other similar disciplines.

Although the EC had published in 1988 a framework directive on flavoring substances known as 88/388/EEC, it was not until 1996 that the European Parliament and the Council of the European Union adopted Regulation EC No. 2232/96 laying down a Community procedure for flavoring substances used or intended for use in or on foodstuffs (European Parliament and Council, 1996). Article 3(1) of this Regulation required that within one year of entry into force, the member states must notify to the EC a list of flavoring substances which may be used in or on foodstuffs marketed in their territory. As a result, the Commission compiled a register of those flavoring substances notified by the member states. The register includes flavoring substances both accepted for use and those subject to restrictions or prohibitions in some member states. In

February 1999, the Register was adopted in accordance with the agreement reached by the member states (EC, 1999) and contains more than 2800 entries.

One amendment has been made to the Register (EC, 2000b) adding a number of flavoring substances that had been omitted in the first publication of the Register and to amend a number of entries in the light of new information.

Although in 1991 the SCF had published its 'Guidelines for the Evaluation of Flavourings for Use in Foodstuffs: 1. Chemically Defined Flavouring Substances,' on December 2, 1999 the SCF expressed its opinion pursuant to Article 4 of Regulation 2232/96 that substances previously evaluated by JECFA, the Council of Europe Category A substances, and the SCF Category 1 substances need not be re-evaluated and were acceptable for use. It also stated that future evaluations should not be duplicative of the JECFA evaluation program. Following the advice of the SCF, the Commission adopted Commission Regulation (EC) No. 1565/2000 laying down the measures necessary for the adoption of an evaluation program (EC, 2000a). This Regulation lays down the data requirements and format for submitting data on substances to be evaluated.

At the present time, Article 5 of Regulation 2232/96 is being carried out. This article requires that the evaluation program adopted pursuant to Article 4 be carried out within five years of adoption of the evaluation program.

9.4.4 The Council of Europe

Founded in 1949, the Council of Europe (CE) is an intergovernmental organization which at the present time is composed of 41 member states representing more than 800 million people (CE, 2000a). The CE was established to promote greater unity and cooperation between the nations of Europe, for the purpose of safeguarding and realizing the ideals and principles which are their common heritage, and facilitating their economic and social progress (CE, 1992). In 1959 the CE entered into a partial agreement in the social and public health field. This partial agreement includes only 18 of the 41 member states. Flavoring substances have been dealt with under this partial agreement. In particular, the CE established the Committee of Experts on Flavoring Substances, referred to as the Committee of Experts, whose function is to provide an opinion on the safety-in-use of flavoring substances in foods. As such, the Committee of Experts published three editions of the so-called 'Blue Book' (CE, 1970, 1974 and 1981). These first three editions are comprised of lists of acceptable flavoring substances used in foods from both natural and artificial sources.

The goals of the CE in publishing the Blue Book are to provide a list of flavoring substances according to their acceptability for use in food, as well as to supplement and revise these lists when necessary, and to recommend to the manufacturers of flavoring substances, the criteria which must be met in order to assist the Committee of Experts to form an opinion on the safety of the flavoring substance in food.

In the first three editions of the Blue Book, flavoring substances for use in foods evaluated by the Committee of Experts were divided into two lists: (i) substances considered from natural sources; and (ii) substances considered from artificial sources. The Committee of Experts established certain subcategories for natural flavoring substances (i.e., admissible flavoring substances, temporarily admissible flavoring substances, flavoring substances not fully evaluated, or flavoring substances for which no or insufficient technological or toxicological information was available to classify them as acceptable natural sources of flavorings). In addition, the Committee of Experts established a subcategory of artificial flavoring substances acceptable for use in foods, in which levels of permitted use in the final food product are stated.

In 1992, the Council of Europe published the fourth edition of the Blue Book (CE, 1992). In the fourth edition, the flavoring substances are not divided into lists of natural source materials and chemically-defined flavoring substances. Instead this edition contains only chemically-defined flavoring substances arranged according to their chemical structure. The system of listing was changed in order to simplify future updating of the list and to facilitate the re-evaluation of flavoring substances (CE, 1992). In addition, it was recognized that grouping chemically-related substances, in particular compounds belonging to homologous series, has the advantage that toxicological data available for the whole group can be considered in the evaluation of individual substances (CE, 1992). Within each chemical group, the substances are listed by their principal name (i.e., the same as that used by FEMA) and by ascending CE number.

In the 1992 edition, flavoring substances are classified into Category A or B based on the information currently available (i.e., Category A – flavoring substances which may be used in foodstuffs, or Category B – flavoring substances for which further information is required before an opinion on their safety-in-use can be determined) (CE, 1992). In addition, where available, an ADI is specified by the Committee of Experts. The ADI is typically the one recommended by JECFA, and, in cases where the substance has not been evaluated by JECFA and where toxicological data are insufficient to establish an ADI, acceptable upper use levels in food (i.e., those that would be considered to result in no risk to health) are specified.

In 2000, the Council of Europe published two Blue Books, one on natural sources of flavorings and another on chemically-defined flavoring substances (CE, 2000a, 2000b) promising that there will be six reports on the natural sources of flavoring substances. These latest two Council of Europe publications contain updated evaluations on 899 chemically-defined flavoring substances, listed according to chemical class and designated as either Category A or B in accordance with the criteria used in the fourth edition to classify substances; and updated evaluations on 101 natural sources of flavorings. In the later book, 'active principles' and 'other chemical components' currently under evaluation by the Committee of Experts are listed.

The assessment of the safety-in-use of a flavoring substance listed in the CE's Blue Books involves the compilation and review of toxicological data from several sources, including published and unpublished data from industry and public institutions (CE, 2000a, 2000b). In forming its opinion, the Committee of Experts relies on essentially the same data and employs much of the same evaluation criteria as JECFA and the FEMA Expert Panel:

- toxicity data in animals or humans, such as data on acute or short-term toxicity including mutagenicity, and long-term toxicity, including carcinogenicity, and reproductive toxicity/teratogenicity where available
- metabolic fate and biotransformation to compounds of known toxicological properties
- levels of use in various dietary components
- natural occurrence in normal food
- comparison of chemical structure with that of compounds of known toxicological and biochemical properties (CE, 1992).

9.5 References

BARLOW, S. M., KOZIANOWSKI, G., WURTZEN, G. and SCHLATTER, J. 2001. Threshold of toxicology concern for chemical substances present in the diet. *Food Chem. Toxicol.* 39: 893–905.

BOCK, S. and ATKINS, F. 1990. Patterns of food hypersensitivity during sixteen years of double-blind, placebo controlled food challenges. *J. Pediatr.* 117: 561–7.

CE. 1970. *Natural Flavouring Substances, their Sources, and Added Artificial Flavouring Substances.* First edition. Council of Europe, Maisonneuve, France.

CE. 1974. *Natural Flavouring Substances, their Sources, and Added Artificial Flavouring Substances.* Second edition. Council of Europe, Maisonneuve, France.

CE. 1981. *Flavoring Substances and Natural Sources of Flavourings.* Third Edition. Council of Europe, Maisonneuve, France.

CE. 1992. *Flavouring Substances and Natural Sources of Flavourings. Volume I. Chemically-Defined Flavouring Substances*, Fourth Edition. Council of Europe, Maisonneuve, France.

CE. 2000a. *Natural Sources of Flavourings Report No. 1*, Council of Europe, Strasbourg, Cedex, France.

CE. 2000b. *Chemically-defined Flavouring Substances.* Council of Europe, Strasbourg, Cedex, France.

CRAMER, G.M., FORD, R.A. and HALL, R.L. 1978. Estimation of toxic hazard – A decision tree approach. *Food Cosmet Toxicol* 16: 255–76.

EUROPEAN COMMISSION. 1999. Commission Decision of 17 February 1999 adopting a register of flavouring substances in or on foodstuffs. OJ L84, 27.3.1999, p. 1.

EUROPEAN COMMISSION. 2000a. Commission Regulation (EC) No. 1565/2000 of

18 July 2000 laying down the measures necessary for the adoption of an evaluation programme in application of Regulation (EC) No. 2232/96 of the European Parliament and of the Council. OJ L180, 19.7.2000, p. 8.

EUROPEAN COMMISSION. 2000b. Commission Decision of 18 July 2000 amending Decision 1999/217/EC adopting a register of flavouring substances used in or on foodstuffs. OJ L197, 3.8.2000, p. 53.

EUROPEAN COUNCIL. 1988. Council Directive of 22 June 1988 on the approximation of the laws of the member states relating to flavourings for use in foodstuffs and to source materials for their production (88/388/EEC). OJ L184, 15.7.1988, p. 61.

EUROPEAN PARLIAMENT AND COUNCIL. 1996. Regulation (EC) No. 2232/96 of the European Parliament and of the Council of 28 October 1996 laying down a Community procedure for flavouring substances used or intended for use in or on foodstuffs. OJ L299, 23.11.1996, p.1.

FAO. 1995. Report of the FAO Technical Consultation on Food Allergies. Rome, Nov. 13–14, Food and Agric Org., Rome.

FAO/WHO. 2000. Safety aspects of genetically modified foods of plant origin. Report of a Joint FAO/WHO Expert Consultation on Foods Derived from Biotechnology. Food and Agriculture Organization of the United Nations and World Health Organization. WHO, Geneva, Switzerland.

FDA. 1982. Toxicological Principles for the Safety Assessment of Direct Food Additives and Color Additives Used in Food. Redbook. US Food and Drug Administration, Bureau of Foods, Washington, DC.

FEDERAL REGISTER. 1993. Food additives; Threshold of regulation for substances used in food-contact articles. Fed Reg 58(195): 52719–52729.

FEDERAL REGISTER. 1995. Food additives; Threshold of regulation for substances used in food-contact articles. Fed Reg 60(136): 36582–36596.

FEDERAL REGISTER. 1997. 21 CFR Parts 170, 184, 186, and 570. (Docket No. 97N-0103). Substances Generally Recognized as Safe. 62(74): 18938–18964.

GOLD, L.S., SLONE, T.H. and BERNSTEIN, L. 1989. Summary of carcinogenic potency and positivity for 492 rodent carcinogens in the carcinogenic potency database. *Environ Health Perspectives* 79: 259–72.

HALL, R.L., FORD, R.A. and HALLAGAN, J.B. 1999. Comparison of two methods to assess the intake of flavoring substances. *Food Additives and Contaminants* 16(11): 481–95.

HALLAGAN, J.B. and HALL, R.L. 1995a. FEMA GRAS – A GRAS assessment program for flavor ingredients. *Reg Toxicol Pharmacol* 21(3): 422–30.

HALLAGAN, J.B. and HALL, R.L. 1995b. Safety Assessment of Flavor Ingredients Produced by Genetically Modified Organisms. In *Genetically Modified Foods* edited by K-H. Engel, G.R. Takeoka and R. Teranish, American Chemical Society, Washington DC, pp. 59–69.

INTERNATIONAL FOOD BIOTECHNOLOGY COUNCIL (IFBC). 1990. Biotechnologies and food: assuring the safety of foods produced by genetic modification. *Reg Toxi Pharmacol* 12(Supp.): S1–S196.

INTERNATIONAL FOOD BIOTECHNOLOGY COUNCIL (IFBC) AND THE INTERNATIONAL LIFE SCIENCES INSTITUTE (ILSI) ALLERGY AND IMMUNOLOGY INSTITUTE (AII). 1996. Allergenicity of Foods Produced by Genetic Modification. *Critical Reviews in Food Science and Nutrition*: S1–S181.

IOFI. 1995. *European Inquiry on Volume of Use*. International Organization of the Flavor Industry.

JECFA. 1998. Toxicological evaluation of certain food additives and contaminants. The 44th Meeting of the Joint FAO/WHO Expert Committee on Food Additives. WHO Technical Report Series 884.

JECFA. 1999. Evaluation of certain food additives and contaminants. The 49th Report of the Joint FAO/WHO Expert Committee on Food Additives. WHO Technical Report Series 884.

MUNRO, I.C., KENNEPOHL, E., ERICKSON, R.E., PORTOGHESE, P.S., WAGNER, B.M., EASTERDAY, O.D. and MANLEY, C.H. 1993. Safety assessment of ingested heterocyclic amines: Initial report. *Reg Toxicol Pharmacol* 17(2): S1–S109.

MUNRO, I.C., FORD, R.A., KENNEPOHL, E. and SPRENGER, J.G. 1996. Correlation of structural class with no-observed-effect levels: A proposal for establishing a threshold of concern. *Food Chem Toxicol* 34: 829–67.

MUNRO, I.C., SHUBIK, P. and HALL, R. 1998. Principles for the safety evaluation of flavoring substances. *Food Chem Toxicol* 36: 529–40.

MUNRO, I.C., KENNEPOHL, E. and KROES, R. 1999. A procedure for the safety evaluation of flavoring substances. *Food Chem Toxicol* 37: 207–32.

RICHLING, E., HARING, D., HERDERICH, M. and SCHREIER P. 1998. Determination of Heterocyclic Aromatic Amines (HAA) in Commercially Available Meat Products and Fish by High Performance Liquid Chromatography – Electrospray Tandem Mass Spectrometry (HPLC-ESI-MS-MS). *Chromatographia* 48: 258–61.

RULIS, A.M. 1989. Establishing a threshold of concern. In *Risk Assessment in Setting National Priorities* (J.J. Bonin and D.E. Stevenson, eds.), Volume 7, pp. 271–8. Plenum Press, New York.

SCF. 1991. Guidelines for the Evaluation of Flavorings for Use in Foodstuffs: I. Chemically Defined Flavoring Substances. Commission of the European Communities, Scientific Committee for Food, Brussels, Belgium.

SCF. 1999. Opinion on a programme for the evaluation of flavouring substances (expressed on 2 December 1999). SCF/CS/FLAV/TASKE/11 Final 6/12/99. Annex I to the minutes of the 119th Plenary Meeting.

SLOAN, A. 1986. A perspective on popular perceptions of adverse reaction to foods. *J. Allergy Clin. Immunol.* 78: 140–59.

SMITH, R. L., DOULL, J., FERON, J. I., GOODMAN, J. L., MUNRO, I. C., NEWBERNE, P. M., PORTOGHESE, P. S., WADDELL, W. J., WAGNER, B. M., ADAMS, J. B. and MCGOWAN, M. M. 2001. GRAS Flavoring Substances 20. *Food Technology* 55(12): 34–6, 38, 40, 42, 44–55.

STOFBERG, J. and KIRSCHMAN, J. 1985. The consumption ratio of flavoring materials: A mechanism for setting priorities for safety evaluation. *Food Chem Toxicol* 23: 857–60.

STOFBERG, J. and GRUNDSCHOBER, F. 1987. Consumption ratio and food predominance of flavoring materials. *Perfumer Flavorist* 12: 27–68.
TAYLOR, S. 1985. Food Allergies. *Food Technol* 39: 98–102.
TAYLOR, S. and DORMEDY, E. 1998. The role of flavoring substances in food allergy and intolerance. *Advances in Food and Nutrition Research* 42: 1–44.
WHO. 1987. *Principles for the Safety Assessment of Food Additives and Contaminants in Food*. WHO Environmental Health Criteria 70. World Health Organization, Geneva, Switzerland.
WHO. 1996. Food Additive Series 35, WHO Technical Report of the 44th meeting, Annex 5.
WOODS, L.A. and DOULL, J. 1991. GRAS evaluation of flavoring substances by the Expert Panel of FEMA. *Reg Toxicol Pharmacol* 14: 48–58.

10

Sweeteners

G. von Rymon Lipinski, Nutrinova Nutrition Specialities and Food Ingredients GmbH, Frankfurt

10.1 Introduction

Sweeteners are food additives by definition in almost all countries and need approval for use by food laws, food regulations, decrees or certificates released and issued by the national government or other responsible governmental institutions. Generally three prerequisites must be fulfilled to obtain an approval for use:

- safety for the consumer under normal and foreseeable use conditions[1]
- technological justification
- absence of consumer deception.

Safety studies have to include the wide variety of different test systems commonly used today to exclude risks originating from consumption of these products. This applies in particular to sweeteners, as they may be consumed by the general population for prolonged periods. As other sweet-tasting products would be available as an alternative, safety demands on sweeteners are especially high. Although diabetics were the most important sweetener consumers in the past, improved possibilities to manufacture high-quality sweet-tasting products containing sweeteners has resulted in increased consumption of sweeteners in general over the last decades.

Sweetness free, or at least virtually free, from calories establishes the need for intense sweeteners while suitability for diabetics, absence of carcinogenicity with functionality similar to sucrose and related carbohydrates form the basis for application of bulk sweeteners.

10.2 Definitions

Sweet-tasting substances can be grouped together on the basis of different characteristics. Many sweet-tasting substances are fully or at least partly metabolised. Sucrose and similar carbohydrates have to be used in substantial concentrations to provide some sweetness. They are completely absorbed and fully metabolised into carbon dioxide and water and therefore provide approximately 4 kcal or 17 kJ per gram. Their contribution to the calorie content of foods and beverages can be substantial.

Bulk sweeteners which can serve as sugar substitutes are less rapidly absorbed than sucrose and related carbohydrates. Therefore their contribution to the calorie content of the diet is below that of the sucrose level. Values for energy utilisation given in the literature differ, depending on the test system used. Whereas in the European Union a standard value of 2.4 kcal or 10 kJ/g is used[2] other countries like the USA use substance-specific values. Due to their calorific value sugar substitutes offer limited possibilities for energy reduction only.

The most striking difference between intense sweeteners and the other types of sweet-tasting substances is the high difference in sweetness intensity. Intense sweeteners are at least ten times sweeter than sucrose. Most are several hundred times sweeter than sucrose and sometimes even exceed the factor of 1000 times higher sweetness than sucrose. As a consequence, the intensely sweet products have to be used in minute quantities only which would not contribute substantially to the energy content of the diet, should they be metabolised and utilised for energy. In addition, some of the intense sweeteners are not metabolised, and are excreted completely unchanged. Therefore they cannot contribute to the energy content of foods and beverages at all.

Because of their non-metabolism or insignificant contribution to the calorie content of a diet, intense sweeteners seem to be an ideal means to lower calorie intake in sweet-tasting foods and beverages. Such a simple approach, however, would not take into consideration that function and properties of bulk sweeteners determine characteristics of many sweet-tasting products, e.g. texture, appearance and shelf stability amongst others.[3]

10.3 Functionality and uses

10.3.1 Intense sweetener functionality

Intense sweeteners are characterised by a high sweetness intensity on a weight basis. Sweetness intensity values differ; for general comparisons standard sweetness intensity values are often used with sucrose being the standard with a sweetness intensity of 1. Sweetness intensities depend on a number of factors, e.g. concentration and presence of flavours or taste components. They are therefore not a suitable tool for calculation of use concentrations except for very preliminary approaches.

Taste characteristics in general determine applicability of intense sweeteners. A time-intensity profile of sweetness perception similar to sucrose is desirable, and a delay in sweetness onset or a lingering sweetness are generally perceived as less pleasant. Side-tastes like bitter, liquorice or metallic taste are disadvantages which limit the applicability of some sweeteners.

Synergistic effects occurring in sweetener blends have often been used in countries approving saccharin and cyclamate. With the availability of new sweeteners synergism has found renewed interest. Synergism can be broken into two basic aspects: quantitative synergism and qualitative taste improvement. Quantitative synergism is a pronounced increase in perceived sweetness occurring in certain sweetener blends. Use levels can be reduced accordingly, and quantity savings in the range of 30 to 50% are possible for certain sweetener blends. Qualitative synergism, a taste improvement, is another advantage of certain synergistic sweetener blends. Such quality improvement is especially obvious in blends of sweeteners having different time-intensity profiles, e.g. acesulfame and aspartame.[4] Due to such synergistic effects a more balanced sweetness profile is obtained, and even adaptation of the sweetness profile to flavour profiles is possible to some extent.

Simple applicability of sweeteners is desirable. Good solubility in water generally facilitates application as many foods, and, of course, beverages contain substantial amounts of water. High solubility allows use of the sweeteners in stock solutions, which is advantageous for simple dosing and blending.

Intense sweeteners should be highly stable, as sweetness, being one of the basic tastes, is easily perceived, and sweetness changes may influence the general product characteristics strongly. Sweetness during processing, i.e. the ability of sweeteners to withstand elevated temperature, and stability during prolonged storage are therefore characteristics intense sweeteners should have. It should, in addition, be kept in mind, that the possibility of reactions with other food constituents may bring about changes of product characteristics, especially changes of flavours, should sweeteners react with flavour constituents. Such reactivity is therefore undesirable.

10.3.2 Bulk sweetener functionality

Bulk sweeteners have a sweetening power roughly similar to sucrose and are, in contrast to intense sweeteners, used in similar quantities as carbohydrate sweeteners.[3] Most of these bulk sweeteners are slightly less sweet than sucrose. Their sweetness intensities vary between less than half of to approximately the same as sucrose. When consumed in solid form or in products where they are present as solids, they may give a more or less pronounced cooling sensation. Generally their tastes are perceived as less round than the sugar taste but they harmonise well with many flavours. Combination with intense sweeteners does not only bring the sweetness intensity to higher levels but also often rounds the rather 'flat' sweetness compared to sucrose.

Bulk sweeteners are distinguished by high stability at elevated temperatures. They do not undergo caramelisation reactions like sucrose. They do not react with common food constituents either. As a result products containing bulk sweeteners instead of carbohydrates may be more pale than sugar-based products and may have a slightly different flavour. Such differences can, however, be compensated for by slight modifications of recipes, if necessary.

As some bulk sweeteners are hygroscopic, different packaging materials may be necessary for products containing these sweeteners.

10.3.3 Uses

Sweet-tasting carbohydrates like sucrose or glucose are multifunctional food ingredients. Sweet taste is one of the key properties for food applications, but functional properties are often very important and must not be underestimated. Among the functional properties of these carbohydrates, bulking is one of the most important. It is obvious that properties of hard and soft candies, chewing gums, chocolate, and sweet bakery products with carbohydrate contents ranging from one fourth or one third to almost one hundred percent are strongly influenced by sucrose or other sweet-tasting carbohydrates commonly used in their production. Bulk sweeteners can replace carbohydrates in many functions, and blending of bulk sweeteners having different properties allows adaptation to the required product characteristics.

Humectant properties of sucrose and other carbohydrates may be important for certain products, as water activity can be controlled by the level of humectants. Replacement of the carbohydrates by sweeteners may therefore influence the water activity of products and require other technologies, other packaging materials or additional steps to control microorganisms.

High levels of soluble carbohydrates may have an influence on the texture characteristics of products, too. Gelling of pectins, for example, depends on the level of sucrose, which has to be high to get the typical gel structure of jams and marmalades.

Browning caused by Maillard-type reactions between carbohydrates and amino acids or caramelisation contribute to appearance, taste and flavour of many products. Carbohydrate sweeteners are important for these reactions, and other types of sweetening agents may not react in a similar way, therefore rendering products more pale.

Several other functional characteristics of sweet carbohydrates may have importance for certain products, too, like flavour enhancement. Therefore, one of the important steps in using the appropriate sweetener or sweetener system is the basic decision whether functionality of the sweeteners used is appropriate for the product to be sweetened and to select an appropriate sweetener or rather a sweetener system.

As bulk sweeteners provide the functionality of carbohydrate sweeteners they are mainly used in applications in which normally high amounts of sugar would be used. These are, in particular, confectionery products, chewing gum and fine bakery wares. They may also serve as carriers for table-top sweeteners.

Intense sweeteners have low functionality besides their sweet taste. Therefore intense sweeteners cannot be used as the only sweetening agents whenever at least one of the mentioned functions is important for a product. Combinations of intense and bulk sweeteners will come close to sucrose and other sweet carbohydrates in functionality and taste, and can therefore be considered as an interesting alternative to sugar in applications requiring functional properties. They are often used in addition to bulk sweeteners in the typical application of these.

Many sweet-tasting foods and beverages, however, do not require the functionality of sucrose and sweet carbohydrates. These products are the typical fields of application of intense sweeteners. As bulk sweeteners are used for taste reasons rather than functionality these products offer possibilities to reduce calories without sacrificing any important product characteristic. Intense sweeteners are used as the sole sweetening agents in beverages, table-top sweeteners like powder or tablets, desserts and dairy products besides a variety of further areas of lesser importance.

10.4 The available sweeteners

A large number of sweet-tasting substances has been described in the scientific literature, most of them in the course of the last 20 to 30 years. The number of substances having an intense sweetness is by far higher than the number of products with a sweetening power similar to sucrose which could therefore be used as bulking sugar substitutes. Only a small number of these has found

Table 10.1 Available intense sweeteners

Europe and USA	
E/INS 950	Acesulfame K
E/INS 951	Aspartame
E/INS 954	Saccharin*
Europe	
E/INS 952	Cyclamate*
E/ - 959	Neohesperidin dihydrochalcone
E/INS 957	Thaumatin***
USA	
E/INS 955	Sucralose **
Others	
-/INS 956	Alitame
	Stevioside

* includes some salts
** due for approval in the EU
*** in the USA available as a flavour enhancer

Table 10.2 Available bulk sweeteners

E/INS 593	Isomalt
E/INS 966	Lactitol
E/INS 965	Maltitol and maltitol syrup
E/INS 421	Mannitol
E/INS 420	Sorbitol and sorbitol syrup
E/INS 967	Xylitol

practical application owing to the very demanding requirements on safety and technical characteristics like stability under processing and storage conditions in a wide variety of foods and beverages.

The available intense sweeteners belong to very different structural classes of sweeteners (Table 10.1). They were normally discovered by chance. All internationally important sweeteners are produced synthetically and only two less important products are isolated from plants.

A variety of bulk sweeteners is now available (Table 10.2). They belong to the class of sugar alcohols. Some monosaccharide alcohols have been used for decades, and disaccharide alcohols have broadened the range of available functional characteristics.

10.5 Sweetener safety testing

10.5.1 General

Although the extent and type of safety studies necessary to obtain food additive approvals, including sweetener approvals, is, as a rule, not defined precisely in food additive legislation, guidelines have been published by different bodies to provide interested companies and institutions with guidance for such testing. Examples of these rules are OECD guidelines for toxicological tests, the Principles for the Safety Assessment of Food Additives and Contaminants in Food published by the World Health Organisation,[5] and the 'Red Book' issued by the US Food and Drug Administration, the FDA.[6] It should be kept in mind, however, that such guidelines can always only give a framework and a program for the safety testing of food additives has to be developed on the basis of the properties of the substance in question. Generally, a full range of safety studies as proposed in the guidelines is a prerequisite for a successful food additive petition.

A full set of studies normally includes short-term and long-term animal studies on chronic effects and potential carcinogenicity, studies on reproductive and developmental toxicity, genotoxicity, kinetics and metabolism, pharmacological properties and special studies depending on the characteristics of the substance and observation in the standard set of studies. Human clinical studies may be necessary for substances which are metabolised and may interfere with functions of the human body.

10.5.2 Toxicology
Absence of carcinogenity, genotoxicity, developmental and reproductive toxicity and of chronic toxicity effects at low exposure levels are indispensable prerequisites for food additive approvals. All substances approved in the European Union or the USA or deemed generally recognised as safe (GRAS) in the USA fulfil this requirement.

10.5.3 Physiology
Metabolism via normal metabolic pathways or fast excretion without metabolism are desirable characteristics. Some intense sweeteners are excreted unchanged while others are metabolised. Bulk sweetener absorption is lower and slower than for carbohydrates and results in reduced caloric availability which is partly due to metabolites formed by intestinal bacteria. Such metabolites and osmotic effects of not fully absorbed bulk sweeteners can cause laxative effects. Generally, the calorific value of bulk sweeteners is lower than for carbohydrates. Intense and bulk sweeteners are, as far as they are metabolised, not dependent on insulin. They are therefore acceptable for diabetics as part of a suitable diet.

10.6 Case study: acesulfame K

10.6.1 Product
The extent of safety studies necessary to obtain food additive approval can be demonstrated by the studies carried out on acesulfame K (trade name Sunett®), one of the sweeteners developed in course of the last 25 years,[7] which has been endorsed for food use by the Joint Expert Committee on Food Additives (JECFA) of the WHO and FAO and the Scientific Committee for Foods (SCF) of the EU and has meanwhile been approved in more than 100 countries. This program shows the wide range of studies necessary.

10.6.2 Kinetics and biotransformation
Studies on kinetics and biotransformation were carried out in rats, dogs, pigs and men. In this study a ^{14}C label was introduced at position 6 on the acesulfame K ring system. Half-lives were approximately 4 h in rats, 1.3 h in dogs and 2.5 h in man. No indication for accumulation which would have prevented the use as a food additive was found even after repeated dosing. No formation of metabolites was found either. Therefore no species-specific metabolism had to be taken into consideration for toxicity studies. Additional investigations into the distribution in pregnant rats and excretion with milk indicated that radioactivity did not pass into the faeces and that excretion with milk was limited to a minute proportion of the ingested quantity. Therefore no high exposure to fetuses or weanlings can be expected.[7]

10.6.3 Toxicity
The acute toxicity of acesulfame K is so low it can be regarded as virtually non-toxic. The oral LD_{50} was determined to be approximately 7.4 g/kg of body weight which corresponds to human doses by magnitudes higher than those which could be expected for inadvertent overdosage.[7]

In subchronic toxicity studies levels up to 10% in the diet were administered to animals, and on the basis of the results of the studies 3% in the diet were given to rats, mice, and dogs in several long-term studies aimed at detecting chronic toxicity in dogs and rats, or carcinogenicity in rats and mice. No dose-related effects were observed in all these studies, and a no-observed-effect level (NOEL) of 3% in the diet (equivalent to 1500–3000 mg/kg of body weight in rats or 900 mg/kg of body weight in dogs) was determined.[7]

10.6.4 Genotoxicity
A full range of studies aimed at detecting genotoxic effects was carried out. In none of the studies, using different genotoxicity mechanisms, was there any indication of genotoxicity.[7]

10.6.5 Reproductive toxicity
Reproductive toxicity of acesulfame K was studied in test systems aimed at detecting teratogenicity, oral embryotoxicity and in a multigeneration study. No teratogenicity, no embryotoxicity, and no effects on reproduction, development of the fetuses and lactation performance were found.[7]

10.6.6 Pharmacology
In a screening system used for potential new drugs, acesulfame K was inert in all test systems except where potassium has an effect, like a transient influence on blood pressure after injection or reduction of digoxin toxicity.[7]

10.6.7 Influence on diabetes
Neither in the screening system nor in a streptozotocin-induced diabetes in rats did acesulfame K show any influence on insulin release or blood glucose levels. Although after ingestion of extremely high concentrations of acesulfame K a temporary and transient increase in insulin secretion was reported without influence on the blood glucose level, single doses corresponding to one can of acesulfame K in sweetened beverages did neither stimulate insulin secretion nor have an influence on blood glucose. Data generated in these studies show that acesulfame K is safe for diabetics.[7]

10.6.8 Stability

In order to exclude formation of reaction or decomposition products which would have had an influence on safety considerations, extensive stability studies were carried out. Only under extreme conditions, i.e. acid pH, prolonged storage or high temperatures, could formation of hydrolysis products be detected. Although these products would not even be detected in an acid soft drink after storage periods well exceeding normal levels, and great overspiking was necessary to detect traces, these products, i.e. acetoacetic acid derivatives, were studied in a range of toxicity, genotoxicity and metabolism studies to exclude any potential risk originating from these compounds.[7] The extent of these studies fulfilled FDA Red Book requirements for the expected intake. International agencies and national health authorities concluded after evaluation of the data, that they do not bring about a safety risk.

As acesulfame K contains cyclic nitrogen, a nitrosation study was carried out which did not show any nitrosation under normal use conditions. In model reactions with food constituents or substances modelling groups of food constituents, no reactions of acesulfame K with these components was detected.[7]

10.6.9 International acceptance

Data on the safety studies were submitted to international agencies like the Joint Expert Committee for Food Additives of the WHO and FAO (JECFA), and the Scientific Committee on Food (SCF) of the EC. Both committees endorsed acesulfame K as a food additive. Initial acceptance was based on an NOEL of 900 mg/kg in dogs which were considered to be the most sensitive species. Therefore Acceptable Daily Intake (ADI) values of 0–9 mg/kg of body weight were allocated.[8,9] Evidence that rats would be an appropriate model for risk assessment was the reason for JECFA to change the ADI to 0–15 mg/kg of body weight on the basis of a no-effect level of 1500–3000 mg/kg in rats.[10] Countries allocating their own ADI values like the USA and Canada have come to the same conclusion. The SCF still retains its 0–9 mg/kg ADI.[11]

10.7 Other sweeteners

10.7.1 Alitame

Alitame, L-α-aspartyl-N-(2,2,4,4-tetramethyl-3-thietanyl)-D-alanine amide, has undergone a series of safety studies. While most of the studies did not show adverse effects and no indications for carcinogenicity were found, a dose-dependent increase in liver weights was found at levels above 100 mg/kg which was identified as the no-effect level. While JECFA has allocated an ADI of 0–1 mg/kg[12] only a few countries, but neither the European Union nor the USA, have approved alitame.

10.7.2 Aspartame

Aspartame, N-α-L-aspartyl-L-phenylalanine methyl ester, trade names NutraSweet®, and Aspartil®, is a dipeptide derivative. Like dipeptides aspartame is metabolised into the constituents, i.e. amino acids and methanol. Therefore studies into the metabolic behaviour and the fate of metabolites were carried out. Levels of blood aspartate and glutamate were measured after intake of high aspartame doses. Changes were transient and allegations of influences of high aspartame levels on brain function could never be verified.

Special attention was paid to the potential influence on phenylketonurics (PKU), as aspartame contains phenylalanine. As persons suffering from PKU should avoid uncontrolled intake of phenylalanine-containing food constituents or food additives, most countries require a warning on aspartame-sweetened products unless the aspartame level brought about by constituents of these products will exceed the aspartame levels.[13,14] Evaluations of aspartame were carried out by JECFA, and an ADI of 0–40 mg/kg of body weight was allocated.[15] The SCF allocated the same level,[10] whereas the FDA published a value of 50 mg/kg.[16]

The stability of aspartame in liquid media is highly dependent on the pH level and, besides hydrolysis into constituents, formation of a diketopiperazine (DKP) 5-benzyl-3,6-dioxo-2-piperazine acetic acid is possible, a component which may also form during heating and for which a limit of up to 1.5% has been set in the specifications. Intake of this component may be substantial. Therefore a full range of studies was carried out on this product, too. No negative results were obtained either, and an ADI of 0–7.5 mg/kg of body weight has been allocated by international agencies.[15] In order to ascertain intake below the acceptable levels some countries have introduced maximum levels for DKP in soft drinks although excess intake above the acceptable limits seems highly unlikely.

10.7.3 Cyclamate

Cyclamates are the salts of the cyclohexylsulfamic acid. In 1969 cyclamates were blamed for being carcinogenic following a study in which an increase in bladder tumours was observed in a study on a cyclamate-saccharin blend. Following publication of the study the FDA and a number of other health authorities banned cyclamate from use in foods and beverages. On the basis of additional studies it has meanwhile been accepted that cyclamate is not carcinogenic.

Whereas cyclamate as such is not metabolised in the human body, bacteria in the intestinal tract may degrade cyclamates to cyclohexylamine. The degradation rates vary widely and may occasionally exceed 50%, but are generally limited to a few percent of the total intake, should any metabolism occur at all. Changes within the same individual within fairly short periods have been observed.

The discussion on the acceptability of cyclamates has focused on pharmacological and toxicological effects observed in studies with cyclohexylamine.

238 Food chemical safety

Some cardiovascular effects observed after high doses seem less important than testicular atrophy observed in juvenile rats. It seems possible that the latter effect is species-specific. Respective investigations are being carried out.

The ADI value for cyclamate of 0–11 mg/kg as allocated by JECFA[17] is based on the no-effect level for cyclohexylamine, a 63% availability of cyclohexylamine for conversion due to absorption of 37% and an average conversion rate of 30%. The SCF has meanwhile reduced the ADI to 0–7 mg/kg assuming a higher conversion rate.[18]

Some countries approve cyclamate and the EU Sweetener Directive lists cyclamate for a number of applications, while others still ban cyclamates. Attempts to get a re-approval in the USA have remained unsuccessful so far.

10.7.4 Neohesperidin dihydrochalcone

A series of studies on neohesperidin dihydrochalcone and some structurally related compounds is available. Neohesperidin dihydrochalcone is metabolised in the human body apparently through activity of intestinal bacteria. Most of the toxicological studies were carried out in the 1960s and 1970s and limited additional information has since become available. No signs of carcinogenicity were observed and slight changes of physiological parameters were found at a level of 5% in the diet. A no-effect level of 1% in the diet was identified which resulted in allocation of an ADI of 0–5 mg/kg by the SCF,[10] but apparently no evaluation has been made by JECFA. Accordingly neohesperidin dihydrochalcone is approved in the EU but not listed in the Codex Alimentarius draft General Standard for Food Additives. It is not approved in the USA either.

10.7.5 Saccharin

Saccharin is 1,2-benzisothiazol-3(2H)-on-1,1-dioxide, often also called o-benzoic acid sulfimide. Discovered in 1878, it is the oldest available intense sweetener.

The toxicology of saccharin has been controversial for the last 15 years. Although it has been used for more than 100 years, toxicological studies have been carried out continuously using the increasingly sophisticated methods and the rising standards of toxicology. After completion of a study showing an increased incidence of bladder tumours in rats in 1977 the Canadian Health Protection Branch (HPB) and the FDA banned saccharin. A public outcry prevented the FDA from implementing the ban, but a warning label for potential carcinogenic effect was introduced in the USA which has meanwhile been removed. Investigations following the study with probably the largest number of animals ever used in a safety study confirmed the former observations, but showed a very steep increase in bladder tumours at high concentrations.[19] A statistical risk evaluation for the normal human intake gave borders between 1 : 10 000 and 1 : 100 millions for humans and could not provide any better insight into the potential risk for humans.[20]

Mechanistic studies carried out in the course of recent years indicated that sodium saccharin may be involved in the formation of specific proteins occurring in rat urine which may contribute to the formation of bladder tumours. Most scientists and regulatory authorities now share the view that the incidence of bladder tumours in male rats after saccharin ingestion is species-specific and therefore probably without any significance for humans. Accordingly, the former temporary ADI of 0–2.5 mg/kg of body weight has been raised to 0–5 mg/kg by JECFA[21] and the SCF.[22]

During the discussions of a potential carcinogenicity of saccharin, contaminants, especially o-toluene sulphonamide, were blamed for the toxicological effects observed in these studies. The studies carried out on such contaminants which in former times occurred in concentrations well exceeding 1000 ppm, did, however, not reveal any such effects, and the observed effects today are believed to be saccharin-specific. In order to minimise risks for consumers, low maximum levels for these products were introduced into the specifications.

10.7.6 Stevioside

Stevioside is a constituent of the leaves of the plant *Stevia rebaudiana* (Bertoni), a glycoside of the diterpene derivative steviol. Most available studies were carried out on crude extracts of the leaves or preparations of varying purity, which raises questions about the value of such studies for safety assessment. In the human body a substantial part of stevioside is metabolised to steviol. The available long-term data seem inconclusive and not to cover all relevant aspects either. Results of some of these studies and mutagenicity of the metabolite steviol in at least two genotoxicity test systems, however, are of concern. Accordingly, the SCF concluded recently that the substance is not acceptable as a sweetener on the basis of the available data[23] and that the plant containing the sweetener, *Stevia rebaudiana*, is not acceptable for human consumption either on the basis of the available information.[24] Following this conclusion the EU Commission denied requests to approve stevioside and also banned the sale of *Stevia* leaves. Stevioside has not been endorsed by JECFA and not been approved in the USA while some countries growing the *Stevia* plant or having less stringent rules for food additives of plant origin permit its use.

10.7.7 Sucralose

Sucralose, 1,6-dichloro-1,6-dideoxy-β-D-fructofuranosyl-4-chloro-4-deoxy-α-D-galactopyranoside, formerly also called trichlorogalactosucrose, has been developed and studied since the late 1970s. Sucralose has been studied extensively using the relevant modern test systems. In the human body sucralose is not utilised and enzyme adaptation to sucralose is highly unlikely. No carcinogenicity was found in long-term studies and an NOEL of 1500 mg/kg was established, resulting in allocation of an ADI of 0–15 mg/kg by JECFA[25] and the SCF.[26]

Although sucralose is highly stable, studies on the breakdown products 1,6-dichloro fructose and 4-chloro galactose were also carried out.

Sucralose is approved in many countries worldwide and has also been endorsed by the SCF. A listing in the EU Sweetener Directive is expected in 2002.

10.7.8 Thaumatin

Other than the sweeteners discussed so far thaumatin is a polypeptide consisting of amino acids commonly found in food proteins. It is quickly and completely digested like proteins and did, after demonstration of its metabolic characteristics, only require a rather limited set of safety data. In contrast to the other intense sweeteners the ADI of thaumatin is 'not specified', as for substances of similar composition.[27] It is approved in many countries but, owing to its flavour enhancing properties, is often used as a flavour enhancer rather than a sweetener.

10.7.9 Bulk sweeteners

The toxicological profiles of bulk sweeteners are similar. Acute toxicity values, expressed as LD_{50} are normally around 20 g/kg and therefore so high that consumption of amounts coming close to these seems virtually impossible. In long-term studies concentrations as high as 10% in the diet cause only non-specific, if any effects, that generally 'not specified' ADI values have been allocated, e.g. for isomalt.[28] Laxative effects can occur after consumption of large amounts although adaptation seems possible.

10.7.10 Sweetener blends

In many sweetener blends, especially with intense sweeteners, synergistic sweetness enhancement can be observed (Fig. 10.1). The lack of any substantial toxicity and the high safety margin between life-time no-effect levels and human exposure as well as the lack of a common target organ for effects of intense sweeteners exclude any appreciable risk of negative combination effects. In contrast, the synergistic taste enhancement results in lower consumption of intense sweeteners. Accordingly, provisions of the EU Sweetener Directive rather encourage use of blends of intense sweeteners. Use of such blends is also possible in virtually all countries approving intense sweeteners.

10.7.11 Intake considerations

The intake of food additives in general and of intense sweeteners in particular has been discussed in the course of the last few years. Interest has focused on intense sweeteners, although the database for sweeteners and possibilities to exclude excess intake above the acceptable levels on the basis of calculations and estimates are much better than for most other components.

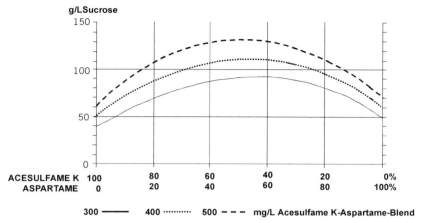

Fig. 10.1. Synergistic sweetness enhancement in acesulfame K–aspartame blends.

A first simple consideration can be based on the standard sweetness intensities. Sweetness intensities of intense sweeteners are multiplied by the acceptable daily intake level and the standard body weight of 60 kg as suggested by JECFA for intake calculations giving values which may be called 'acceptable daily sucrose equivalent'. Whenever such values are well above the average daily sucrose intake the risk of excess intake above the ADI is small. This statement is based on the fact that hardly one single sweetener will be used in all products sweetened and that, in addition, intense sweeteners lack the functionality of sucrose and other bulk sweeteners which is essential for a number of products like confectionery and baked goods. The substantial proportion of the average daily sucrose consumption from these products would mostly be replaced by bulk sweeteners and to a minute proportion by intense sweeteners, as they only round the sweetness of these products and contribute only a small proportion of the total sweetness of these product groups. Using this simple calculation a high safety margin between the acceptable daily sucrose equivalent and reasonable intake estimates can be shown for almost all sweeteners.

Intake estimates and calculations have been performed repeatedly for intense sweeteners for which probably the most extensive database among food additives exists. All studies and all calculations starting from reasonable assumptions indicate that only a minute proportion of consumers may come close to the ADI which may only seldom be exceeded by persons having food habits substantially different from the majority of the population. The best available data originate from a biomarker study on acesulfame and saccharin in which even the highest consumers among children consumed only a fraction of the ADI.[29] Several intake studies were carried out on aspartame with the uniform result that no appreciable risk to exceed the ADI was found.[14]

As bulk sweeteners have 'not specified' ADI values, a similar range of intake studies is not available, and laxative properties more or less exclude excessive intake.

10.8 Regulatory status

10.8.1 General

The applicability of bulk and intense sweeteners is determined by their regulatory status. Bulk and intense sweeteners require food additive approval in many countries.

An important step and for some countries even a prerequisite before food additive approval is endorsement for food use by international scientific bodies like the JECFA or the European SCF. These committees evaluate the safety data, identify a no-observed-effect level and allocate an ADI, usually by applying a safety factor of 100 to the ADI. While numerical values have been allocated for all intense sweeteners, the ADIs for bulk sweeteners are normally 'not specified' as any numerical limitation would not be reasonable for these substances.

The Sweetener Directive of the European Union and a later amendment[30] list the approved sweeteners and use conditions for all EU member states (Table 10.3). Bulk sweeteners are generally approved for use following the *quantum satis* principle, i.e. following good manufacturing practice and at levels not higher than is necessary to achieve the intended purpose. For intense sweeteners generally maximum use levels are listed, depending on the specific application, with the exception of table-top sweeteners in which they can be used following good manufacturing practice. Only sweeteners conforming to specific purity criteria are permitted which are set out in a Directive laying down special purity criteria.[31]

In the USA the available bulk sweeteners are listed under different provisions, like food additive, interim status, GRAS (Generally Recognised As Safe) or GRAS by self-determination or self-affirmation of the manufacturers. Intense sweeteners require food additive approval which includes a listing of the approved fields of use or may be a listing as a general purpose sweetener.[32]

Beyond national approvals the Codex Alimentarius is developing a General Standard for Food Additives which will be the applicable basis for international trade. All and only sweeteners endorsed for food use by JECFA are/or will be listed in this standard. For bulk sweeteners the standard has progressed to permanent while the part dealing with intense sweeteners is still under discussion.

Generally the approval of intense sweeteners following good manufacturing practice should be acceptable whenever an estimated intake well below the ADI can be demonstrated for technologically reasonable and most probable use levels. Intense sweetener approvals in the USA are based on such considerations. Too strong a sweetness for excess use levels and aftertastes becoming perceptible at excess concentrations will render foods containing such sweetener levels unpalatable and therefore necessarily limit the use to reasonable concentrations.

For bulk sweeteners, which are often an important and sometimes the most important ingredient, any limitation would be unreasonable and often even render their use impossible. In the EU consumers are alerted of potential laxative effects by labelling of products containing more than 10% by weight of sugar alcohols.

Table 10.3 Approved use levels for intense sweeteners in some categories of sugar-free or calorie-reduced products in the European Union (values in mg/kg or mg/l)

	Acesulfame K	Aspartame	Cyclamate	Neohesperidin DC	Saccharin	Thaumatin
Beverages	350	600	400	30/50*	80/100*	
Desserts	350	1000	250	50	100	
Sweets and confections	500/1000/ 2500*	1000/2000/ 6000*	500/2500*	100/150/ 400*	300/500/ 3000*	50
Chewing gum	2000	5500	1500	400	1200	
Ice cream	800	800	250	50	100	50
Jams and marmalades	1000	1000	1000	50	200	50
Canned fruit	350	1000	1000	50	200	
Fine bakery wares	1000	1700	1600	150	170	
Dietary supplements	350/500 2000*	600/2000/ 5500*	400/500/ 1250*	50/100/ 400*	80/500/ 1200*	

*Different levels for different sub-categories.

10.8.2 Purity criteria

Only sweeteners conforming to certain purity criteria are permitted for use in foods. Such purity specifications are normally based on the specifications of the substance used in safety evaluations as byproducts and impurities may affect the toxicological profile of a substance.

Purity criteria are published by JECFA as well as by national authorities. The JECFA specifications[33] may be endorsed by the Codex Alimentarius as advisory specifications which means that they are equivalent to a Codex Alimentarius standard. In the EU a directive laying down special criteria of purity for sweeteners for use in foodstuffs with subsequent amendments sets the purity standards,[31] while in the USA criteria may be listed in conjunction with the approval. Normally a monograph of the Food Chemicals Codex[34] is considered the applicable basis for assessment of purity.

All purity specifications normally list identity tests, an assay, further criteria like loss on drying, general maximum levels for heavy metals and substance-specific limits of byproducts.

While the above EU Directive lists the criteria only, the JECFA specifications refer to analytical methods in a Guide to Specifications[35] or describe the analytical method in the specification monograph itself. Similarly, the Food Chemicals Codex lists and describes the necessary analytical methods for purity control.

10.9 Analytical methods

As in many countries where maximum levels are set for the use of intense sweeteners in foods, analytical methods for identification and quantitative determination are required. Although a wide variety of methods is described, the European Standards for sweetener analysis deserve special attention. They were based on the most widely used analytical methods available and also demonstrated applicability and reliability in collaborative studies in several laboratories.

Two main standards were developed, one for the sweeteners showing UV absorption, i.e. acesulfame, aspartame and saccharin,[36] and a separate standard for cyclamate[37] which does not show a substantial UV absorption. Both standards use HPLC and propose specific sample preparation procedures. HPLC is also the suitable technique for the quantitative determination of sucralose.[38]

10.10 Outlook

The extensive safety studies necessary for food additive approval and the high cost of such studies are more or less prohibitive for the development of new sweeteners. Although some intensely sweet plant constituents have been considered for development, apparently only one new sweetener has undergone a full safety evaluation program and may have a chance of being approved in the foreseeable future, the aspartame derivative neotame, N-[N-(3,3-dimethylbutyl)-α-L-aspartyl]-L-phenyl alanine methyl ester.[39] Petitions for international evaluation and national approvals have been filed and are apparently under evaluation. Among the bulk sweeteners erythritol is a newcomer which is distinguished by a lower calorific value than other bulk sweeteners.

While the market for intense sweeteners is substantial it seems questionable whether a newcomer could expand the total market size instead of cannibalising markets of existing sweeteners. Similar considerations apply for bulk sweeteners with their specific fields of applications and higher cost than carbohydrates. Any new development would have to compete with the established sweeteners and new products would have to earn the substantial cost of development first which would be very difficult. It seems therefore unlikely that more than very few new developments, if any more than those mentioned before, will be seen in the foreseeable future.

10.11 Summary

Food additives in general and sweeteners in particular are extensively tested for their safety before approvals are granted. Safety studies follow some general guidelines and rules, but are specifically adapted to the properties of the single products. For the common intense sweeteners the acceptable daily intake levels as allocated by international agencies and national authorities are in most cases fully sufficient to cover common food habits.

Intense and bulk sweeteners are endorsed by international agencies and approved in a large number of countries. Acesulfame K, aspartame and saccharin are available as sweeteners in the EU and Europe while sucralose is approved in the USA and due for approval in Europe and cyclamate, neohesperidin dihydrochalcone and thaumatin are available in Europe. As bulk sweeteners isomalt, lactitol, maltitol, mannitol, sorbitol and xylitol are commonly available.

10.12 Bibliography

BIRCH G G (Ed.), *LFRA Ingredients Handbook – Sweeteners*, Leatherhead, Leatherhead Publishing, 2000.
CORTI A (Ed.), *Low-calorie Sweeteners: Present and Future*, Basle, Karger, 1999.
MAYER D G and KEMPER F H (Eds), *Acesulfame K*, New York, Marcel Dekker 1991.
O'BRIEN NABORS I (Ed), *Alternative Sweeteners*, New York, Marcel Dekker, 3rd Edn, 2001.
STEGINK L D and FILER L J (Eds), *Aspartame – Physiology and Biochemistry*, New York, Marcel Dekker, 1984.

TSCHANZ C, BUTCHKO H H, STARGEL W W and KOTSONIS F N (Eds), *The Clinical Evaluation of a Food Additive – Assessment of Aspartame*, Boca Raton, CRC Press, 1996.

10.13 References

1. Council Directive No. 89/107/EEC on the approximation of the laws of the Member States concerning additives for use in foods intended for human consumption, *Official Journal of the European Communities* 1989 L40 27–32.
2. Council Directive No. 90/496/EEC on nutrition labelling for foodstuffs, *Official Journal of the European Communities* 1990 L276 40–44.
3. VON RYMON LIPINSKI, G-W, 'Which sweetener for which product?', *Food Marketing Technol* 1991 5(2) 5, 8–11.
4. VON RYMON LIPINSKI, G-W, 'Multiple sweeteners', *Food Marketing Technol* 1990 4(5) 20–22.
5. *Principles for the Safety Assessment of Food Additives and Contaminants in Food*, Environmental Health Criteria 70, Geneva, World Health Organization, 1987.
6. *Toxicological Principles for the Safety Assessment of Direct Food Additives and Color Additives Used in Food*, Washington, US Food and Drug Administration, 1982 and draft revision 1993.
7. MAYER D G and KEMPER F H (Eds), *Acesulfame K*, New York, Marcel Dekker, 1991.
8. *Toxicological evaluation of certain food additives and contaminants*, WHO Food Additives Series No. 18, 12–14, Geneva, World Health Organization, 1983.
9. *Report of the Scientific Committee for Foods No. 16, Sweeteners*, Luxembourg, Commission of the European Communities, 1985.
10. *Toxicological evaluation of certain food additives and contaminants*, WHO Food Additives Series No. 28, 183–218, Geneva, World Health Organization, 1993.
11. SCIENTIFIC COMMITTEE ON FOOD, *Re-evaluation of Acesulfame K*, SCF/CS/ADD/EDUL/194, Brussels, European Commission, 2000.
12. *Toxicological evaluation of certain food additives and contaminants*, WHO Food Additives Series No. 37, 9–11, Geneva, World Health Organization, 1996.
13. STEGINK L D and FILER L J (Eds), *Aspartame – Physiology and Biochemistry*, New York, Marcel Dekker, 1984.
14. TSCHANZ C, BUTCHKO H H, STARGEL W W and KOTSONIS F N (Eds), *The Clinical Evaluation of a Food Additive – Assessment of Aspartame*, Boca Raton, CRC Press, 1996.
15. *Toxicological evaluation of certain food additives*, WHO Food Additives Series No. 16, 28–32, Geneva, World Health Organization, 1981.
16. *Fed Reg* 1981 46(142) 38285–38308.

17. *Toxicological evaluation of certain food additives*, WHO Food Additives Series No. 17, 66–81, Geneva, World Health Organization, 1981.
18. SCIENTIFIC COMMITTEE ON FOOD, *Revised opinion on cyclamic acid and its sodium and calcium salts*, SCF/CS/ADD/EDUL/192, Brussels, European Commission, 2000.
19. SCHOENIG G P, GOLDENTHAL E I, GEIL R G, FRIT C H, RICHTER W R and CARLBORG F W, Evaluation of the dose response and in utero exposure to saccharin in the rat, *Food Chem Toxicol* 1985 23 475–90.
20. Saccharin – Current status, *Food Chem Toxicol* 1985 23 543–6.
21. *Toxicological evaluation of certain food additives and contaminants*, WHO Food Additives Series No. 32, 105–33, Geneva, World Health Organization, 1993.
22. SCIENTIFIC COMMITTEE ON FOOD, *Opinion on Saccharin and its sodium, potassium and calcium salts*, SCF/ADD/EDUL/148, Brussels, European Commission, 1997.
23. SCIENTIFIC COMMITTEE ON FOOD, *Opinion on Stevioside as a sweetener*, SCF/CS/ADD/EDUL/167, Brussels, European Commission, 1999.
24. SCIENTIFIC COMMITTEE ON FOOD, *Opinion on Stevia Rebaudiana Bertoni plants and leaves*, SCF/NF/STE/3, Brussels, European Commission, 1999.
25. *Toxicological evaluation of certain food additives and contaminants*, WHO Food Additives Series No. 28, 219–28, Geneva, World Health Organization, 1991.
26. SCIENTIFIC COMMITTEE ON FOOD, *Opinion of the Scientific Committee on Food on sucralose*, SCF/CS/ADDS/EDUL/190, Brussels, European Commission, 2000.
27. *Toxicological evaluation of certain food additives and contaminants*, WHO Food Additives Series No. 20, 239–50, Geneva, World Health Organization, 1984.
28. *Toxicological evaluation of certain food additives and contaminants*, WHO Food Additives Series No. 20, 207–37, Geneva, World Health Organization, 1984.
29. WILSON L A, WILKINSON K, CREWS H A, DAVIES A M, DICK C S and DUMSDAY V L, Urinary monitoring of saccharin and acesulfame K as biomarkers of exposure to food additives, *Food Additives Contaminants* 1999 16(6) 227–38.
30. European Parliament and Council Directive No. 94/35/EC on sweeteners for use in foodstuffs, *Official Journal of the European Communities* 1994 L237 3–12 and European Parliament and Council Directive No. 96/83/EC amending directive 94/35/EC on sweeteners for use in foodstuffs, *Official Journal of the European Communities* 1996 L48 16–19.
31. Commission Directive No. 95/31/EC laying down special purity criteria concerning sweeteners for use in foodstuffs, *Official Journal of the European Communities* 1995 L178 1–19 with subsequent amendments.
32. *Code of Federal Regulations* Title 21 Part 172, Washington, National Archives and Records Administration.

33. *Compendium of Food Additive Specifications*, FAO Food and Nutrition Paper No. 52, Rome, FAO, 1992 with several additions.
34. *Food Chemicals Codex*, 4th Edn. Washington, National Academy Press, 1996 with supplements.
35. *Guide to Specifications*, FAO Food and Nutrition Paper No. 5 Rev. 2, Rome, FAO, 1991.
36. *CEN Standard 12856*, 1999.
37. *CEN Standard 12857*, 1999.
38. QUINLAN M E and JENNER M R, Analysis and stability of the sweetener sucralose in beverages, *J Food Sci* 1990 55(1) 244–6.
39. WITT J, Discovery and development of neotame, in Corti A. (Ed.), *Low-calorie Sweeteners: Present and Future*, Basle, Karger, 1999.

11

Food additives, other than colours and sweeteners

Brian Whitehouse, Consultant, UK

11.1 Introduction: classifiying the range of additives

Classification of substances often results in some unsatisfactory results, and food additives are no exception. In the evaluation of food additives, functional use is a helpful way of categorising additives. Functional use and level of incorporation are important considerations in assessing the likely intake by consumers. Use and intake level are clearly linked, as the functional effect depends on the amount added. The level of use may be self-limiting by organoleptic considerations. However, whilst it is logical to have functional use as a classification, many additives have more than one functional use, and some could be described as having multiple uses. International documentation differs in its treatment of functional uses. The European Council legislation has three broad classifications:

- Sweeteners (Directive 94/35/EC)
- Colours (Directive 94/36/EC)
- Additives, other than Sweeteners and Colours (Directive 95/2/EC) – Miscellaneous Additives' Directive

In the latter Directive, except for antioxidants and preservatives, individual uses are not listed in either the basic Directive or in the Directives on purity criteria (specifications). In the EU, the 'functional use' of the main additives permitted for use in processed foods are described in Directive 89/107/EC. There are 23 listed which are shown in Table 11.1. In this chapter we are concerned with 17 categories listed in the Directive 95/2/EC 'on food additives other than colours and sweeteners' (these are indicated by an asterisk in Table 11.1).

Table 11.1 EU food additive categories

EU categories of food additives	Definition
Acid*	Substance which increases the acidity of a foodstuff and/or imparts a sour taste to it
Acidity regulator*	Substance which alters or controls the acidity or alkalinity of a foodstuff
Anti-caking agent*	Substance which reduces the tendency of individual particles of a foodstuff to adhere to one another
Anti-foaming agent*	Substance which prevents or reduces foaming
Antioxidant	Substance which prolongs the shelf life of foodstuffs by protecting them against deterioration caused by oxidation, such as fat rancidity and colour changes
Bulking agent*	Substance which contributes to volume of a foodstuff without contributing significantly to its available energy value
Carriers including carrier solvents	These are used to dissolve, dilute or otherwise physically modify a food additive without altering its technological function (and without exerting any technological effect themselves) in order to facilitate its handling, application or use
Colour	Substance which adds or restores colour in a food, and includes natural constituents of foodstuffs and natural sources which are not consumed as foodstuffs as such and not normally used as characteristic ingredients of food
Emulsifier*	Substance which makes it possible to form or maintain a homogenous mixture of two or more immiscible phases such as oil and water in a foodstuff
Emulsifying salt*	Substance which converts proteins contained in cheese into a dispersed form and thereby brings about homogenous distribution of fat and other components
Firming agent*	Substance which makes or keeps tissues of fruit or vegetables firm or crisp, or interacts with gelling agents to produce or strengthen a gel
Flavour enhancer*	Substance which enhances the existing taste and/or odour of a foodstuff
Flour treatment agent (FTA)	Substance (other than emulsifiers) which are added to flour or dough to improve baking quality
Glazing agent	Substance which, when applied to the external surface of a foodstuff, imparts a shiny appearance or provides a protective coating
Humectant*	Substance which prevents foodstuffs drying out by counteracting the effect of an atmosphere having a low degree of humidity, or promote the dissolution of a powder in an aqueous medium
Modified starch	Substance obtained by one or more chemical treatments of edible starches, which may have undergone a physical or enzymatic treatment, and may be acid or alkali thinned or bleached
Packaging gas*	Gas, other than air, introduced into a container before, during or after the placing of a foodstuff in that container

Table 11.1 *continued*

EU categories of food additives	Definition
Propellant gas*	Gas, other than air, which expels a foodstuff from a container
Raising agent*	Substance or combinations of substances which liberate gas and thereby increase the volume of a dough or a batter
Sequestrant*	Substance which forms chemical complexes with metallic ions
'Stabilizers'*	Substance which makes it possible to maintain the physico-chemical state of a foodstuff; stabilizers include substances which enable maintenance of a homogenous dispersion of two or more immiscible substances in a foodstuff and include also substances which stabilize, retain or intensify an existing colour of a foodstuff
Sweetener	Substance which imparts a sweet taste to foodstuffs
Thickener	Substance which increases the viscosity of a foodstuff

* Indicates additive discussed in this chapter. Other additives discussed in this chapter but not defined above are adjuvants, foam stabilizers and foaming agents. These are described in Table 11.2. For historical reasons 'Modified starch' is an exceptional category, because all the other categories are functional uses, whereas modified starches are a group of substances.

The EU system for categorising additives by functional use contains some inconsistencies. As an example, no raising agents are listed but sodium bicarbonate, recognised as a raising agent, is listed in the category: acidity regulators. 'Acid' and 'Acidity regulator' are both 'acidity regulators' but are listed in two separate categories. Acetic acid is an acidity regulator and a preservative but is only listed in the former category. In contrast, polyols are in both the Sweeteners and the Miscellaneous Additives Directives, because of their multifunctionality as a humectant and bulking agent as well as sweetener. There are several additives which do not fall into any of the 23 EU categories, such as colour adjuvants, foaming agents and foam stabilizers.

The International Numbering System (INS) for food additives of the Codex Alimentarius (Volume 1A 1995) lists as 'functional classes' what the EU calls 'categories' and what JECFA calls 'functional uses'. The INS was based on the EU procedure for E numbers allocated for ease of identification of additives permitted for use in the European Union. Mostly the two numbering systems use the identical number for the same substance, even if the name of the substance may follow a different convention, such as pyrophosphate/diphosphate, glycerol triacetate/triacetin. Over the years, for carotenes and tocopherols, deviations in the numbering have developed. With a few exceptions, the list of functional classes of the INS is identical with the EU list of categories. The exceptions are that the INS includes Colour retention agent and Foaming agent, which the EU list excludes, and the INS does not include Packaging gas or Sequestrant as functional classes. However, the INS table of functional classes includes sub-classes where, for example, 'sequestrant' can be found as a sub-class of emulsifying salt.

Table 11.2 EU, JECFA and INS classification of additives

EU categories of food additives	JECFA functional uses (includes)	INS functional classes/sub-class
Acid	Acids	Acid/acidifier
Acidity regulator	Acidity regulators (alkali, buffer, buffering agent, neutralizing agent)	Acidity regulator/acid, alkali, base, buffer, buffering agent, pH adjusting agent
	Adjuvants (density adjustment for flavouring oils in beverages, diluent for colours and other food additives, encapsulating agent, excipient)	
	Adsorbants (decolourizing agent)	
Anti-caking agent	Anti-caking agent (drying or dusting agent)	Anti-caking agent/antistick agent, drying agent, dusting agent, dusting powder, release agent
Anti-foaming agent	Anti-foaming agent (defoaming agent)	Anti-foaming agent
Antioxidant	Antioxidant	Antioxidant/antioxidant synergist, sequestrant
Bulking agent	Bulking agents (component of chewing gum base, filler)	Bulking agent/filler
	Carrier solvents (carrier, carrier for flavour)	
	Clouding agent (cloud producing agent)	
Colour	Colours (decorative pigment, food colour, surface colourant)	Colour
Emulsifier	Emulsifiers (antispattering agent, dispersing agent, plasticizer, suspension agent)	Emulsifier/plasticizer, dispersing agent, surface active agent, surfactant, wetting agent
Emulsifying salt		Emulsifying salt/melding salt, sequestrant
	Enzyme preparations* (animal, microbial and plant origins)	
	Extraction solvents*	
	Filtering aids* (clarifying agent, precipitation agent)	
Firming agent	Firming agents	Firming agent
Flavour enhancer	Flavour enhancers (salt substitute, seasoning agent)	Flavour enhancer/flavour modifier, tenderizer
	Flavouring agent (flavouring adjunct, flavour)	

Flour treatment agent	Flour treatment agents (bleaching agent, dough conditioner, dough strengthening agent, flour bleaching agent, oxidizing agent)	Flour treatment agent/bleaching agent, dough improver, flour improver
	Foaming agents	Foaming agent/whipping agent, aerating agent
	Freezing agents* (cryogenic freezant, liquid freezant)	
Gelling agent	Gelling agents	Gelling agents
Glazing agent	Glazing agents (film coating, protective coating, surface finishing/treatment agent)	Glazing agent/coating, sealing agent, polish
Humectant	Humectants (moisture retaining agent, wetting agent)	Humectant/moisture/water retention agent, wetting agent
Modified starch		
	Nutrient supplements	
Preservative	Preservatives (antibrowning agent, antimicrobial preservative, antimould/antirope agent, fumigant, fungistatic agent, sterilizing agent)	Preservative/antimicrobial preservative, antimycotic agent, bacteriophage control agent, chemosterilant/wine maturing agent, disinfection agent
Packaging gas		
Propellant gas	Propellants (packing/packaging gas)	Propellant
Raising agent	Raising agents (leavening agent)	Raising agent/leavening agent
Sequestrant	Sequestrants	
Stabilizer	Stabilizers (colloidal/emulsion stabilizers)	Stabilizer/binder, firming agent, moisture/water retention agent, foam stabilizer
Sweetener	Sweeteners (sweetening agent)	Sweetener/artificial sweetener, nutritive sweetener
	Synergists (antimicrobial synergist, synergist and solubilizers for antioxidants and flavours)	
Thickener	Thickeners (binder, binding agent, texturizing agent)	Thickener/texturizer, bodying agent

* Generally regarded as processing aids rather than additives.

The Joint FAO/WHO Expert Committee on Food Additives (JECFA) includes 'Functional Use' in the specifications published in Food and Nutrition Paper 52 and its Addenda. JECFA includes an indication of the functional use or uses, as part of its specifications of purity of additives. JECFA has developed these as part of the description of the additive, and although JECFA provides advice on specifications to the Codex Committee on Food Additives and Contaminants, the uses quoted in those specifications are not necessarily the same as the functions given for the same substance in the Codex INS system. Frequently more than one use is listed and these uses often refer to countries outside the EU, and may not include the reason for use listed by the EU. The various categories, uses or classes of additive used by the EU, INS and JECFA are listed in Table 11.2. With minor exceptions, the JECFA functional uses marked with an asterisk are generally regarded as processing aids and not additives and therefore outside the scope of this volume.

11.2 The regulatory background

Food additives in the EU are regulated by a Framework Directive 89/107, which among other things defines food additives and processing aids (to which the Directive does not apply). Annex I lists the categories referred to above. (For reasons which are not obvious, these are not listed in any recognizable order!). In annex II, the very important 'three legs' on which food additives stand are listed:

- technological need
- no hazard to the health of the consumer
- no deception of the consumer.

Firstly, the case of need is described as 'demonstrable advantages of benefit to the consumer' under four categories:

- preservation of nutritional quality
- provision of constituents for foods for special dietary needs
- enhancement of keeping quality
- as aids to manufacture, processing, preparation, treatment, packing transport or storage of food.

Secondly, to ensure that there is no hazard to health, food additives must be subjected to appropriate toxicological testing and evaluation, taking into account cumulative, synergistic or potentiating effects of use and possible intolerance. This process results in establishing an 'Acceptable Daily Intake' (ADI) (see next section). This screening process includes potential re-evaluation to be carried out in the light of changing conditions of use or new scientific information. Food additives must also comply with the approved criteria of purity (i.e. the approved specification). Sources for specifications for some of the main additives used in foods are set out in Appendix 1 (at the end of this chapter). Approval of an

Food additives, other than colours and sweeteners 255

additive for use involves establishing a specification defining the additive and governing purity, specifying the foodstuffs to which the additives may be added, limiting its use to the lowest level needed to achieve the desired effect, and taking into account the ADI, including special groups of consumers. The third 'leg', relates to enhancing keeping quality or stability of a food, or improving organoleptic qualities, without changing the nature, substance or quality of the food as to deceive consumers.

11.3 Acceptable daily intake (ADI)

ADI is defined as an estimate of the amount of an additive in a food or beverage, expressed on a body weight basis, that can be ingested daily over a lifetime without appreciable risk. The ADI is listed in units of mg/kg of body weight. When a food additive is toxicologically tested and evaluated, a 'no-effect level' is established, usually in experimental animals, meaning that at that dose level, no adverse effects were to be found. To account for differences in effect on animals and humans, and to allow for the fact that an additive may be consumed over a lifetime, a safety factor is then applied, typically 100. Thus, if a no-effect level of 10 g per kg of body weight is established, this value is divided by 100 giving 0.1 g per kg body weight which will be called the ADI for that substance. This implies that a person weighing 60 kg could safely consume 6 g of this additive per day for a lifetime. Where the no-effect level of an additive is very high and the likely consumption is such that no hazard can be foreseen, instead of giving a very high value to the ADI, it is usual for the phrase, ADI 'Not specified' to be used. This term applies to 122 of the 240 additives being considered in this chapter (see Table 11.3). Of the functional uses of the additives dealt with in this chapter, preservatives, which by their very function, are biologically active, have the largest proportion of substances with a numerical ADI.

The EU Directive 95/2/EC which covers the additives in this chapter limits the amount of certain additives to particular uses or quantities, usually based on the ADI. It also has special limits for foods intended for products for young children. For example, modified starches listed in Part 3 of Annex VI, although ADI 'Not specified', are restricted to 50 g/kg in weaning foods for infants and young children. In the past a series of different and sometimes confusing expressions such as ADI 'Not limited' or 'Acceptable' were used instead of 'ADI Not specified', and older texts must, therefore, be read with caution! Appendices 2 and 3 list the ADI allocated by the Scientific Committee for Food in the EU and by JECFA at an international level to the additives in this chapter.

11.4 JECFA safety evaluation

The process of evaluation starts at a meeting of the Codex Alimentarius Committee on Food Additives and Contaminants (CCFAC). The Committee,

Table 11.3 EU-approved additives with a numerical ADI and with ADI 'not specified'*

Category	Total number	With numerical ADI	ADI 'Not specified'
Acid	13	5	8
Acidity regulator	50	17	33
Anti-caking agent	15	8	7
Anti-foaming agent	2	2	—
Bulking agent	1	—	1
Emulsifier	34	22	12
Emulsifying salt	8	8	—
Firming agent	2	2	—
Flavour enhancer	18	—	18
Flour treatment agent**	6	6	—
Glazing agent	6	1	5
Humectant	3	1	2
Packaging gas	3	—	3
Preservative	46	31	15
Propellant gas	3	—	3
Raising agent	2	1	1
Sequestrant	5	4	1
Stabilizer	7	6	1
Others:			
Colour adjuvants, retention agents	5	2	3
Foaming agents, stabilizers	4	3	1
Total	233	119	114

Notes
* The complete EU list includes 364 additives.
** Although flour treatment agents are a listed category in Directive 87/107, and some have been evaluated by SCF, they are not regulated at present by the EU.

which consists of about 175 delegations from Codex Alimentarius Commission (CAC) member states, meets to consider matters relevant to food additives and contaminants, and any member state may propose that an additive be included on the agenda of JECFA. The member state making the proposal undertakes to ensure that the toxicological, specification and likely consumption data are made available to JECFA. It is likely that an additive manufacturer from that member state will provide this data. It is also likely that the member state government itself has evaluated the product in terms of the 'three legs' on which the additive rests before proposing the JECFA evaluation. These 'legs' are:

- technological need
- no hazard to the health of the consumer
- no deception of the consumer.

If the CCFAC agrees with the proposal, the additive is included on a future JECFA agenda. Prior to the next JECFA meeting, a call for data is published, and any item put forward for inclusion will be in that data call, requesting all available toxicological, specification and consumption data.

The specification is an integral part of the evaluation. It will include manufacturing and raw material details, as well as the functional use, levels of contaminants which are likely to be present if they are of concern, remembering always the product specified must represent the product which has been toxicologically tested. This is why the raw materials and methods of manufacture are required as part of the detail to be supplied. In view of the importance of both raw materials and methods of manufacture to the specification, it is curious that this detail is frequently omitted from the published specification. A change of either may affect the product which may not represent the material toxicologically tested. In this century, openness in relation to things incorporated into food products is clearly demanded by consumers, and there cannot be secret ingredients, or, except for proprietary aspects of a process, methods of manufacture which cannot be briefly described. It is commonplace to do so in specifications for products derived from plant sources, for example, but not currently with synthetic products. The specification may contain criteria governing composition and how this can be measured to ensure that all such additives used are of uniform quality.

The toxicological data will be sent to a WHO expert for evaluation, with a second expert appointed to comment on the report prepared for submission to the Expert Committee. The specification will be written based on information submitted by an FAO expert, and another FAO expert will consider the likely consumption. When the Expert Committee meets, these three groups of people discuss the aspects of the toxicology, specification and likely consumption before an additive is approved and an ADI is allocated.

11.5 Summary

This chapter concludes with three appendices summarising regulations covering additives both at EU and international level. These appendices cover:

- source details for additive specifications (Appendix 1)
- ADI values and source details organised by additive category (Appendix 2) and by additive name (Appendix 3).

Appendix 1: Food additive specifications EU, JECFA, FCC and JSSFA

Category	E No	Sub	Additive name	Directive	JECFA Specn	FCC IV	JSSFA
Preservative	260		Acetic acid	96/77	FNP 52	p 10	p 102
Emulsifier	472	a	Acetic esters of fatty acid glycerides	98/86	FNP 52	p 12	–
Acid	355		Adipic acid	2000/63	FNP 52 Add 7	p 16	p 104
Acidity Regulator	523		Aluminium ammonium sulphate	2000/63	FNP 52	p 20	p 108
Acidity Regulator	522		Aluminium potasssium sulphate	2000/63	FNP 52 Add 8	p 21	p 108
Anti-caking Agent	559		Aluminium silicate (Kaolin)	2001/30	FNP 52	p 208	–
Firming Agent	521		Aluminium sodium sulphate	2000/63		p 21	–
Firming Agent	520		Aluminium sulphate	2000/63	FNP 52	p 22	–
Acidity Regulator	503	(i)	Ammonium carbonate	2000/63	FNP 52	p 25	p 109
Acidity Regulator	503	(ii)	Ammonium hydrogen carbonate	2000/63	FNP 52	p 25	p 109
Acidity Regulator	527		Ammonium hydroxide	2000/63	FNP 52 Add 4	p 26	p 109
Emulsifier	442		Ammonium phosphatides	98/86	FNP 52 Add 8	–	–
FTA	517		Ammonium sulphate	2000/63	FNP 52	p 28	p 111
Packaging Gas	938		Argon	2000/63	FNP 52 Add 7	–	–
Glazing Agent	901		Bees wax, white and yellow	2000/63	FNP 52 Add 1	p 41	p 118
Anti-caking Agent	558		Bentonite	2001/30	FNP 52	p 41	–
Preservative	210		Benzoic acid	96/77	FNP 52 Add 4	p 43	p 119
Preservative	230		Biphenyl	96/77	FNP 52	–	p 161
Preservative	284		Boric acid	96/77	FNP 52	–	–
Flavour Enhancer	634		Calcium 5'-ribonucleotides	2001/30	FNP 52	–	p 134
Preservative	263		Calcium acetate	96/77	FNP 52	p 53	–
Anti-caking Agent	556		Calcium aluminium silicate	2001/30	FNP 52	p 53	–
Preservative	213		Calcium benzoate	96/77	FNP 52 Add 4	–	–
Flavour Enhancer	623		Calcium diglutamate	2001/30	FNP 52	–	–
Emulsifying salt	450	(vii)	Calcium dihydrogen diphosphate	98/86	FNP 52	p 53	p 129
Sequestrant	385		Calcium disodium EDTA	96/77	FNP 52	p 59	p 130
Anti-caking Agent	538		Calcium ferrocyanide	2000/63	–	–	–
Acidity regulator	578		Calcium gluconate	2000/63	FNP 52 Add 6	p 60	p 130

Function	Number		Name	Year	FNP	Page
Flavour Enhancer	629		Calcium guanylate	2001/30	FNP 52	–
Preservative	227		Calcium hydrogen (bi) sulphite	96/77	FNP 52 Add 6	–
Acidity Regulator	352	(ii)	Calcium hydrogen malate	2000/63	FNP 52	–
Acidity Regulator	526		Calcium hydroxide	2000/63	FNP 52	p 61
Flavour Enhancer	633		Calcium inosinate	2001/30	FNP 52	–
Acidity Regulator	327		Calcium lactate	96/77	FNP 52	p 62
Acidity Regulator	352	(i)	Calcium malate	2000/63	FNP 52	–
Acidity Regulator	529		Calcium oxide	2000/63	FNP 52	p 64
Emulsifying salt	452	(iv)	Calcium polyphosphate	98/86	FNP 52	–
Preservative	282		Calcium propionate	96/77	FNP 52 Add 6	p 70
Anti-caking Agent	552		Calcium silicate	2000/63	FNP 52	p 11 1st supp
Preservative	203		Calcium sorbate	96/77	FNP 52 Add 6	p 73
Emulsifier	482		Calcium stearoyl-2-lactylate	98/86	FNP 52 Add 4	p 74
FTA	516		Calcium sulphate	2000/63	FNP 52	p 76
Preservative	226		Calcium sulphite	96/77	FNP 52	–
Acidity Regulator	354		Calcium tartrate	2001/30	FNP 52	–
Glazing Agent	902		Candelilla wax	2000/63	FNP 52	p 77
FTA	927	b	Carbamide	2000/63	FNP 52	p 428
Propellant	290		Carbon dioxide	96/77	FNP 52 Add 5	p 12 1st supp
Glazing Agent	903		Carnauba wax	2000/63	FNP 52 Add 6	p 90
Acid	330		Citric acid	96/77	FNP 52 Add 7	p 102
Emulsifier	472	c	Citric esters of fatty acid glycerides	98/86	FNP 52	–
Stabilizer	468		Crosslinked Na carboxy methyl cellulose	2000/63	–	–
Adjuvant	459		Cyclodextrin, beta-	2000/63	FNP 52 Add 3	p 15 1st supp
FTA	920		Cysteine (L-)	2000/63	–	p 113
Acidity Regulator	333	(ii)	Dicalcium citrate	96/77	FNP 52	p 58
Acidity Regulator	450	(vi)	Dicalcium diphosphate	98/86	FNP 52	p 70
Acidity Regulator	341	(ii)	Dicalcium phosphate	96/77	FNP 52	p 8 1st supp
Acidity Regulator	343	(ii)	Dimagnesium phosphate	2000/63	FNP 52	p 233
Preservative	242		Dimethyl dicarbonate	96/77	FNP 52	p 19 1st supp
Anti-foaming Agent	900		Dimethyl polysiloxane	2000/63	FNP 52	p 123
Flavour Enhancer	628		Dipotassium guanylate	2001/30	FNP 52	p 126
Flavour Enhancer	632		Dipotassium inosinate	2001/30	FNP 52	p 127
Acidity Regulator	340	(ii)	Dipotassium phosphate	96/77	FNP 52	p 38 1st supp

–	–
–	–
p 131	–
p 132	–
p 134	–
p 135	p 136
–	–
p 137	–
p 141	p 141
p 149	–
–	–
p 153	p 155
p 128	p 129
p 133	–
–	–
–	–
–	–

Appendix 1 continued

Category	E No	Sub	Additive name	Directive	JECFA Specn	FCC IV	JSSFA
Stabilizer	336	(ii)	Dipotassium tartrate	96/77	FNP 52	–	–
Flavour Enhancer	635		Disodium 5'-ribonucleotides	2001/30	FNP 52	p 339	p 168
Acidity Regulator	331	(ii)	Disodium citrate	96/77	FNP 52	p 350	–
Raising agent	450	(i)	Disodium diphosphate	98/86	FNP 52 Add 2	p 39 1st supp	p 163
Flavour Enhancer	627		Disodium guanylate	2001/30	FNP 52 Add 2	p 126	p 166
Flavour Enhancer	631		Disodium inosinate	2001/30	FNP 52 Add 2	p 127	p 167
Acidity Regulator	339	(ii)	Disodium phosphate	96/77	FNP 52	p 374	p 166
Stabilizer	335	(ii)	Disodium tartrate	96/77	FNP 52	p 384	p 170
Glazing Agent	469		Enz. hydrolysed carboxy methyl cellulose	2000/63	FNP 52	–	–
Preservative	214		Ethyl p-hydroxybenzoate	96/77	FNP 52 Add 6	p 395	p 177
Foam Stabilizer	570		Fatty acids	2000/63	FNP 52	p 153	–
Colour Retn. Aid	579		Ferrous gluconate	98/86	FNP 52 Add 7	p 154	p 183
Colour Retn. Aid	585		Ferrous lactate	98/86	FNP 52	p 164	–
Acid	297		Fumaric acid	2000/63	FNP 52 Add 7	–	p 201
Acid	574		Gluconic acid	2000/63	FNP 52	p 172	p 203
Acid	575		Glucono-delta-lactone	2000/63	FNP 52 Add 6	p 22 1st supp	p 203
Flavour Enhancer	620		Glutamic acid	2001/30	FNP 52	p 176	p 204
Humectant	422		Glycerol	98/86	FNP 52	p 177	p 205
Emulsifier	445		Glycerol esters of wood rosin	98/86	FNP 52 Add 3	p 424	–
Humectant	1518		Glyceryl triacetate (triacetin)	2000/63	FNP 52 Add 4	p 186	p 206
Flavour Enhancer	640		Glycine and its sodium salt	2000/63	–		
Flavour Enhancer	626		Guanylic acid	2001/30	FNP 52	p 189	–
Packaging Gas	939		Helium	2000/63	FNP 52 Add 7		
Preservative	239		Hexamethylene tetramine	96/77	FNP 52	p 194	p 211
Acid	507		Hydrochloric acid	2000/63	FNP 52		
Flavour Enhancer	630		Inosinic acid	2001/30	FNP 52	p 181	–
Emulsifier	472	b	Lact. esters of fatty acid glycerides	98/86	FNP 52 Add 4	p 270	p 255
Acid	270		Lactic acid	96/77	FNP 52	p 220	p 226
Emulsifier	322		Lecithins	96/77	FNP 52 Add 2		

Function	No.		Name	Year	Document	Page	Other
Preservative	1105		Lysozyme	96/77	FNP 52 Add 1	–	p 231
Acidity Regulator	504	(i)	Magnesium carbonate	2001/30	FNP 52	p 230	–
Flavour Enhancer	625		Magnesium diglutamate	2001/30	FNP 52	–	–
Acidity Regulator	528		Magnesium hydroxide	2000/63	FNP 52	p 232	–Anti-
Acidity Regulator	504	(ii)	Magnesium hydroxide carbonate	2001/30	FNP 52	p 233	p 232
caking Agent	530		Magnesium oxide	2000/63	FNP 52	–	–Anti-
Emulsifier	470	b	Magnesium salts of fatty acids	98/86	FNP 52	p 235	–
caking Agent	553	a(i)	Magnesium silicate	2000/63	FNP 52	p 235	–
Anti-caking Agent	553	(a) (i)	Magnesium silicate (synthetic)	2000/63	FNP 52	–	–
Anti-caking Agent	553	(a) (ii)	Magnesium trisilicate	2000/63	FNP 52	p 237	p 233
Acid	296		Malic acid	2001/30	FNP 52 Add 7	–	–
Acid	353		Metatartaric acid	–	–	–	–
Preservative	218		Methyl p-hydroxybenzoate	96/77	FNP 52 Add 3	p 95	p 241
Emulsifier	460	(i)	Micro crystalline cellulose	98/86	FNP 52 Add 8	p 290	–
Glazing Agent	905		Micro-crystalline wax	2001/30	FNP 52 Add 5	–	–
Emulsifier	472	f	Mix. ac./ tart.acid esters of fatty acid glycerides	98/86	FNP 52 Add 6	p 258	p 205
Emulsifier	471		Mono and diglycerides of fatty acids	98/86	FNP 52	p 119	–
Emulsifier	472	e	Mono/diacet. tart. esters of fatty acid glycerides	98/86	FNP 52 Add 6	p 257	–
Flavour Enhancer	624		Monoammonium glutamate	2001/30	FNP 52	–	–
Acidity Regulator	333	(i)	Monocalcium citrate	96/77	FNP 52	p 9 1st supp	p 128
Acidity Regulator	341	(i)	Monocalcium phosphate	96/77	FNP 52 Add 4	–	–
Acidity Regulator	343	(i)	Monomagnesium phosphate	2000/63	FNP 52	p 259	p 244
Acidity Regulator	332	(i)	Monopotassium citrate	96/77	FNP 52	p 39 1st supp	p 245
Flavour Enhancer	622		Monopotassium glutamate	2001/30	FNP 52	–	p 261
Acidity Regulator	340	(i)	Monopotassium phosphate	96/77	FNP 52	p 260	p 246
Acidity Regulator	331	(i)	Monosodium citrate	96/77	FNP 52	p 374	p 289
Flavour Enhancer	621		Monosodium glutamate	2001/30	FNP 52	–	–
Acidity Regulator	339	(i)	Monosodium phosphate	96/77	FNP 52	–	–
Preservative	912		Montan acid esters	2001/30	–	p 263	–
Emulsifier	470	a	Na, K and Ca salts of fatty acids	98/86	FNP 52	p.266	–
Preservative	235		Natamycin (pimaricin)	96/77	FNP 52		
Preservative	234		Nisin	96/77	FNP 52		

Appendix 1 *continued*

Category	E No	Sub	Additive name	Directive	JECFA Specn	FCC IV	JSSFA
Propellant	941		Nitrogen	2000/63	FNP 52 Add 7	p 268	–
Propellant	942		Nitrous oxide	2000/63	FNP 52 Add 8	p 269	–
Preservative	231		Orthophenyl phenol	96/77	FNP 52	–	p 254
Preservative	914		Oxidized polyethylene wax	2001/30	–	–	–
Packaging Gas	948		Oxygen	2000/63	FNP 52 Add 7	–	–
Sequestrant	451	(ii)	Pentapotassium triphosphate	98/86	FNP 52	p 328	–
Sequestrant	451	(i)	Pentasodium triphosphate	98/86	FNP 52	p 41 1st supp	–
Acid	338		Phosphoric acid	96/77	FNP 52 Add 3	p 38 1st supp	p 255
Bulking Agent	1200		Polydextrose	2000/63	FNP 52 Add 3	p 297	–
Anti-foaming Agent	(1521)		Polyethylene glycol 6000 (1521=INS). No E no.	2000/63	FNP 52	p 301	–
Emulsifier	475		Polyglycerol esters of fatty acids	98/86	FNP 52	p 303	p 205
Emulsifier	476		Polyglycerol polyricinoleate	98/86	FNP 52	–	–
Emulsifier	432		Polyoxyethylene sorbitan monolaurate	98/86	FNP 52	p 306	–
Emulsifier	433		Polyoxyethylene sorbitan monooleate	98/86	FNP 52	p 308	–
Emulsifier	434		Polyoxyethylene sorbitan monopalmitate	98/86	FNP 52	–	–
Emulsifier	435		Polyoxyethylene sorbitan monostearate	98/86	FNP 52	p 307	–
Emulsifier	436		Polyoxyethylene sorbitan tristearate	98/86	FNP 52	p 307	–
Emulsifier	431		Polyoxyethylene stearate	98/86	FNP 52	–	–
Colour Retn. Aid	1202		Polyvinylpolypyrrolidone	Draft	FNP 52	p 309	p 258
Stabilizer	1201		Polyvinylpyrrolidone	Draft	FNP 52	p 310	–
Preservative	261		Potassium acetate	96/77	FNP 52	–	–
Acidity Regulator	357		Potassium adipate	2001/30	FNP 52	–	–
Anti-caking Agent	555		Potassium aluminium silicate (Mica)	2001/30	FNP 52 Add 4	p 313	–
Preservative	212		Potassium benzoate	96/77	FNP 52	p 314	p 260
Acidity Regulator	501	(i)	Potassium carbonate	2000/63	FNP 52	–	–
Anti-caking Agent	536		Potassium ferrocyanide	2000/63	FNP 52	p 317	p 262
Acidity Regulator	577		Potassium gluconate	2000/63	FNP 52 Add 6	–	p 262
Preservative	228		Potassium hydrogen (bi) sulphite	96/77	FNP 52	–	–

Acidity Regulator	501	(ii)	Potassium hydrogen carbonate	2000/63	FNP 52	p 313	—
Acidity Regulator	515	(ii)	Potassium hydrogen sulphate	2000/63	FNP 52	—	—
Acidity Regulator	525		Potassium hydroxide	2000/63	FNP 52	p 319	p 262
Acidity Regulator	351		Potassium malate	2000/63	FNP 52	—	—
Preservative	224		Potassium metabisulphite	96/77	FNP 52 Add 7	p 322	p 266
Preservative	252		Potassium nitrate	96/77	FNP 52 Add 3	p 323	p 264
Preservative	249		Potassium nitrite	96/77	FNP 52 Add 3	p 323	—
Emulsifying salt	452	(ii)	Potassium polyphosphate	98/86	FNP 52	p 328	p 265
Preservative	283		Potassium propionate	96/77	FNP 52	—	—
Stabilizer	337		Potassium sodium tartrate	96/77	FNP 52	—	—
Preservative	202		Potassium sorbate	96/77	FNP 52 Add 6	p 327	p 266
Flavour Enhancer	515	(i)	Potassium sulphate	2000/63	FNP 52	p 328	—
Stabilizer	336	(i)	Potassium tartrate	96/77	FNP 52	p 312	p 259
Emulsifier	460	(ii)	Powdered cellulose	98/86	FNP 52	p 96	p 267
Emulsifier	477		Propan-1,2-diol esters of fatty acids	98/86	FNP 52 Add 5	p 322	p 272
Humectant	1520		Propan-1,2-diol	2000/63	FNP 52 Add 5	p 331	p 271
Preservative	280		Propionic acid	96/77	FNP 52 Add 6	p 115 1st supp	p 271
Preservative	216		Propyl p-hydroxybenzoate	96/77	FNP 52 Add 3	—	p 274
Foaming Agent	999		Quillaia extract	2000/63	FNP 52 Add 8	p 347	p 275
Glazing Agent	904		Shellac	2000/63	FNP 52 Add 8	p 348	p 279
Anti-caking Agent	551		Silicon dioxide	2000/63	FNP 52	p 349	p 281
Preservative	262	(i)	Sodium acetate	96/77	FNP 52	—	p 282
Acidity Regulator	356		Sodium adipate	2001/30	FNP 52	p 353	—
FTA	541		Sodium aluminium phosphate, acidic	2000/63	FNP 52	p 351	—
Anti-caking Agent	554		Sodium aluminium silicate	2001/30	FNP 52	p 355	p 283
Preservative	211		Sodium benzoate	96/77	FNP 52 Add 4	—	—
Emulsifying salt	452	(iii)	Sodium calcium polyphosphate	2000/63	FNP 52	—	—
Acidity Regulator	500	(i)	Sodium carbonate	2000/63	FNP 52	p 357	p 285
Preservative	215		Sodium ethyl p-hydroxybenzoate	96/77	FNP 52	—	—Anti-
caking Agent	535		Sodium ferrocyanide	2000/63	FNP 52	p 363	—
Sequestrant	576		Sodium gluconate	2000/63	FNP 52 Add 6	p 363	p 291
Preservative	262	(ii)	Sodium hydrogen acetate (diacetate)	96/77	FNP 52	—	—
Raising Agent	500	(ii)	Sodium hydrogen carbonate	2000/63	FNP 52	p 355	p 284
Preservative	222		Sodium hydrogen (bi) sulphite	96/77	FNP 52 Add 7	p 356	291

Appendix 1 *continued*

Category	E No	Sub	Additive name	Directive	JECFA Specn	FCC IV	JSSFA
Acidity Regulator	350	(ii)	Sodium hydrogen malate	2000/63	FNP 52	—	—
Acidity Regulator	514	(ii)	Sodium hydrogen sulphate	2000/63	FNP 52	p 356	—
Acidity Regulator	524		Sodium hydroxide	2000/63	FNP 52	p 364	p 292
Acidity Regulator	350	(i)	Sodium malate	2000/63	FNP 52	—	p 295
Preservative	223		Sodium metabisulphite	96/77	FNP 52 Add 7	p 369	p 302
Preservative	219		Sodium methyl p-hydroxybenzoate	96/77	FNP 52	—	—
Preservative	251		Sodium nitrate	96/77	FNP 52 Add 3	p 373	p 297
Preservative	250		Sodium nitrite	96/77	FNP 52 Add 3	p 373	p 297
Preservative	232		Sodium orthophenyl phenol	96/77	FNP 52	—	p 299
Emulsifying salt	452	(i) 2	Sodium polyphosphate insoluble	98/86*	FNP 52 Add 4	p 370	p 301
Sequestrant	452	(i) 1	Sodium polyphosphate soluble	98/86	FNP 52 Add 4	p 375	—
Preservative	281		Sodium propionate	96/77	FNP 52 Add 5	p 377	p 301
Preservative	217		Sodium propyl p-hydroxybenzoate	96/77	FNP 52	—	—
Acidity Regulator	500	(iii)	Sodium sesqui carbonate	2000/63	FNP 52	p 380	—
Emulsifier	481		Sodium stearoyl-2-lactylate	98/86	FNP 52 Add 4	p 380	—
Colour Adjuvant	514	(i)	Sodium sulphate	2000/63	FNP 52 Add 8	p 383	p 305
Preservative	221		Sodium sulphite	96/77	FNP 52 Add 7	p 383	p 305
Stabilizer	335	(i)	Sodium tartrate	96/77	FNP 52	p 384	—
Preservative	285		Sodium tetraborate (borax)	96/77	FNP 52	—	—
Preservative	200		Sorbic acid	96/77	FNP 52	p 387	p 306
Emulsifier	493		Sorbitan monolaureate	98/86	FNP 52 Add 8	—	p 306
Emulsifier	494		Sorbitan monooleate	98/86	FNP 52	—	p 306
Emulsifier	495		Sorbitan monopalmitate	98/86	FNP 52	—	p 306
Emulsifier	491		Sorbitan monostearate	98/86	FNP 52	p 387	p 306
Emulsifier	492		Sorbitan tristearate	98/86	FNP 52 Add 1	—	p 306
Colour Retn. Aid	512		Stannous chloride	2000/63	FNP 52	p 394	—
FTA	483		Stearyl tartrate	98/86	FNP 52 Add 4	—	—
Acid	363		Succinic acid	2000/63	FNP 52	p 397	p 308
Emulsifier	474		Sucroglycerides	98/86	FNP 52 Add 5	—	—

Emulsifier	444		Sucrose acetate isobutyrate	98/86	FNP 52 Add 4	–
Emulsifier	473		Sucrose esters of fatty acids	98/86	FNP 52 Add 7	p 44 1st supp
Preservative	220		Sulphur dioxide	96/77	FNP 52 Add 6	p 401
Acid	513		Sulphuric acid	2000/63	FNP 52	p 402
Anti-caking Agent	553	(b)	Talc	2001/30	FNP 52 Add 8	p 404
Emulsifier	472	d	Tart. esters of fatty acid glycerides	98/86	FNP 52	–
Acid	334		Tartaric acid (L(+)-)	96/77	FNP 52 Add 7	p 407
Emulsifying salt	450	(v)	Tetrapotassium diphosphate	98/86	FNP 52	p 326
Emulsifying salt	450	(iii)	Tetrasodium diphosphate	98/86	FNP 52	p 378
Emulsifier	479	b	Therm. ox. soya oil/with fatty acid glycerides	98/86	FNP 52 Add 1	–
Acidity Regulator	380		Triammonium citrate	2000/63	FNP 52	–
Acidity Regulator	333	(iii)	Tricalcium citrate	96/77	FNP 52	p 58
Acidity Regulator	341	(iii)	Tricalcium phosphate	96/77	FNP 52	p 10 1st supp
Foam Stabilizer	1505		Triethyl citrate	2000/63	FNP 52	p 426
Acidity Regulator	332	(ii)	Tripotassium citrate	96/77	FNP 52	p 316
Acidity Regulator	340	(iii)	Tripotassium phosphate	96/77	FNP 52	p 325
Acidity Regulator	331	(iii)	Trisodium citrate	96/77	FNP 52	p 360
Emulsifying salt	450	(ii)	Trisodium diphosphate	98/86	FNP 52	p 375
Acidity Regulator	339	(iii)	Trisodium phosphate	96/77	FNP 52	p 375

p 310	
–	
p 311	
p 311	
p 312	
p 265	
p 302	
–	
p 128	
p 323	
p 323	
p 324	
p 324	
–	
p 325	

Notes to Appendix 1

E No = The number allocated under the EU numbering system. Although the INS number is quoted in the specifications published by JECFA, it does not in general use the numbers to identify additives. Association of the name and number would make for more certain identification and indexing, especially when there are different nomenclatures in use. For example Crosscarmellose= INS468 = E468 = Crosslinked sodium carboxy methyl cellulose and E1520 = Propane1,2 diol = INS1520 = Propylene glycol. An index by INS number would be helpful in the other reference books cited (JECFA FNP 52, FCC and JSSFA).

Sub = the sub classification of the E number.

Additive name = The name used in the EU Directive. The JECFA names and the FCC names are often different from the EU name. Thus in FCC, Monosodium phosphate is Sodium phosphate, monobasic, and found together with the other two sodium phosphates. So depending on which list is consulted, the product has to be sought by name under 'S', 'M', 'D', or 'T'! The EU name Diacetyl tartaric and fatty acid esters of glycerol, refers to the INS (CCFAC) name, Mixed tartaric, acetic and fatty acid esters of glycerol, which is the same product as the JECFA name, Tartaric, acetic and fatty acid esters of glycerol, (mixed). Each is numbered (E)472d! Propane 1,2 diol (EU) is propylene glycol (JECFA and INS). The INS list (published by the Codex Alimentarius Volume 1a 1995) is in both alphabetic and numeric order.

Directive = The number of the relevant EU Directive on purity criteria (specification) for the additive listed.

JECFA Spec. The food additive specifications published in the FAO Food and Nutrition Paper 52, (FNP 52) and its Addenda (covering the years 1992 to 2000) are numbered 'Add' from 1 to 8. The intention is to publish a consolidated volume, as a CD ROM and/or on the World Wide Web.

FCC IV = Food Chemical Codex (US), National Academic Press, Fourth edition and its first supplement ('1st supp') with page number (p ..).

JSSFA = Japan's Specifications and Standards for Food Additives 7th edition, 2000 with page number (p ...).

Appendix 2: Acceptable daily intake (ADI) values and references by additive category

Category	E No	Sub	Additive name	SCF ADI	SCF Report Series	JECFA TRS
Acid	270		Lactic acid	NS	25th p 7 et seq	539, p 23, 1973
Acid	296		Malic acid	NS	25th p 7 et seq	445, p 16, 1970
Acid	297		Fumaric acid	0-6	25th p 7 et seq	789, p 18, 1990
Acid	330		Citric acid	NS	25th p 7 et seq	539, p 19, 1973
Acid	334		Tartaric acid (L(+)-)	0-30	25th p 7 et seq	696, p 28, 1983
Acid	338		Phosphoric acid	0-70	25th p 7 et seq	733, p 11, 1986
Acid	353		Metatartaric acid	Acc	25th p 7 et seq	Not evaluated
Acid	355		Adipic acid	0-30	25th p 7 et seq	789, p 18, 1990
Acid	363		Succinic acid	0-30	5th Series	733, p 11, 1986
Acid	507		Hydrochloric acid	NS	25th p 7 et seq	339, p 16, 1965
Acid	513		Sulfuric acid	NS	25th p 7 et seq	733, p 11, 1986
Acid	574		Gluconic acid	NS	25th p 7 et seq	891, p 43, 1998
Acid	575		Glucono-delta-lactone	NS	25th p 7 et seq	891, p 43, 1998
Acidity Regulator	327		Calcium lactate	NS	25th p 7 et seq	539, p 23, 1973
Acidity Regulator	331	(ii)	Disodium citrate	NS	25th p 7 et seq	539, p 19, 1973
Acidity Regulator	331	(i)	Monosodium citrate	NS	25th p 7 et seq	539, p 19, 1973
Acidity Regulator	331	(iii)	Trisodium citrate	NS	25th p 7 et seq	733, p 12, 1986
Acidity Regulator	332	(i)	Monopotassium citrate	NS	25th p 7 et seq	539, p 19, 1973
Acidity Regulator	332	(ii)	Tripotassium citrate	NS	25th p 7 et seq	Not evaluated
Acidity Regulator	333	(ii)	Dicalcium citrate	NS	25th p 7 et seq	539, p 19, 1973
Acidity Regulator	333	(i)	Monocalcium citrate	NS	25th p 7 et seq	539, p 19, 1973
Acidity Regulator	333	(iii)	Tricalcium citrate	NS	25th p 7 et seq	710, p 19, 1984
Acidity Regulator	339	(ii)	Disodium phosphate	0-70	25th p 7 et seq	733, p 11, 1986
Acidity Regulator	339	(i)	Monosodium phosphate	70	25th p 7 et seq	733, p 11, 1986
Acidity Regulator	339	(iii)	Trisodium phosphate	70	25th p 7 et seq	733, p 11, 1986
Acidity Regulator	340	(ii)	Dipotassium phosphate	70	25th p 7 et seq	733, p 11, 1986
Acidity Regulator	340	(i)	Monopotassium phosphate	70	25th p 7 et seq	733, p 11, 1986
Acidity Regulator	340	(iii)	Tripotassium phosphate	70	25th p 7 et seq	733, p 11, 1986

Function	Code		Name	Reference		
Acidity Regulator	341	(ii)	Dicalcium phosphate	70	25th p 7 et seq	733, p 11, 1986
Acidity Regulator	341	(i)	Monocalcium phosphate	70	25th p 7 et seq	733, p 11, 1986
Acidity Regulator	341	(iii)	Tricalcium phosphate	70	25th p 7 et seq	733, p 11, 1986
Acidity Regulator	343	(ii)	Dimagnesium phosphate	70	25th p 7 et seq	733, p 11, 1986
Acidity Regulator	343	(i)	Monomagnesium phosphate	70	25th p 7 et seq	733, p 11, 1986
Acidity Regulator	350	(ii)	Sodium hydrogen malate	NS	25th p 7 et seq	445, p 16, 1970
Acidity Regulator	350	(i)	Sodium malate	NS	25th p 7 et seq	733, p 11, 1986
Acidity Regulator	351		Potassium malate	NS	25th p 7 et seq	733, p 11, 1986
Acidity Regulator	352	(ii)	Calcium hydrogen malate	NS	25th p 7 et seq	445, p 16, 1970
Acidity Regulator	352	(i)	Calcium malate	NS	25th p 7 et seq	445, p 16, 1970
Acidity Regulator	354		Calcium tartrate	30	25th p 7 et seq	696, p 28, 1983
Acidity Regulator	356		Sodium adipate	30	25th p 7 et seq	789, p 18, 1990
Acidity Regulator	357		Potassium adipate	30	25th p 7 et seq	789, p 18, 1990
Acidity Regulator	380		Triammonium citrate	NS	25th p 7 et seq	733, p 11, 1986
Acidity Regulator	450	(vi)	Dicalcium diphosphate	70	25th p 7 et seq	733, p 11, 1986
Acidity Regulator	500	(i)	Sodium carbonate	NS	25th p 7 et seq	733, p 11, 1986
Acidity Regulator	500	(iii)	Sodium sesqui carbonate	NS	25th p 7 et seq	733, p 11, 1986
Acidity Regulator	501	(i)	Potassium carbonate	NS	25th p 7 et seq	733, p 11, 1986
Acidity Regulator	501	(ii)	Potassium hydrogen carbonate	NS	25th p 7 et seq	733, p 11, 1986
Acidity Regulator	503	(i)	Ammonium carbonate	NS	25th p 7 et seq	733, p 11, 1986
Acidity Regulator	503	(ii)	Ammonium hydrogen carbonate	NS	25th p 7 et seq	733, p 11, 1986
Acidity Regulator	504	(i)	Magnesium carbonate	NS	25th p 7 et seq	733, p 11, 1986
Acidity Regulator	504	(ii)	Magnesium hydroxide carbonate	NS	25th p 7 et seq	733, p 11, 1986
Acidity Regulator	514	(ii)	Sodium hydrogen sulphate	NS	25th p 7 et seq	733, p 11, 1986
Acidity Regulator	515	(ii)	Potassium hydrogen sulphate	NS	25th p 7 et seq	733, p 11, 1986
Acidity Regulator	522		Aluminium potasssium sulphate	7	25th p 8	733, p 11, 1986
Acidity Regulator	523		Aluminium ammonium sulphate	7	25th p 8	733, p 11, 1986
Acidity Regulator	524		Sodium hydroxide	NS	25th p 7 et seq	733, p 11, 1986
Acidity Regulator	525		Potassium hydroxide	NS	25th p 7 et seq	733, p 11, 1986
Acidity Regulator	526		Calcium hydroxide	NS	25th p 7 et seq	733, p 11, 1986
Acidity Regulator	527		Ammonium hydroxide	NS	25th p 7 et seq	733, p 11, 1986
Acidity Regulator	528		Magnesium hydroxide	NS	25th p 7 et seq	733, p 11, 1986
Acidity Regulator	529		Calcium oxide	NS	25th p 7 et seq	733, p 11, 1986
Acidity Regulator	577		Potassium gluconate	NS	25th p 7 et seq	891, p 42, 1998

Appendix 2 *continued*

Category	E No	Sub	Additive name	SCF ADI	SCF Report Series	JECFA TRS
Acidity regulator	578		Calcium gluconate	NS	25th p 7 et seq	891, p 46, 1998
Adjuvant	459		Cyclodextrin-beta	5	41st Series	859, p 28, 1995
Anti-caking Agent	530		Magnesium oxide	NS	5th Series	733, p 11, 1986
Anti-caking Agent	535		Sodium ferrocyanide	0.025	25th p 18	557, p 23, 1974
Anti-caking Agent	536		Potassium ferrocyanide	0.025	25th p 18	557, p 23, 1974
Anti-caking Agent	538		Calcium ferrocyanide	0.025	25th p 8	Not evaluated
Anti-caking Agent	551		Silicon dioxide	NS	24th Series	733, p 11, 1986
Anti-caking Agent	552		Calcium silicate	NS	25th p 7 et seq	733, p 11, 1986
Anti-caking Agent	553	a(i)	Magnesium silicate	NS	25th p 7 et seq	733, p 11, 1986
Anti-caking Agent	553	(a) (i)	Magnesium silicate (synthetic)	NS	25th p 7 et seq	733, p 11, 1986
Anti-caking Agent	553	(a) (ii)	Magnesium trisilicate	NS	25th p 7 et seq	733, p 11, 1986
Anti-caking Agent	553	(b)	Talc	NS	5th Series	751, p 33, 1986
Anti-caking Agent	554		Sodium aluminium silicate	7	25th p 8 et seq	733, p 11, 1986
Anti-caking Agent	555		Potassium aluminium silicate (Mica)	7	25th p 8 et seq	733, p 11, 1986
Anti-caking Agent	556		Calcium aluminium silicate	7	25th p 8 et seq	733, p 11, 1986
Anti-caking Agent	558		Bentonite	7	25th p 8 et seq	599, p 11, 1976
Anti-caking Agent	559		Aluminium silicate (Kaolin)	7	25th p 8	828, p 8, 1992
Anti-foaming Agent	900		Dimethyl polysiloxane	1.5	25th p 18	648, p 31, 1980
Anti-foaming Agent	1521		Polyethylene glycol 6000 (1521=INS). No E no.)	Acc (0-10J)	36th	648, p 17, 1980
Bulking Agent	1200		Polydextrose	NS	26th Series	759, p 31, 1987
Colour Adjuvant	514	(i)	Sodium sulphate	NS	25th p 7 et seq	896, p 239, 1999
Colour Retn. Aid	512		Stannous chloride	Acc	25th p 8 et seq	683, p 32, 1982
Colour Retn. Aid	579		Ferrous gluconate	0.8 (J)	25th p 8	759, p 29, 1987
Colour Retn. Aid	585		Ferrous lactate	0.8 (J)	25th p 8	530, p 23, 1973
Colour Retn. Aid	1202		Polyvinylpolypyrrolidone	NS	26th Series	696, p 26, 1983
Emulsifier	322		Lecithins	NS	24th Series	539, p 20, 1973
Emulsifier	431		Polyoxyethylene stearate	25(J)	7th Series	281, p 140, 1964
Emulsifier	432		Polyoxyethylene sorbitan	10	7th Series	281, p 140, 1964

Function	Code	Sub	Name	Amount	Series	Reference
Emulsifier	433		Polyoxyethylene sorbitan monooleate monolaurate	10	7th Series	281, p 140, 1964
Emulsifier	434		Polyoxyethylene sorbitan monopalmitate	10	7th Series	281, p 140, 1964
Emulsifier	435		Polyoxyethylene sorbitan monostearate	10	7th Series	281, p 140, 1964
Emulsifier	436		Polyoxyethylene sorbitan tristearate	10	7th Series	281, p 140, 1964
Emulsifier	442		Ammonium phosphatides	30	7th p 44	17th 1973
Emulsifier	444		Sucrose acetate isobutyrate	10	32nd Series	868, p 43, 1996
Emulsifier	445		Glycerol esters of wood rosin	12.5	26th p 11	868, p 16, 1996
Emulsifier	460	(i)	Micro crystalline cellulose	NS	34th Series	884, p14,1997
Emulsifier	460	(ii)	Powdered cellulose	NS	7th p 9 etseq	599, p 12, 1976
Emulsifier	470	b	Magnesium salts of fatty acids	NS	25th p 7 et seq	733, p 11, 1986
Emulsifier	470	a	Na, K and Ca salts of fatty acids	NS	25th p 7 et seq	733, p 11, 1986 1988
Emulsifier	471		Mono and diglycerides of fatty acids	NS	24th Series	539, p 20, 1973
Emulsifier	472	a	Acetic esters of fatty acid glycerides	NS	24th p 21	539, p 20, 1973
Emulsifier	472	c	Citric esters of fatty acid glycerides	NS	24th Series	789, p 9, 1989
Emulsifier	472	b	Lact. esters of fatty acid glycerides	NS	7th p 43	539, p 20, 1973
Emulsifier	472	f	Mix. ac./ tart.acid esters of fatty acid glycerides	NS	32nd Series	539, p 20, 1973
Emulsifier	472	e	Mono/diacet. tart. esters of fatty acid glycerides	NS	43rd Series	539, p 19, 1973
Emulsifier	472	d	Tart. esters of fatty acid glycerides	NS	7th p 21	539, p 20, 1973
Emulsifier	481		Sodium stearoyl-2-lactylate	20	7th p 9 etseq	539, p 37, 1973
Emulsifier	482		Calcium stearoyl-2-lactylate	20	7th p 9 et seq	539, p 37, 1973
Emulsifier	491		Sorbitan monostearate	25	7th, p 34	281, p 110, 1964
Emulsifier	492		Sorbitan tristearate	25	7th, p 35	281, p 110, 1964
Emulsifier	493		Sorbitan monolaureate	5	7th p 35	281, p 110, 1964
Emulsifier	494		Sorbitan monooleate	5	7th p 35	281, p 110, 1964
Emulsifier	495		Sorbitan monopalmitate	25	7th, p 34	281, p 110, 1964
Emulsifier	473		Sucrose esters of fatty acids	20	83rd mtg	884, p 16, 1997
Emulsifier	474		Sucroglycerides	20	83rd mtg	859, p 12, 1995
Emulsifier	475		Polyglycerol esters of fatty acids	25	7th p 9 et seq	789, p 13, 1990
Emulsifier	476		Polyglycerol polyricinoleate	7.5	8th Series	539, p 20, 1973
Emulsifier	477		Propan- 1,2-diol esters of fatty acids	25	7th Series	539, p 24, 1973
Emulsifier	479	b	Therm. ox. soya oil / with fatty acid glycerides	25	21st p 51	828, p 8, 1992
Emulsifying salt	450	(vii)	Calcium dihydrogen diphosphate	70	25th p 7 et seq	733, p 11, 1986
Emulsifying salt	450	(v)	Tetrapotassium diphosphate	70	25th p 7 et seq	733, p 11, 1986

Appendix 2 *continued*

Category	E No	Sub	Additive name	SCF ADI	SCF Report Series	JECFA TRS
Emulsifying salt	450	(iii)	Tetrasodium diphosphate	70	25th p 7 et seq	733, p 11, 1986
Emulsifying salt	450	(ii)	Trisodium diphosphate	70	25th p 7 et seq	733, p 11, 1986
Emulsifying salt	452	(iv)	Calcium polyphosphate	70	25th p 7 et seq	733, p 11, 1986
Emulsifying salt	452	(ii)	Potassium polyphosphate	70	25th p 7 et seq	733, p 11, 1986
Emulsifying salt	452	(iii)	Sodium calcium polyphosphate	70	25th p 7 et seq	733, p 11, 1986
Emulsifying salt	452	(i) 2	Sodium polyphosphate insoluble	70	25th p 7 et seq	733, p 11, 1986
Firming Agent	520		Aluminium sulphate	7	25th p 8 et seq	733, p 11, 1986
Firming Agent	521		Aluminium sodium sulphate	7	25th p 8 et seq	Withdr.55th2000
Flavour Enhancer	515	(i)	Potassium sulphate	NS	25th p 7 et seq	733, p 11, 1986
Flavour Enhancer	620		Glutamic acid	NS	25th p 7 et seq	759, p 29, 1987
Flavour Enhancer	621		Monosodium glutamate	NS	25th p 7 et seq	759, p 29, 1987
Flavour Enhancer	622		Monopotassium glutamate	NS	25th p 7 et seq	759, p 29, 1987
Flavour Enhancer	623		Calcium diglutamate	NS	25th p 7 et seq	733, p 11, 1986
Flavour Enhancer	624		Monoammonium glutamate	NS	25th p 7 et seq	759, p 29, 1987
Flavour Enhancer	625		Magnesium diglutamate	NS	25th p 7 et seq	759, p 29, 1987
Flavour Enhancer	626		Guanylic acid	NS	25th p 7 et seq	723, p 25, 1986
Flavour Enhancer	627		Disodium guanylate	NS	25th p 7 et seq	733, p 25, 1986
Flavour Enhancer	628		Dipotassium guanylate	NS	25th p 7 et seq	837, p 13, 1993
Flavour Enhancer	629		Calcium guanylate	NS	25th p 7 et seq	733, p 25, 1986
Flavour Enhancer	630		Inosinic acid	NS	25th p 7 et seq	733, p 25, 1986
Flavour Enhancer	631		Disodium inosinate	NS	25th p 7 et seq	733, p 25, 1986
Flavour Enhancer	632		Dipotassium inosinate	NS	25th p 7 et seq	837, p 13, 1993
Flavour Enhancer	633		Calcium inosinate	NS	25th p 7 et seq	733, p 25, 1986
Flavour Enhancer	634		Calcium 5'-ribonucleotides	NS	25th p 7 et seq	723, p 25, 1986
Flavour Enhancer	635		Disodium 5'-ribonucleotides	NS	25th p 7 et seq	723, p 25, 1986
Flavour Enhancer	640		Glycine and its sodium salt	Acc	25th p 16	Not evaluated
Foam Stabilizer	1505		Triethyl citrate	20	26th Series	710, p 19, 1984
Foam Stabilizer	570		Fatty acids	NS	25th p 7 et seq	733, p 11, 1986
Foaming Agent	999		Quillaia extract	5	7th p 10 et seq	733, p 41, 1985

Category	No.	Name		Value	Series	Reference
FTA	483	Stearyl tartrate		20	25th Series	N.I. 55th 2000
FTA	517	Ammonium sulphate		NS	25th p 7 et seq	733, p 11, 1986
FTA	541	Sodium aluminium phosphate, acidic		7	25th p 7 et seq	733, p 11, 1986
FTA	920	L-Cysteine		Acc	25th p 7 et seq	Not evaluated
FTA	927	Carbamide		Acc	81st mtg 12.91	837, p 28, 1993
FTA	516	Calcium sulphate		NS	25th p 7 et seq	733, p 11, 1986
Glazing Agent	469	Enz. hydrolysed carboxy methyl cellulose		NS	32nd Series	891, p 42, 1998
Glazing Agent	904	Shellac		Acc	26th Series	828, p 27, 1992
Glazing Agent	905	Micro-crystalline wax		20	37th Series	828, p 24, 1992
Glazing Agent	901	Bees wax, white and yellow		Acc	25th p 8 et seq	828, p 24, 1992
Glazing Agent	902	Candelilla wax		Acc(t)	26th p 8	828, p 24, 1992
Glazing Agent	903	Carnauba wax		Acc	26th p 8	828, p 24, 1992
Humectant	422	Glycerol		NS	11th Series	599, p 13, 1976
Humectant	1518	Glyceryl triacetate (triacetin)		NS	26th Series?	648, p 17, 1980
Humectant	1520	Propan-1,2-diol		25	11thSeries	539, p 24, 1973
Packaging Gas	938	Argon		Acc	25th p 7 et seq	Not evaluated
Packaging Gas	939	Helium		Acc	25th Series	Not evaluated
Packaging Gas	948	Oxygen		Acc	25th p 17	733, p 39, 1988
Preservative	200	Sorbic acid		25	35th Series	539, p 18, 1973
Preservative	202	Potassium sorbate		25	5th Series	539, p 18, 1973
Preservative	203	Calcium sorbate		25	5th Series	539, p 18, 1973
Preservative	210	Benzoic acid		5(t)	35th p 29	868, p 41, 1996
Preservative	211	Sodium benzoate		5(t)	35th p 29	868, p 41, 1996
Preservative	212	Potassium benzoate		5(t)	35th p 29	868, p 41, 1996
Preservative	213	Calcium benzoate		5	35th p 29	868, p 41, 1996
Preservative	214	Ethyl p-hydroxybenzoate		10(t)	35th Series	539, p 17, 1973
Preservative	215	Sodium ethyl p-hydroxybenzoate		10	35th Series	2001
Preservative	216	Propyl p-hydroxybenzoate		10	35th Series	539, p 17, 1973
Preservative	217	Sodium propyl p-hydroxybenzoate		10	35th Series	2001
Preservative	218	Methyl p-hydroxybenzoate		10	35th Series	539, p 17, 1973
Preservative	219	Sodium methyl p-hydroxybenzoate		10	35th Series	2001
Preservative	220	Sulphur dioxide		0.7	5th Series	891, p 30, 1998
Preservative	221	Sodium sulphite		0.7	5th Series	891, p 30, 1998
Preservative	222	Sodium hydrogen (bi) sulphite		0.7	5th Series	891, p 30, 1998

Appendix 2 *continued*

Category	E No	Sub	Additive name	SCF ADI	SCF Report Series	JECFA TRS
Preservative	223		Sodium metabisulphite	0.7	5th Series	891, p 30, 1998
Preservative	224		Potassium metabisulphite	0.7	5th Series	891, p 30, 1998
Preservative	226		Calcium sulphite	0.7	35th Series	891, p 30, 1998
Preservative	227		Calcium hydrogen (bi) sulphite	0.7	5th Series	891, p 30, 1998
Preservative	228		Potassium hydrogen (bi) sulphite	0.7	5th Series	891, p 30, 1998
Preservative	230		Biphenyl	0.05(J)	Not evaluated?	309, p 25, 1964
Preservative	231		Orthophenyl phenol	02.(J)	Not evaluated	309, p 25, 1964
Preservative	232		Sodium orthophenyl phenol	0.2(J)	Not evaluated	309, p 25, 1964
Preservative	234		Nisin	0.83	26th Series	430, p 33, 1990
Preservative	235		Natamycin (pimaricin)	0.3	9th Series	599, p 16, 1976
Preservative	239		Hexamethylene tetramine	0.15	4th Series	539, p 17, 1973
Preservative	242		Dimethyl dicarbonate	Acc	26th p 9	806, p 23, 1991
Preservative	249		Potassium nitrite	0.06	1st Series	859, p 29, 1995
Preservative	250		Sodium nitrite	0.06	26th Series	859, p 29, 1995
Preservative	251		Sodium nitrate	3.7	26th Series	859, p 29, 1995
Preservative	252		Potassium nitrate	3.7	1st Series	859, p 29, 1995
Preservative	260		Acetic acid	NS	25th p 7 et seq	733, p 11, 1986
Preservative	261		Potassium acetate	NS	25th p 7 et seq	733, p 11, 1986
Preservative	262	(i)	Sodium acetate	NS	25th p 7 et seq	733, p 11, 1986
Preservative	262	(ii)	Sodium hydrogen acetate (diacetate)	NS	25th p 7 et seq	733, p 11, 1986
Preservative	263		Calcium acetate	NS	25th p 7 et seq	733, p 11, 1986
Preservative	280		Propionic acid	NS	26th Series	539, p 23, 1973
Preservative	281		Sodium propionate	NS	5th Series	539, p 23, 1973
Preservative	282		Calcium propionate	NS	5th Series	539, p 23, 1973
Preservative	283		Potassium propionate	Not limited	1st Series	539, p 23, 1973
Preservative	284		Boric acid	Acc	26th p 10	228, p 37, 1962
Preservative	285		Sodium tetraborate (borax)	Acc	26th Series	228, p 37, 1961
Preservative	912		Montan acid esters	Acc	26th p 8 Partial	Not evaluated

Function	Number	Name		ADI	Series	TRS
Preservative	914	Oxidized polyethylene wax		Acc	26th Series	Not evaluated
Preservative	1105	Lysozyme		Acc		828, p 8, 1992
Propellant	290	Carbon dioxide		Acc	32nd Series	733, p 39, 1988
Propellant	941	Nitrogen		Acc	25th p 17	733, p 39, 1988
Propellant	942	Nitrous oxide		Acc	25th p 17	733, p 40, 1985
Raising agent	450	Disodium diphosphate	(i)	70	25th p 7 et seq	733, p 11, 1986
Raising Agent	500	Sodium hydrogen carbonate	(ii)	NS	25th p 7 et seq	733, p 11, 1986
Sequestrant	385	Calcium disodium EDTA		2.5	4th Series	539, p 18, 1973
Sequestrant	451	Pentapotassium triphosphate	(ii)	70	25th p 7 et seq	733, p 11, 1986
Sequestrant	452	Sodium polyphosphate soluble	(i) 1	70	25th p 7 et seq	733, p 11, 1986
Sequestrant	451	Pentasodium triphosphate	(i)	70	25th p 7 et seq	733, p 11, 1986
Sequestrant	576	Sodium gluconate		NS	25th p 7 et seq	891, p 42, 1998
Stabilizer	335	Disodium tartrate	(ii)	30	25th p 7 et seq	696, p 28, 1983
Stabilizer	335	Sodium tartrate	(i)	30	25th p 7 et seq	733, p 11, 1986
Stabilizer	336	Dipotassium tartrate	(ii)	30	25th p 7 et seq	696, p 28, 1983
Stabilizer	336	Potassium tartrate	(i)	30	25th p 7 et seq	733, p 11, 1986
Stabilizer	337	Potassium sodium tartrate		30	25th p 7 et seq	733, p 11, 1986
Stabilizer	468	Crosslinked Na carboxy methyl cellulose		NS	35th	Not evaluated
Stabilizer	1201	Polyvinylpyrrolidone		50	26th Series	751, p 30, 1986

Notes to Appendix 2

SCF ADI Acceptable Daily Intake allocated by the Scientific Committee on Food, formerly known as the Scientific Committee *for* Food. (The SCF has often considered and then adopted JECFA ADI values.) Numbers in this column are the numerical ADI in mg per kg body weight, (mg/kg b.w.) expressed as 0–X, because 0 mg/kg b.w. is also acceptable!
NS = Not specified. This implies that the no-effect level and conditions of use have been assessed to be such as to cause no toxicological concern. It does *not* mean that no ADI could be allocated, because of, for instance, lack of submitted evidence.
Acc = Acceptable (an earlier usage implying the same as Not specified, bearing in mind the way in which the additive is used and the proposed use level e.g. Boric acid is shown as Acceptable, but only if its use is restricted to preservation of caviar).
NL = Not limited; an early version of Not specified.
Not evaluated = Not evaluated by the SCF. Associated with an ADI, this implies that JECFA opinion has been taken without further consideration.
(J) = JECFA ADI.
(t) = Temporary ADI, which will be reviewed after a specified period of time, when it may either be confirmed or the use of the additive prohibited.
SCF Report Series = the number of the series of reports of SCF opinions. (Details of the SCF opinions can be found on the web at http://europa.eu.int/comm/food/fs/sc/scf/reports_en.html.)
JECFA TRS = The issue number of the most recent World Health Organization (WHO) Technical Report Series (TRS) on the additive in question. TRS number, page, and year of publication. (The JECFA references may be found on the International Life Sciences Institute (ISI) website http://jecfa.ilsi.org/search.cfm. Although it claims to be up to 1997, in reality some of the reported evaluations are incorrectly quoted.) The JECFA reports numbering system is somewhat Byzantine, relying on the JECFA meeting number. The date of publication of the TRS report may be one or more years after the meeting. The WHO publishes the Technical report series which summarizes the meeting including the toxicological discussions, and the Food Additive Series (FAS) which includes the toxicological monographs.

Appendix 3: Acceptable daily intake (ADI) values and references by additive name

Category	E No	Sub	Additive name	SCF ADI	SCF Report Series	JECFA TRS
Preservative	260		Acetic acid	NS	25th p 7 et seq	733, p 11, 1986
Emulsifier	472	a	Acetic esters of fatty acid glycerides	NS	24th p 21	539, p 20, 1973
Acid	355		Adipic acid	30	25th p 7 et seq	789, p 18, 1090
Acidity Regulator	523		Aluminium ammonium sulphate	7	25th p 8	733, p 11, 1986
Acidity Regulator	522		Aluminium potassium sulphate	7	25th p 8	733, p 11, 1986
Anti-caking Agent	559		Aluminium silicate (Kaolin)	7	25th p 8	828, p 8, 1992
Firming Agent	521		Aluminium sodium sulphate	7	25th p 8 et seq	Withdr.55th2000
Firming Agent	520		Aluminium sulphate	7	25th p 8 et seq	733, p 11, 1986
Acidity Regulator	503	(i)	Ammonium carbonate	NS	25th p 7 et seq	733, p 11, 1986
Acidity Regulator	503	(ii)	Ammonium hydrogen carbonate	NS	25th p 7 et seq	733, p 11, 1986
Acidity Regulator	527		Ammonium hydroxide	NS	25th p 7 et seq	733, p 11, 1986
Emulsifier	442		Ammonium phosphatides	30	7th p 44	17th 1973
FTA	517		Ammonium sulphate	NS	25th p 7 et seq	733, p 11, 1986
Packaging Gas	938		Argon	Acc	25th p 7 et seq	Not evaluated
Glazing Agent	901		Bees wax, white and yellow	Acc	25th p 8 et seq	828, p 24, 1992
Anti-caking Agent	558		Bentonite	7	25th p 8 et seq	599, p 11, 1976
Preservative	210		Benzoic acid	5(t)	35th p 29	868, p 41, 1996
Preservative	230		Biphenyl	0.05(J)	Not evaluated?	309, p 25, 1964
Preservative	284		Boric acid	Acc	26th p 10	228, p 37, 1962
Flavour Enhancer	634		Calcium 5'-ribonucleotides	NS	25th p 7 et seq	723, p 25, 1986
Preservative	263		Calcium acetate	NS	25th p 7 et seq	733, p 11, 1986
Anti-caking Agent	556		Calcium aluminium silicate	7	25th p 8 et seq	733, p 11, 1986
Preservative	213		Calcium benzoate	5	35th p 29	868, p 41, 1996
Flavour Enhancer	623		Calcium diglutamate	NS	25th p 7 et seq	733, p 11, 1986
Emulsifying salt	450	(vii)	Calcium dihydrogen diphosphate	70	25th p 7 et seq	733, p 11, 1986
Sequestrant	385		Calcium disodium EDTA	2.5	4th Series	539, p 18, 1973
Anti-caking Agent	538		Calcium ferrocyanide	0.025	25th p 8	Not evaluated
Acidity regulator	578		Calcium gluconate	NS	25th p 7 et seq	891, p 46, 1998

Function	No.		Name	Level	Meeting	Reference
Flavour Enhancer	629		Calcium guanylate	NS	25th p 7 et seq	733, p 25, 1986
Preservative	227		Calcium hydrogen (bi) sulphite	0.7	5th Series	891, p 30, 1998
Acidity Regulator	352	(ii)	Calcium hydrogen malate	NS	25th p 7 et seq	445, p 16, 1970
Acidity Regulator	526		Calcium hydroxide	NS	25th p 7 et seq	733, p 11, 1986
Flavour Enhancer	633		Calcium inosinate	NS	25th p 7 et seq	733, p 25, 1986
Acidity Regulator	327		Calcium lactate	NS	25th p 7 et seq	539, p 23, 1973
Acidity Regulator	352	(i)	Calcium malate	NS	25th p 7 et seq	445, p 16, 1970
Acidity Regulator	529		Calcium oxide	NS	25th p 7 et seq	733, p 11, 1986
Emulsifying salt	452	(iv)	Calcium polyphosphate	70	25th p 7 et seq	733, p 11, 1986
Preservative	282		Calcium propionate	NS	5th Series	539, p 23, 1973
Anti-caking Agent	552		Calcium silicate	NS	25th p 7 et seq	733, p 11, 1986
Preservative	203		Calcium sorbate	25	5th Series	539, p 18, 1973
Emulsifier	482		Calcium stearoyl-2-lactylate	20	7th p 9 et seq	539, p 37, 1973
FTA	516		Calcium sulphate	NS	25th p 7 et seq	733, p 11, 1986
Preservative	226		Calcium sulphite	0.7	35th Series	891, p 30, 1998
Acidity Regulator	354		Calcium tartrate	30	25th p 7 et seq	696, p 28, 1983
Glazing Agent	902		Candelilla wax	Acc(t)	26th p 8	828, p 24, 1992
FTA	927		Carbamide	Acc	81st mtg 12.91	837, p 28, 1993
Propellant	290		Carbon dioxide	Acc	32nd Series	733, p 39, 1988
Glazing Agent	903		Carnauba wax	Acc	26th p 8	828, p 24, 1992
Acid	330		Citric acid	NS	25th p 7 et seq	539, p 19, 1973
Emulsifier	472	c	Citric esters of fatty acid glycerides	NS	24th Series	789, p 9, 1989
Stabilizer	468		Crosslinked Na carboxy methyl cellulose	NS	35th	Not evaluated
Adjuvant	459		Cyclodextrin-beta	5	41st Series	859, p 28, 1995
Acidity Regulator	333	(ii)	Dicalcium citrate	NS	25th p 7 et seq	539, p 19, 1973
Acidity Regulator	450	(vi)	Dicalcium diphosphate	70	25th p 7 et seq	733, p 11, 1986
Acidity Regulator	341	(ii)	Dicalcium phosphate	70	25th p 7 et seq	733, p 11, 1986
Acidity Regulator	343	(ii)	Dimagnesium phosphate	70	25th p 7 et seq	733, p 11, 1986
Preservative	242		Dimethyl dicarbonate	Acc	26th p 9	806, p 23, 1991
Anti-foaming Agent	900		Dimethyl polysiloxane	1.5	25th p 18	648, p 31, 1980
Flavour Enhancer	628		Dipotassium guanylate	NS	25th p 7 et seq	837, p 13, 1993
Flavour Enhancer	632		Dipotassium inosinate	NS	25th p 7 et seq	837, p 13, 1993
Acidity Regulator	340	(ii)	Dipotassium phosphate	70	25th p 7 et seq	733, p 11, 1986
Stabilizer	336	(ii)	Dipotassium tartrate	30	25th p 7 et seq	696, p 28, 1983

Appendix 3 *continued*

Category	E No	Sub	Additive name	SCF ADI	SCF Report Series	JECFA TRS
Flavour Enhancer	635		Disodium 5'-ribonucleotides	NS	25th p 7 et seq	723, p 25, 1986
Acidity Regulator	331	(ii)	Disodium citrate	NS	25th p 7 et seq	539, p 19, 1973
Raising agent	450	(i)	Disodium diphosphate	70	25th p 7 et seq	733, p 11, 1986
Flavour Enhancer	627		Disodium guanylate	NS	25th p 7 et seq	733, p 25, 1986
Flavour Enhancer	631		Disodium inosinate	NS	25th p 7 et seq	733, p 25, 1986
Acidity Regulator	339	(ii)	Disodium phosphate	70	25th p 7 et seq	733, p 11, 1986
Stabilizer	335	(ii)	Disodium tartrate	30	25th p 7 et seq	696, p 28, 1983
Glazing Agent	469		Enz. hydrolysed carboxy methyl cellulose	NS	32nd Series	891, p 42, 1998
Preservative	214		Ethyl p-hydroxybenzoate	10(t)	35th Series	539, p 17, 1973
Foam Stabilizer	570		Fatty acids	NS	25th p 7 et seq	733, p 11, 1986
Colour Retn. Aid	579		Ferrous gluconate	0.8 (J)	25th p 8	759, p 29, 1987
Colour Retn. Aid	585		Ferrous lactate	0.8 (J)	25th p 8	530, p 23, 1973
Acid	297		Fumaric acid	6	25th p 7 et seq	789, p 18, 1990
Acid	574		Gluconic acid	NS	25th p 7 et seq	891, p 43, 1998
Acid	575		Glucono-delta-lactone	NS	25th p 7 et seq	891, p 43, 1998
Flavour Enhancer	620		Glutamic acid	NS	25th p 7 et seq	759, p 29, 1987
Humectant	422		Glycerol	NS	11th Series	599, p 13, 1976
Emulsifier	445		Glycerol esters of wood rosin	12.5	26th p 11	868, p 16, 1996
Humectant	1518		Glyceryl triacetate (triacetin)	NS	26th Series?	648, p 17, 1980
Flavour Enhancer	640		Glycine and its sodium salt	Acc	25th p 16	Not evaluated
Flavour Enhancer	626		Guanylic acid	NS	25th p 7 et seq	723, p 25, 1986
Packaging Gas	939		Helium	Acc	25th Series	Not evaluated
Preservative	239		Hexamethylene tetramine	0.15	4th Series	539, p 17, 1973
Acid	507		Hydrochloric acid	NS	25th p 7 et seq	339, p 16, 1965
Flavour Enhancer	630		Inosinic acid	NS	25th p 7 et seq	733, p 25, 1986
Emulsifier	472	b	Lact. esters of fatty acid glycerides	NS	7th p 43	539, p 20, 1973
Acid	270		Lactic acid	NS	25th p 7 et seq	539, p 23, 1973
FTA	920		L-Cysteine	Acc	25th p 7 et seq	Not evaluated
Emulsifier	322		Lecithins	NS	24th Series	539, p 20, 1973

Function	Code	Sub	Name	ADI	Reference	Source
Preservative	1105		Lysozyme	Acc	25th p 7 et seq	828, p 8, 1992
Acidity Regulator	504	(i)	Magnesium carbonate	NS	25th p 7 et seq	733, p 11, 1986
Flavour Enhancer	625		Magnesium diglutamate	NS	25th p 7 et seq	759, p 29, 1987
Acidity Regulator	528		Magnesium hydroxide	NS	25th p 7 et seq	733, p 11, 1986
Acidity Regulator	504	(ii)	Magnesium hydroxide carbonate	NS	25th p 7 et seq	733, p 11, 1986
Anti-caking Agent	530		Magnesium oxide	NS	5th Series	733, p 11, 1986
Emulsifier	470	b	Magnesium salts of fatty acids	NS	25th p 7 et seq	733, p 11, 1986
Anti-caking Agent	553	a(i)	Magnesium silicate	NS	25th p 7 et seq	733, p 11, 1986
Anti-caking Agent	553	(a) (i)	Magnesium silicate (synthetic)	NS	25th p 7 et seq	733, p 11, 1986
Anti-caking Agent	553	(a) (ii)	Magnesium trisilicate	NS	25th p 7 et seq	733, p 11, 1986
Acid	296		Malic acid	NS	25th p 7 et seq	445, p 16, 1970
Acid	353		Metatartaric acid	Acc	25th p 7 et seq	Not evaluated
Preservative	218		Methyl p-hydroxybenzoate	10	35th Series	539, p 17, 1973
Emulsifier	460	(i)	Micro crystalline cellulose	NS	34th Series	884, p 14, 1997
Glazing Agent	905		Micro-crystalline wax	20	37th Series	828, p 24, 1992
Emulsifier	472	f	Mix. ac. tart.acid esters of fatty acid glycerides	NS	32nd Series	539, p 20, 1973
Emulsifier	471		Mono and diglycerides of fatty acids	NS	24th Series	539, p 20, 1973
Emulsifier	472	e	Mono diacet. tart. esters of fatty acid glycerides	NS	43rd Series	539, p 19, 1973
Flavour Enhancer	624		Monoammonium glutamate	NS	25th p 7 et seq	759, p 29, 1987
Acidity Regulator	333	(i)	Monocalcium citrate	NS	25th p 7 et seq	539, p 19, 1973
Acidity Regulator	341	(i)	Monocalcium phosphate	70	25th p 7 et seq	733, p 11, 1986
Acidity Regulator	343	(i)	Monomagnesium phosphate	70	25th p 7 et seq	733, p 11, 1986
Acidity Regulator	332	(i)	Monopotassium citrate	NS	25th p 7 et seq	539, p 19, 1973
Flavour Enhancer	622		Monopotassium glutamate	NS	25th p 7 et seq	759, p 29, 1987
Acidity Regulator	340	(i)	Monopotassium phosphate	70	25th p 7 et seq	733, p 11, 1986
Acidity Regulator	331	(i)	Monosodium citrate	NS	25th p 7 et seq	539, p 19, 1973
Flavour Enhancer	621		Monosodium glutamate	NS	25th p 7 et seq	759, p 29, 1987
Acidity Regulator	339	(i)	Monosodium phosphate	70	25th p 7 et seq	733, p 11, 1986
Preservative	912		Montan acid esters	Acc	26th p 8 Partial	Not evaluated
Emulsifier	470	a	Na, K and Ca salts of fatty acids	NS	25th p 7 et seq	733, p 11, 1986 1988
Preservative	235		Natamycin (pimaricin)	0.3	9th Series	599, p 16, 1976
Preservative	234		Nisin	0.83	26th Series	430, p 33, 1990
Propellant	941		Nitrogen	Acc	25th p 17	733, p 39, 1988

Appendix 3 *continued*

Category	E No	Sub	Additive name	SCF ADI	SCF Report Series	JECFA TRS
Propellant	942		Nitrous oxide	Acc	25th p 17	733, p 40, 1985
Preservative	231		Orthophenyl phenol	02.(J)	Not evaluated	309, p 25, 1964
Preservative	914		Oxidized polyethylene wax	Acc	26th Series	Not evaluated
Packaging Gas	948		Oxygen	Acc	25th p 17	733, p 39, 1988
Sequestrant	451	(ii)	Pentapotassium triphosphate	70	25th p 7 et seq	733, p 11, 1986
Sequestrant	451	(i)	Pentasodium triphosphate	70	25th p 7 et seq	733, p 11, 1986
Acid	338		Phosphoric acid	70	25th p 7 et seq	733, p 11, 1986
Bulking Agent	1200		Polydextrose	NS	26th Series	759, p 31, 1987
Anti-foaming Agent	1521		Polyethylene glycol 6000 (1521=INS). No E no.)	Acc (0-10J)	36th	648, p 17, 1980
Emulsifier	475		Polyglycerol esters of fatty acids	25	7th p 9 et seq	789, p 13, 1990
Emulsifier	476		Polyglycerol polyricinoleate	7.5	8th Series	539, p 20, 1973
Emulsifier	432		Polyoxyethylene sorbitan monolaurate	10	7th Series	281, p 140, 1964
Emulsifier	433		Polyoxyethylene sorbitan monooleate	10	7th Series	281, p 140, 1964
Emulsifier	434		Polyoxyethylene sorbitan monopalmitate	10	7th Series	281, p 140, 1964
Emulsifier	435		Polyoxyethylene sorbitan monostearate	10	7th Series	281, p 140, 1964
Emulsifier	436		Polyoxyethylene sorbitan tristearate	10	7th Series	281, p 140, 1964
Emulsifier	431		Polyoxyethylene stearate	25(J)	7th Series	281, p 140, 1964
Colour Retn. Aid	1202		Polyvinylpolypyrrolidone	NS	26th Series	696, p 26, 1983
Stabilizer	1201		Polyvinylpyrrolidone	50	26th Series	751, p 30, 1986
Preservative	261		Potassium acetate	NS	25th p 7 et seq	733, p 11, 1986
Acidity Regulator	357		Potassium adipate	30	25th p 7 et seq	789, p 18, 1990
Anti-caking Agent	555		Potassium aluminium silicate (Mica)	7	25th p 8 et seq	733, p 11, 1986
Preservative	212		Potassium benzoate	5(t)	35th p 29	868, p 41, 1996
Acidity Regulator	501	(i)	Potassium carbonate	NS	25th p 7 et seq	733, p 11, 1986
Anti-caking Agent	536		Potassium ferrocyanide	0.025	25th p 18	557, p 23, 1974
Acidity Regulator	577		Potassium gluconate	NS	25th p 7 et seq	891, p 42, 1998
Preservative	228		Potassium hydrogen (bi) sulphite	0.7	5th Series	891, p 30, 1998
Acidity Regulator	501	(ii)	Potassium hydrogen carbonate	NS	25th p 7 et seq	733, p 11, 1986

Function	Number	Name		Limit	Series	Reference
Acidity Regulator	515	Potassium hydrogen sulphate	(ii)	NS	25th p 7 et seq	733, p 11, 1986
Acidity Regulator	525	Potassium hydroxide		NS	25th p 7 et seq	733, p 11, 1986
Acidity Regulator	351	Potassium malate		NS	25th p 7 et seq	733, p 11, 1986
Preservative	224	Potassium metabisulphite		0.7	5th Series	891, p 30, 1998
Preservative	252	Potassium nitrate		3.7	1st Series	859, p 29, 1995
Preservative	249	Potassium nitrite		0.06	1st Series	859, p 29, 1995
Emulsifying salt	452	Potassium polyphosphate	(ii)	70	25th p 7 et seq	733, p 11, 1986
Preservative	283	Potassium propionate		Not limited	1st Series	539, p 23, 1973
Stabilizer	337	Potassium sodium tartrate		30	25th p 7 et seq	733, p 11, 1986
Preservative	202	Potassium sorbate		25	5th Series	539, p 18, 1973
Flavour Enhancer	515	Potassium sulphate	(i)	NS	25th p 7 et seq	733, p 11, 1986
Stabilizer	336	Potassium tartrate	(i)	30	25th p 7 et seq	733, p 11, 1986
Emulsifier	460	Powdered cellulose	(ii)	NS	7th p 9 etseq	599, p 12, 1976
Emulsifier	477	Propan- 1,2-diol esters of fatty acids		25	7th Series	539, p 24, 1973
Humectant	1520	Propan-1,2-diol		25	11thSeries	539, p 24, 1973
Preservative	280	Propionic acid		NS	26th Series	539, p 23, 1973
Preservative	216	Propyl p-hydroxybenzoate		10	35th Series	539, p 17, 1973
Foaming Agent	999	Quillaia extract		5	7th p 10 et seq	733, p 41, 1985
Glazing Agent	904	Shellac		Acc	26th Series	828, p 27, 1992
Anti-caking Agent	551	Silicon dioxide		NS	24th Series	733, p 11, 1986
Preservative	262	Sodium acetate	(i)	NS	25th p 7 et seq	733, p 11, 1986
Acidity Regulator	356	Sodium adipate		30	25th p 7 et seq	733, p 11, 1986
FTA	541	Sodium aluminium phosphate, acidic		7	25th p 7 et seq	789, p 18, 1990
Anti-caking Agent	554	Sodium aluminium silicate		7	25th p 7 et seq	733, p 11, 1986
Preservative	211	Sodium benzoate		5(t)	25th p 8 et seq	733, p 11, 1986
Emulsifying salt	452	Sodium calcium polyphosphate	(iii)	70	35th p 29	868, p 41, 1996
Acidity Regulator	500	Sodium carbonate	(i)	NS	25th p 7 et seq	733, p 11, 1986
Preservative	215	Sodium ethyl p-hydroxybenzoate		10	25th p 7 et seq	733, p 11, 1986
Anti-caking Agent	535	Sodium ferrocyanide		0.025	35th Series	2001
Sequestrant	576	Sodium gluconate		NS	25th p 18	557, p 23, 1974
Preservative	222	Sodium hydrogen (bi) sulphite		NS	25th p 7 et seq	891, p 42, 1998
Preservative	262	Sodium hydrogen acetate (diacetate)	(ii)	0.7	5th Series	891, p 30, 1998
Raising Agent	500	Sodium hydrogen carbonate	(ii)	NS	25th p 7 et seq	733, p 11, 1986

Appendix 3 continued

Category	E No	Sub	Additive name	SCF ADI	SCF Report Series	JECFA TRS
Acidity Regulator	350	(ii)	Sodium hydrogen malate	NS	25th p 7 et seq	445, p 16, 1970
Acidity Regulator	514	(ii)	Sodium hydrogen sulphate	NS	25th p 7 et seq	733, p 11, 1986
Acidity Regulator	524		Sodium hydroxide	NS	25th p 7 et seq	733, p 11, 1986
Acidity Regulator	350	(i)	Sodium malate	NS	25th p 7 et seq	733, p 11, 1986
Preservative	223		Sodium metabisulphite	0.7	5th Series	891, p 30, 1998
Preservative	219		Sodium methyl p-hydroxybenzoate	10	35th Series	2001
Preservative	251		Sodium nitrate	3.7	26th Series	859, p 29, 1995
Preservative	250		Sodium nitrite	0.06	26th Series	859, p 29, 1995
Preservative	232		Sodium orthophenyl phenol	0.2(J)	Not evaluated	309, p 25, 1964
Emulsifying salt	452	(i) 2	Sodium polyphosphate insoluble	70	25th p 7 et seq	733, p 11, 1986
Sequestrant	452	(i) 1	Sodium polyphosphate soluble	70	25th p 7 et seq	733, p 11, 1986
Preservative	281		Sodium propionate	NS	5th Series	539, p 23, 1973
Preservative	217		Sodium propyl p-hydroxybenzoate	10	35th Series	2001
Acidity Regulator	500	(iii)	Sodium sesqui carbonate	NS	25th p 7 et seq	733, p 11, 1986
Emulsifier	481		Sodium stearoyl-2-lactylate	20	7th p 9 etseq	539, p 37, 1973
Colour Adjuvant	514	(i)	Sodium sulphate	NS	25th p 7 et seq	896, p 239, 1999
Preservative	221		Sodium sulphite	0.7	5th Series	891, p 30, 1998
Stabilizer	335	(i)	Sodium tartrate	30	25th p 7 et seq	733, p 11, 1986
Preservative	285		Sodium tetraborate (borax)	Acc	26th Series	228, p 37, 1961
Preservative	200		Sorbic acid	25	35th Series	539, p 18, 1973
Emulsifier	493		Sorbitan monolaureate	5	7th p 35	281, p 110, 1964
Emulsifier	494		Sorbitan monooleate	5	7th p 35	281, p 110, 1964
Emulsifier	495		Sorbitan monopalmitate	25	7th, p 34	281, p 110, 1964
Emulsifier	491		Sorbitan monostearate	25	7th, p 34	281, p 110, 1964
Emulsifier	492		Sorbitan tristearate	25	7th, p 35	281, p 110, 1964
Colour Retn. Aid	512		Stannous chloride	Acc	25th p 8 et seq	683, p 32, 1982
FTA	483		Stearyl tartrate	20	25th Series	N.I. 55th 2000
Acid	363		Succinic acid	30	5th Series	733, p 11, 1986
Emulsifier	474		Sucroglycerides	20	83rd mtg	859, p 12, 1995

Category	Number		Name	ADI	SCF Report	JECFA Reference
Emulsifier	444		Sucrose acetate isobutyrate	10	32nd Series	868, p 43, 1996
Emulsifier	473		Sucrose esters of fatty acids	20	83rd mtg	884, p 16, 1997
Acid	513		Sulfuric acid	NS	25th p 7 et seq	733, p 11, 1986
Preservative	220		Sulphur dioxide	0.7	5th Series	891, p 30, 1998
Anti-caking Agent	553	(b)	Talc	NS	5th Series	751, p 33, 1986
Emulsifier	472	d	Tart. esters of fatty acid glycerides	NS	7th p 21	539, p 20, 1973
Acid	334		Tartaric acid (L(+)-)	30	25th p 7 et seq	696, p 28, 1983
Emulsifying salt	450	(v)	Tetrapotassium diphosphate	70	25th p 7 et seq	733, p 11, 1986
Emulsifying salt	450	(iii)	Tetrasodium diphosphate	70	25th p 7 et seq	733, p 11, 1986
Emulsifier	479	b	Therm. ox. soya oil with fatty acid glycerides	25	21st p 51	828, p 8, 1992
Acidity Regulator	380		Triammonium citrate	NS	25th p 7 et seq	733, p 11, 1986
Acidity Regulator	333	(iii)	Tricalcium citrate	NS	25th p 7 et seq	710, p 19, 1984
Acidity Regulator	341	(iii)	Tricalcium phosphate	70	25th p 7 et seq	733, p 11, 1986
Foam Stabilizer	1505		Triethyl citrate	20	26th Series	710, p 19, 1984
Acidity Regulator	332	(ii)	Tripotassium citrate	NS	25th p 7 et seq	Not evaluated
Acidity Regulator	340	(iii)	Tripotassium phosphate	70	25th p 7 et seq	733, p 11, 1986
Acidity Regulator	331	(iii)	Trisodium citrate	NS	25th p 7 et seq	733, p 12, 1986
Emulsifying salt	450	(i)	Trisodium diphosphate	70	25th p 7 et seq	733, p 11, 1986
Acidity Regulator	339	(iii)	Trisodium phosphate	70	25th p 7 et seq	733, p 11, 1986

Notes to Appendix 3

SCF ADI Acceptable Daily Intake allocated by the Scientific Committee on Food, formerly known as the Scientific Committee *for* Food. (The SCF has often considered and then adopted JECFA ADI values.) Numbers in this column are the numerical ADI in mg per kg body weight, (mg/kg b.w.) expressed as 0–X, because 0 mg/kg b.w. is also acceptable!
NS = Not specified. This implies that the no-effect level and conditions of use have been assessed to be such as to cause no toxicological concern. It does *not* mean that no ADI could be allocated, because of, for instance, lack of submitted evidence.
Acc = Acceptable (an earlier usage implying the same as Not specified, bearing in mind the way in which the additive is used and the proposed use level e.g. Boric acid is shown as Acceptable, but only if its use is restricted to preservation of caviar).
NL = Not limited; an early version of Not specified.
Not evaluated = Not evaluated by the SCF. Associated with an ADI, this implies that JECFA opinion has been taken without further consideration.
(J) = JECFA ADI.
(t) = Temporary ADI, which will be reviewed after a specified period of time, when it may either be confirmed or the use of the additive prohibited.
SCF Report Series = the number of the series of reports of SCF opinions. (Details of the SCF opinions can be found on the web at http://europa.eu.int/comm/food/fs/sc/scf/reports_en.html.)
JECFA TRS = The issue number of the most recent World Health Organization (WHO) Technical Report Series (TRS) on the additive in question. TRS number, page, and year of publication. (The JECFA references may be found on the International Life Sciences Institute (ISI) website http://jecfa.ilsi.org/search.cfm. Although it claims to be up to 1997, in reality some of the reported evaluations are incorrectly quoted.) The JECFA reports numbering system is somewhat Byzantine, relying on the JECFA meeting number. The date of publication of the TRS report may be one or more years after the meeting. The WHO publishes the Technical report series which summarizes the meeting including the toxicological discussions, and the Food Additive Series (FAS) which includes the toxicological monographs.

11.6 Bibliography

Code of Federal Regulations 21. Food and Drug Administration, Parts 100 to 199. Revised annually. Published by the Office of Federal Register, National Archives and Records Administration.

Compendium of Food Additives Specifications. FAO Food and Nutrition Paper 52 and addenda 1 to 8. Published by the FAO, Vialle delle Terme di Caracalla, 00100 Rome. ISBN 92-5-104508-9. The compendium and addenda 1 to 7 are on the JECFA web site: http://www.fao.org/WAICENT/FAOINFO/ECONOMIC/ESN/Jecfa/Jecfa.htm

EU Directives. Details of all EU Directives on food and related subjects, including food additives, can be found in the publication issued three time per annum by the Confederation of the Food and Drink Industries of the EU (CIAA), Avenue des Arts 43, B-1040 Brussels. E-mail: ciaa@ciaa.be; web site: http://www.ciaa.be

Food Chemicals Codex, 4th edition. (1996) (Plus supplements) National Academy of Sciences. National Academy Press, Washington DC 20055. ISBN 0-30905-05394.

JECFA Guide to Specifications. FAO Food and Nutrition Paper 5 Rev. 2. Published by the FAO, Vialle delle Terme di Caracalla, 00100 Rome. ISBN 92-5-102991-1.

JECFA Summary of Evaluations Performed. Published by the International Life Sciences International Press. ISBN 0-944398 41-3. http://jecfa.ilsi.org

Joint FAO/WHO Food Standards Programme, *Codex Alimentarius Commission Procedural Manual*, Eleventh edition, FAO and WHO, Rome 2000, ISBN 92-5-1004402-3. (The tenth edition is on the Codex www-pages at http://fao.org/docrep/W5975E/W5975E00.htm)

Scientific Committee on Food. Reports available at: http://europa.eu.int/comm/food/fs/sc/scf/index_en.html

12

The regulation of antioxidants in food

K. Míková, Prague Institute of Chemical Technology

12.1 Introduction

In countries like the United States, Canada, Australia, many European countries and certain others food regulations have existed for many decades. Many countries have yet to formulate national food policies, responding appropriately to their health situation and economy, or, where these policies have been formulated, they often do not reflect appropriately the true nature and extent of current or emerging food safety problems. The advances in science and technology are important stimuli for modification of the laws.[1]

Because food is essential to life and can be improperly prepared or handled, it can threaten life. The purveyor of food therefore has a duty to provide safe and wholesome products to every customer. Given the fundamental importance of food, it is appropriate for any government to define and enforce this ethical obligation and thereby protect what many would consider the right of every individual to safe and wholesome food.

From the legal point of view, antioxidants are substances which prolong the shelf-life of foodstuffs by protecting them against deterioration caused by oxidation, such as fat rancidity, colour changes and loss of nutrient value. Hundreds of compounds, both natural and synthesised, have been reported to possess antioxidant properties. Their use in food, however, is limited by certain obvious requirements, not the least of which is adequate proof of safety.

Owing to differences in their molecular structure, the various antioxidants exhibit substantial differences in effectiveness when used with different types of foodstuffs and when used under different processing and handling conditions.[2] The problem of selecting the optimum antioxidant or combination of antioxidants is further complicated by the difficulty of predicting how the added antioxidant will function in the presence of pro-oxidants and antioxidants already present in the food or produced in the course of processing.

12.2 Toxicological aspects

Since food additives are subjected to the most stringent toxicological testing procedures, only a few synthetic antioxidants have been used in foods for any length of time. Antioxidants are extensively tested for the absence of carcinogenity and other toxic effects in themselves, in their oxidised forms, and in their reaction products with food constituents, for their effectiveness at low concentrations, and for the absence of the ability to impart an unpleasant flavour to the food in which they are used.

Table 12.1 presents the most common antioxidants permitted for use in food products.

The use of antioxidants in food products is governed by regulatory laws of the individual country or by internal standards. Even though many natural and synthetic compounds have antioxidant properties, only a few of them have been accepted as 'generally recognised as safe (GRAS)' substances for use in food products by international bodies such as the Joint FAO/WHO Expert Committee on Food Additives (JECFA) and the European Community's Scientific Committee for Food (SCF).

Toxicological studies are crucial in determining the safety of an antioxidant and also in determining the acceptable daily intake (ADI) levels. ADIs for widely used antioxidants such as BHA, BHT and gallates have changed over the years mainly because of their toxicological effects in various species.[3] Table 12.2 presents the ADIs allocated by JECFA.

The safety of the antioxidant must be established. According to Lehman et al.,[4] an antioxidant is considered safe if it fulfils two conditions: its LD_{50} must not be less than 1000 mg/kg body weight, and the antioxidant should not have any significant effect on the growth of the experimental animal in long-term

Table 12.1 Antioxidants conventionally permitted in foods

ascorbic acid, sodium, calcium salts	glycine
ascorbyl palmitate and stearate	gum guaiac
anoxomer	lecithin
butylated hydroxyanisole (BHA)	ionox-100
butylated hydroxytoluene (BHT)	polyphosphates
tert-butyl hydroquinone[a] (TBHQ)	propyl, octyl, and dodecyl gallates
citric acid, stearyl, and isopropyl esters	tartaric acid
erythorbic acid and sodium salt	thiodipropionic acid, dilauryl and
ethoxyquin	distearyl esters
ethylenediaminetetraacetic acid (EDTA) and calcium disodium salt	tocopherols
	trihydroxy butyrophenone

[a] Not permitted for use in European Economic Community countries

Table 12.2 ADIs of some antioxidants permitted in foods

Antioxidant	ADI (mg/kg bw)
propyl gallate	0–2.5
BHA	0–0.5
BHT	0–0.125
TBHQ	0–0.2
tocopherols	0.15–2.0
gum guaiac	0–2.5
ethoxyquin	0–0.06
phosphates	0–70.0
EDTA	2.5
tartaric acid	0–30.0
citric acid	not limited
lecithin	not limited
ascorbic acid	not limited
sulphites (as sulphur dioxide)	0–0.7
ascorbyl palmitate or ascorbyl stearate (or the sum of both)	0–1.25

bw = body weight

studies at a level 100 times greater than that proposed for human consumption. Approval of an antioxidant for food use also requires extensive toxicological studies of its possible mutagenic, teratogenic and carcinogenic effects.

New toxicological data on some of the synthetic antioxidants cautioned against their use. In the recent past, natural antioxidants attracted the attention of many food manufacturers as a result of the necessity to produce healthy foods. Numerous antioxidative efficacious compounds that are found in animal or plant tissues and that are also available as synthetic molecules are used in several food applications. Herbs and spices occupy a special position in foods as traditional food ingredients and hence are appropriately used directly for their antioxidant characteristics. If they are applied to foods, they do not need to be declared as antioxidants.

Antioxidants should satisfy several requirements before being accepted for incorporation into food products.[5] The antioxidant should be soluble in fats; it should not impart a foreign colour, odour or flavour to the fat even on long storage; it should be effective for a least one year at a temperature of between 25 and 30°C; it should be stable to heat processing and protect the finished product (carry-through effect); it should be easy to incorporate; and it should be effective at low concentrations.

Antioxidants can be added directly to vegetable oils, melted animal fats or other fat-containing or polyphenol-containing systems. In some cases, however, better results are achieved when the antioxidant is administered in a diluent (e.g. propylene glycol or volatile solvents). Food products can also be sprayed with, or dipped in solutions or suspensions of, antioxidants, or they can be packed in films containing antioxidants. The use of antioxidants is possible only if it is technologically substantiated and indispensable.

12.3 The *Codex Alimentarius*

The *Codex Alimentarius* is a collection of internationally adopted food standards presented in a uniform manner.[6] These food standards aim to protect the consumer's health and ensure fair practices in the food trade. The *Codex Alimentarius* also includes provisions of an advisory nature in the form of codes of practice, guidelines and other recommended measures. Codex standards contain requirements for food including provisions for food additives.

The standards and limits adopted by the *Codex Alimentarius* Commission are intended for formal acceptance by governments in accordance with its general principles. *Codex Alimentarius* permits only those antioxidants which have been evaluated by the Joint FAO/WHO Expert Committee on Food Additives (JECFA) for use in foods. Antioxidants may be used only in foods standardised by Codex. The antioxidant provisions of Codex Commodity Standards are included in and superseded by the provision of this Standard. Food categories or individual foods where the use of additives are not allowed or are restricted are defined by this Standard. The primary objective of establishing permitted levels of use of antioxidants in various food groups is to ensure that the intake does not exceed the acceptable daily intake (ADI).

The antioxidants covered by the Standard and their maximum levels of use are based in part on the food additive provisions of previously established Codex Commodity Standards, or upon the request of governments after subjecting the requested maximum levels to an appropriate method which would verify the compatibility of a proposed maximum level with the ADI.

Adequate information should be given about the manner in which the antioxidant is to be used in food. This information may be given on the label or in the documents relating to the sale. All antioxidants must be characterised by the name. The name should be specific and not generic and should indicate the true nature of the food additive. Where a name has been established for a food additive in a Codex list of additives, that name should be used. In other cases, the common or usual name should be listed or, where none exists, an appropriate descriptive name should be used. As an alternative to the declaration of the specific names the International Numbering System for Food Additives (INS) has been prepared by the Codex Committee on Food Additives and Contaminants for the purpose of providing an agreed international numerical system for identifying food additives. It has been based on the restricted system already introduced successfully within the EC. The INS is intended as an identification system approved for use in the member countries. The antioxidants reviewed in *Codex Alimentarius* are listed in Table 12.3.

In the *Codex Alimentarius* a number of further sequestrants and antioxidant synergists are issued which also pertain to the group of antioxidants. The main representatives of sequestrants are acetates, citrates, tartrates, phosphates, orthophosphates, diphosphates, triphosphates, and polyphosphates, esters of glycerol, sorbitol and sorbitol syrup, gluconates, and 1,4-heptonolactone. Lactates and tartaric acid are considered as antioxidant synergists. The spectrum

Table 12.3 List of antioxidants

INS	Name
220	sulphur dioxide
221	sodium sulphite
222	sodium hydrogen sulphite
223	sodium metabisulphite
224	potassium metabisulphite
225	potassium sulphite
226	calcium sulphite
227	calcium hydrogen sulphite
228	potassium bisulphite
300	ascorbic acid (L-)
301	sodium ascorbate
302	calcium ascorbate
303	potassium ascorbate
304	ascorbyl palmitate
305	ascorbyl stearate
306	mixed tocopherols concentrate
307	alpha-tocopherol
308	synthetic gamma-tocopherol
309	synthetic delta-tocopherol
310	propyl gallate
311	octyl gallate
312	dodecyl gallate
313	ethyl gallate
314	guaiac resin (gum guaiac)
315	isoascorbic acid (erythorbic acid)
316	sodium isoascorbate
317	potassium isoascorbate
318	calcium isoascorbate
319	tertiary butylhydroquinone (TBHQ)
320	butylated hydroxyanisole (BHA)
321	butylated hydroxytoluene (BHT)
322	lecithins
323	anoxomer
324	ethoxyquin
330	citric acid
384	isopropyl citrates
385	calcium disodium ethylenediaminetetraacetate
386	disodium ethylenediaminetetraacetate
387	oxystearin
388	thiodipropionic acid
389	dilauryl thiodipropionate
390	distearyl thiodipropionate
391	phytic acid
512	stannous chloride
539	sodium thiosulphate
1102	glucose oxidase

288 Food chemical safety

of antioxidants cited in the *Codex Alimentarius* is very comprehensive. Individual countries set down their own diversified regulation of antioxidants with regard to local customs and habits, agricultural practices and potentials, economic considerations and technological advances.

12.4 The regulation of antioxidants in the EU

EU regulation of antioxidants is stipulated by European Parliament and Council Directive No. 95/2/EC of 20 February 1995 on food additives other than colours and sweeteners.[7]

Only antioxidants which satisfy the requirements laid down by the Scientific Committee for Food may be used in foodstuffs. Having regard to the most recent scientific and toxicological information on these substances, some of them are to be permitted only for certain foodstuffs and under certain conditions of use. On the other hand, a wide range of natural antioxidants may be added to all foodstuffs, with the exception of formulae for special purposes (such as for infants), following the *quantum satis* principle.

The substances which can be used generally as antioxidants in food processing are listed in Table 12.4.

Antioxidants listed in Table 12.4 are permitted in foodstuffs with the exception of unprocessed foodstuffs, butter, pasteurised and sterilised milk, unflavoured live fermented milk products, and natural unflavoured buttermilk. Within the meaning of the directive, the term 'unprocessed' means not having undergone any treatment resulting in a substantial change in the original state of the foodstuffs (except, for instance, cutting, dividing, cleaning, peeling). In some foodstuffs only a limited amount of antioxidants from Table 12.4 may be used. These are listed in Table 12.5. Conditionally permitted antioxidants and related foodstuffs are listed in Table 12.6.

Table 12.4 Antioxidants generally permitted in foods

E number	Name	Maximum level
E 300	ascorbic acid	
E 304	esters of fatty acids with ascorbic acid (i) ascorbyl palmitate (ii) ascorbyl stearate	*quantum satis*
E 306	tocopherol-rich extract	
E 307	α-tocopherol	
E 308	γ-tocopherol	
E 309	β-tocopherol	
E 322	lecithins	
E 330	citric acid	
E 334	tartaric acid	

The regulation of antioxidants in food 289

Table 12.5 Foodstuffs in which a limited amount of antioxidants generally permitted in foods may be used

Foodstuff	Antioxidant	Maximum level
Cocoa and chocolate products	E 330 citric acid	0.5%
	E 334 tartaric acid	0.5%
Non-emulsified oils and fats of animal or vegetable origin (except virgin oils and olive oils)	E 322 lecithins	$30\,g\,l^{-1}$
Refined olive oil, including olive pomace oil	E 307 α-tocopherol	$200\,mg\,l^{-1}$

There is a special position for sulphur dioxide and sulphites (Table 12.7), which are classified as preservatives but may be used as antioxidants when the risk of oxidation is greater than the risk of microbial spoilage. Maximum levels are expressed as sulphur dioxide in mg/kg or mg/l as appropriate and relate to the total quantity available from all sources. Maximum levels range from 15 mg SO_2/kg to 2000 mg SO_2/kg depending on the foodstuff. Sulphur dioxide and sulphites may be used only for a limited number of foods and these are stipulated in the directive, e.g. dried vegetables and fruits, peeled and processed potatoes and white vegetables, sugars, fruit juices and concentrates, beer, wines, mustard, jams, jellies, pectin, gelatine, starches, toppings, dry biscuits, etc. If the sulphur dioxide content is no more than 10 mg/kg or 10 mg/l it is not considered to be present.

The application of antioxidants in foods for infants and young children and for particular nutritional needs is regulated by Directive 89/398/EEC. Antioxidants permitted in weaning foods for infants and young children in good health are listed in Table 12.8.

Antioxidants should be labelled on the retail package with the specific chemical name or with the EC number. The legislation of member states of the EU is influenced by the decision taken within the EC. Some food standards are fully based on EC Directives and some are still based on national considerations. There may be differences between European states, for instance, the utilisation of ascorbic acid as antioxidant for egg products is permitted in France but prohibited in Germany. These differences concern usually the utilisation of antioxidants in various food commodities. The specification of antioxidants mentioned in EC Directives are respected by all member states. But it is still generally required that individual countries of the European Union as well as the central organisation should be approached. The requirements appearing in the EC Directives on additives must be applied by the member states. This means in the first place that for those categories of additives for which a Community positive list exists, member states may not authorise any additives which do not appear on the positive list.

These demands directly or indirectly affect other countries wishing to trade in Europe. In recent years other European countries have also assumed the regulation of additives from the EC Directive to be compatible in multinational

Table 12.6 Antioxidants which are conditionally permitted in foods

E Number	Name	Food	Maximum level (mg kg^{-1})
E 310 E 311 E 312 E 320 E 321	propyl gallate octyl gallate dodecylgallate BHA BHT	fats and oils (for heat treated foodstuffs) frying oil and frying fat (excluding olive pomace oil); lard; fish oil; beef, poultry and sheep fat	200* (gallates and BHA individually or in combination) 100* (BHT) both expressed on fat
		cake mixes cereal-based snack foods milk powder dehydrated soups and broths sauces	200 (gallates and BHA, individually or in combination)
		dehydrated meat processed nuts seasonings and condiments pre-cooked cereals	expressed on fat
		dehydrated potatoes	25 (gallates and BHA, individually or in combination)
		chewing gum dietary supplements	400 (gallates, BHT and BHA, individually or in combination)
E 315 E 316	erythorbic acid sodium erythorbate	semi-preserved and preserved meat products	500 expressed as erythorbic acid
		preserved and semi-preserved fish products; frozen and deep frozen fish	1500 expressed as erythorbic acid
E 385	calcium disodium ethylene diamine tetra-acetate (EDTA)	emulsified sauces canned and bottled pulses, legumes, mushrooms and artichokes	75 250
		canned and bottled fish, crustaceans and molluscs	75
		frozen crustaceans minarine	75 100
E 512	stannous chloride	canned and bottled white asparagus	25 as Sn

* When a combination of gallates, BHA and BHT are used, the individual levels must be reduced proportionally

Table 12.7 Sulphur dioxide and sulphites

E Number	Name
E 220	sulphur dioxide
E 221	sodium sulphite
E 222	sodium hydrogen sulphite
E 223	sodium metabisulphite
E 224	potassium metabisulphite
E 226	calcium sulphite
E 227	calcium hydrogen sulphite
E 228	potassium hydrogen sulphite

Table 12.8 Antioxidants permitted in foodstuffs for infants and young children

E Number	Name	Foodstuff	Maximum level
E 330	citric acid	weaning foods	quantum satis
E 300	L-ascorbic acid	fruit- and vegetable-based drinks, juices and baby foods	0.3 g kg^{-1}
E 301	sodium L-ascorbate		
E 302	calcium L-ascorbate	fat-containing cereal-based foods including biscuits	0.2 g kg^{-1}
E 304	L-ascorbyl palmitate	fat-containing cereals, biscuits, rusks, and baby foods	*100 mg kg^{-1} individually or in combination
E 306	tocopherol-rich extract		
E 307	α-tocopherol		
E 308	γ-tocopherol		
E 309	β-tocopherol		
E 322	lecithins	biscuits and rusks, cereal-based foods, baby foods	10 g kg^{-1}

* 10 mg/kg for follow-on formulae for infants in good health.

markets. The countries of the Eastern bloc have developed new guidelines too, through association with the EU.

The use of food additives is restricted in all European countries by national order concerning food additives, only those antioxidants specially mentioned are allowed to be used. The antioxidants may only be used with the foods mentioned and in the amounts specified. The veterinary service and national food agencies or other national authorities have supervisory powers.

12.5 The regulation of antioxidants in the USA

In the United States antioxidant use is subject to regulation under the Federal Food, Drug, and Cosmetic Act.[8] Antioxidants for food products are also regulated under the Meat Inspection Act, the Poultry Inspection Act, and various state laws. Antioxidants permitted for use in foods are divided into two groups:

1) The following antioxidants are restricted to use in the foodstuffs indicated:
   BHA
   BHT
   Ethoxyquin
   Gallates, dodecyl-, propyl-, octyl-
   Glycine
   Lecithin
   Resin guaiac
   TBHQ
   Tocopherols, α-tocopherol
2) Compositional standards where these exist may also restrict the use of antioxidants. Where no standard exists, antioxidants listed under 'non-standardised products' may be used subject to the conditions imposed. These include:
   Anoxomer
   Ascorbic acid, calcium ascorbate, sodium ascorbate
   Ascorbyl palmitate, ascorbyl stearate
   BHA
   BHT
   Erythorbic acid
   4-Hydroxymethyl-2,6-di-*tert*-butylphenol
   Propyl gallate
   Stannous chloride
   TBHQ (tertiarybutylhydroquinone)
   THBP (trihydroxybutyrophenone)
   Thiodipropionic acid, dilaurylthiodipropionic acid
   Tocopherols, tocopherol acetate

In general, the total concentration of authorised antioxidants added singly or in combination must not exceed 0.02% by weight based on the fat content of the food. Certain exceptions exist in the case of standardised foods and products covered by special regulations. Under the Meat Inspection Act, concentrations up to 0.01% are permitted for single antioxidants based on fat content, with a combined total of no more than 0.02%. The tocopherols and the major acid synergists are unregulated.

Antioxidants that may be used in foods that are subject to a standard of identity are laid down in the relevant standard (Table 12.9). In the case of food that is not subject to a standard of identity, additives must be used in accordance with the conditions or limits of use specified in regulations.[9] For some types of foodstuffs which have standards of identity, the use of antioxidants is not permitted. No antioxidants may be used, for instance, in dairy products, ices, and egg products.

For non-standardised products tocopherols, tocopherol acetate, ascorbic acid and its sodium and calcium salts, ascorbyl palmitate, ascorbyl stearate, erythorbic acid and stannous chloride (max. 0.0015% calculated as tin) may be used according to good manufacturing practice.

Table 12.9 Standardised products in which a limited number of antioxidants may be used

Foodstuff	Antioxidant	Maximum level (mg kg^{-1})
non-alcoholic beverages (from dry mixes)	BHA	2
	BHT	prohibited
dry mixes for beverages	BHA	90
	BHT	prohibited
chewing gum	BHA	1000
	BHT	
	propyl gallate	
animal fat, rendered	BHA	100 singly or
animal fat plus vegetable fat, rendered	BHT	200 combined
	glycine	
	propyl gallate	TBHQ should not be used
	guaiac resin	in combination with
	TBHQ	glycine, propyl gallate or guaiac resin
	tocopherols	300
margarine	propyl-, octyl-, or dodecyl gallates	200
	BHA	TBHQ should not be used
	BHT	in combination with
	ascorbyl palmitate	gallates, ascorbyl
	ascorbyl stearate	palmitate or stearate
	TBHQ	
shrimp, frozen raw breaded	BHA	200 (total content)
	BHT	
	ascorbic acid	
	erythorbic acid	
	ascorbyl palmitate	
	calcium and sodium ascorbates	
	tocopherols	
fruit butters, jams, jellies, preserves	ascorbic acid	1000
fruit, glazed, diced, dry	BHA	32
fruit nectars	ascorbic acid	150
potato granules	BHA	10
potato flakes	BHT	50
bacon, pump-cured	α-tocopherol	500
meats, dried	BHA	100 singly or combined
	BHT	TBHQ and propyl
	propyl gallate	gallate should not be
	TBHQ	used in combination
	tocopherols	30 not to be used in combination with other antioxidants
meats, restructured	tocopherols	30

Table 12.9 *continued*

Foodstuff	Antioxidant	Maximum level (mg kg^{-1})
sausage, dry	BHA	30 singly
	BHT	60 combined
	propyl gallate	TBHQ and propyl gallate should not be used in combination
	TBHQ	
	tocopherols	30 not to be used with other antioxidants
sausages, fresh Italian sausage products, fresh beef patties pizza topping, raw and cooked meatballs, raw and cooked poultry and poultry products	BHA	100 singly
	BHT	200 combined
	propyl gallate	TBHQ and propyl gallate should not be used in combination
	TBHQ	
poultry and poultry products	tocopherols	300 (200 combined with other antioxidants except TBHQ)
	lecithins	*quantum satis*
breakfast cereal, dry	BHA	50
	BHT	
chilli powder, paprika	ethoxyquin	100
desserts (from dry mix)	BHA	2
emulsion stabilisers	BHA	200
	BHT	
flavouring substances	BHA	5000 based on oil content
mixes for desserts	BHA	90
yeast (active, dry)	BHA	1000

Propyl gallate, BHA, BHT, TBHQ, THBP, 4-hydroxymethyl-2,6-di-*tert*-butylphenol, thiodipropionic acid and dilaurylthiodipropionic acid may be used provided that the total antioxidant content does not exceed 200 mg kg^{-1} of the fat or oil content when used according to good manufacturing practice. Anoxomer may be used provided that the total antioxidant content does not exceed 5000 mg kg^{-1} of the fat or oil content of the food.

Petitions for the use of a new food additive may be submitted to the Food and Drug Administration in accordance with the form specified in the regulations.

There are some differences in the regulation of US and EU antioxidants. The restrictions on synthetic antioxidants are more strict in the EU, e.g. TBHQ, THBP, anoxomer, ethoxyquin, guaiac resin and derivates of thiodipropionic acid are not permitted there. On the other hand, sulphur dioxide and sulphites, citric and tartaric acids and their salts and salts of EDTA are not listed as permitted antioxidants in the US.

12.6 The regulation of antioxidants in Australia

In Australia each State and Territory used to have their own food legislation.[8] However, in 1987 the National Health and Medical Research Council produced the Food Standards Code to bring harmonisation of food legislation throughout the country. The Imported Food Control Act requires that all imported foods comply with the Food Standards Code. In August 1991, the new National Food Authority (NFA) was established. NFA is now responsible for setting all food standards in Australia.

The following are considered as permitted antioxidants in all States:

Gallates (propyl-, octyl-, and dodecyl-, or any mixture thereof)
BHA
TBHQ
Lecithins (including phospholipids from natural sources)
Tocopherols – tocopherols can be used with or without citric acid, malic acid, tartaric acid, lactic acid (singly or in combination)
Ascorbic acid and its sodium salt
Erythorbic acid and its sodium salt
Ascorbyl palmitate
BHT (only for walnut kernels, pecan nut kernels, vitamins A and D).

For the group of fats and oils the antioxidants listed in Table 12.10 are used. For fish and fish products (including prawns and shrimps), fruit and vegetable products (including raw peeled potatoes) and meat and meat products (corned, cured, pickled or salted and cooked) only ascorbic acid, erythorbic acid and their sodium salts may be used.

A mixed food containing one or more foods in which antioxidants are permitted may contain antioxidants in not greater amounts than are specifically allowed in the quantity of food or foods containing the antioxidant used in the preparation of the mixed food.

Table 12.10 Permitted antioxidants for fats and oils

Group	Name	Maximum level (%)
essential oils	gallates	0.1
	BHA	0.1
	TBHQ	0.1
	tocopherols	
	lecithin	
	ascorbyl palmitate	
fats and oils, other table spreads	gallates	0.01
	BHA	0.02
	TBHQ	0.02
	tocopherol	
	lecithin	
	ascorbyl palmitate	

12.7 The regulation of antioxidants in Japan

There are certain legal requirements and standards which cover the manufacture, processing, storage and quality of some foodstuffs.[10] In general, these requirements do not cover composition in any detail except in the case of milk products.

There exists a list of permitted additives. This list is concerned only with chemical synthetics (substances obtained by a chemical reaction other than degradation). It means that the substances on the list are those which either do not occur naturally or are not obtained from natural sources. Of the substances which are not on the list it is not always possible to decide whether these may be used in food. The antioxidants and foodstuffs in which a limited amount of antioxidant is permitted are given in Table 12.11. Only the above-mentioned foods may contain antioxidants, except α-tocopherol which may be generally used in foods as an antioxidant.

12.8 Future trends

Synthetic antioxidants such as BHA, BHT and gallates were introduced in the 1940s. In recent years, there has been an enormous demand for natural antioxidants mainly because of adverse toxicological reports on many synthetic compounds. Thus, most of the recent investigations have been targeted towards identification of novel antioxidants from natural sources. Plant phenolic compounds such as flavonoids, sterols, lignanphenols, and various terpene-related compounds are potent antioxidants. Since the antioxidant activities of natural extracts and compounds have been determined by a wide range of methods and varying endpoints, it has also become increasingly difficult to make a realistic assessment of the efficacy of various natural antioxidants. There is an urgent need for standardisation of evaluation methods in order to obtain meaningful information.[3]

Considerable efforts have also been made toward the development of novel compounds with superior antioxidant properties. Some attempts were also made to introduce new synthetic polymeric compounds which are non-absorbable and non-toxic. These are generally hydroxyaromatic polymers with various alkyl and alkoxyl substitutions. Such compounds are usually very large molecules and their absorption from the intestinal tract is practically nil. In addition to their reportedly high antioxidant activity, they are non-volatile under deep-fat frying conditions, which result in nearly quantitative carry-through to the fried items, but they have not yet received FDA approval.

Synthetic analogues or derivatives of α-tocopherol which have better antioxidant properties can be introduced. Many natural antioxidants such as flavonols, flavones, tea leaf catechins, rosemary antioxidants and spice extracts have been reported to be more active than BHA, BHT or the tocopherols in model systems. The food applications of these compounds need to be explored further.

Table 12.11 Japanese restrictions on the use of antioxidants

Antioxidant	Limitation or restriction	Maximum permitted level (mg kg^{-1})
butylated hydroxyanisole (BHA)[a]	butter	200
	fats and oils	200
	frozen fish, shellfish, and whale meat (for dipping solution)	1000
	mashed potato (dried)	200
	salted fish and shellfish	200
	dried fish and shellfish	200
butylated hydroxytoluene (BHT)[b]	butter	200
	chewing gum	750
	fats and oils	200
	frozen fish, shellfish, and whale meat (for dipping solution)	1000
	mashed potato (dried)	200
	salted fish and shellfish	200
	dried fish and shellfish	200
isopropyl citrate (as monopropyl citrate)	fats and oils	100
EDTA CaNa$_2$, EDTA Na$_2$ (as EDTA CaNa$_2$)[c]	canned or bottled soft drinks	35
	canned or bottled food (except for soft drinks)	250
erythorbic acid	only for antioxidant use	
sodium erythorbate	only for antioxidant use	
nordihydroguaiaretic acid	butter	100
	fats and oils	100
propyl gallate	butter	100
	fats and oils	100
resin guaiac	butter	1000
	fats and oils	1000
(±) α-tocopherol	only for antioxidant use	

[a] if used in combination with BHT, total amount of both antioxidants must not exceed permitted level.
[b] if used in combination with BHA, total amount of both antioxidants, except for chewing gum, must not exceed permitted level.
[c] to be chelated with calcium ions before addition to the food.

The toxicological effects of food antioxidants have been the focus of controversy in recent years. Toxicological studies are mainly carried out to establish the no-effect level for an ADI for humans.

The laws must be modified to ensure the integration of these issues in determining the criteria for safety. Food safety will not only be an issue of protection, but also one of maintenance and promotion of health. It also means that the regulatory and judicial system must be prepared to accept this expanded role for food in the matrix of national life and the international market. Major efforts are being made at the international level to obtain maximum coordination in regulation of principles for control and acceptance of new additives.

12.9 Sources of further information and advice

The selection of the most suitable antioxidant depends on the character of food and the targets which should be attained. Naturally occuring fats and oils contain indigenous antioxidants that protect the unsaturated lipids from free-radical destruction in their native vegetable and animal sources. On the other hand, fats and oils exist in a commingled fashion with reactive substances which cause their rapid decomposition. Intensity of oxidative alterations is also influenced by the shelf-life of products and storage conditions. All these facts should be considered when deciding whether any and if so what antioxidant will be used.

Further restriction is made by legislative regulation of antioxidants in foods. The most detailed information about permitted and recommended antioxidants may be seen in *Codex Alimentarius*. It is modernised on the basis of the actual knowledge of WHO. However various countries regulate the use of antioxidants by their own national legislation (see above). The divergences result from tradition, the composition of local diet and the boarding custom practices. The national regulations are usually established by the Ministry of Health or another state authority.

Lists of permitted antioxidants and foods in which antioxidants may be used are presented in national directives dealing with additives and contaminants and in duty tariffs. They are published in the special bulletins issued by the Ministries, European Parliament, FDA or analogous institutions and are publicly available. Nowadays they are often published on the Internet. These national regulations must be respected in international trade. They can be used or misused in the restraints on food exports and imports.

The food producer has full responsibility for the choice of suitable antioxidant and should obtain all information about the antioxidant from the data sheet (product information) that declares its safety. In most countries the antioxidants used in the product must be labelled on the package. Customers prefer foods with a minimum of additives as indicated by E number, and so the trend is directed towards mixtures of spices that contain antioxidants that do not need to be declared.

12.10 References

1   MIDDLEKAUF R D, SHUBIK P, *International Food Regulation Handbook*, New York, Marcel Dekker, 1996.

2. NAWAR W W, *Lipids in Food Chemistry*, Fenema O R (ed.), New York, Marcel Dekker, 1985.
3. MADHAVI D L, DESHPANDE S S, SALUNKHE D K (ed.), *Food Antioxidants*, New York, Marcel Dekker, 1996.
4. LEHMAN A J, FITZHUGH O G, NELSON A A, WOODARD G, *Adv. Food Res*, 1951 **197** 3.
5. COPPEN P P, *Rancidity in Foods*, Allen J C and Hamilton R J (eds), London, Elsevier Applied Science, 1989, 83.
6. *Codex Alimentarius* General Requirements, 1995 Vol. **1A**.
7. *Official Journal of the European Communities*, 1995 **38** L 61.
8. *Overseas Food Legislation I*, 2nd ed., Leatherhead, The British Food Association, 1994.
9. *Code of Federal Regulations* Title 21, Washington, Food and Drugs Department of Health & Human Services USA, 1993.
10. *Japan's Legal Requirements for Food Additives*, 1981.

Index

absolute flavourings 208
Acceptable Daily Intakes (ADIs) 61,
      63–4, 254, 255, 256
   by additive category 266–73
   by additive name 274–81
   antioxidants 284, 285
   flavourings 210, 216, 223
   risk evaluation 75
   risk management 77
   sweeteners 236, 242
acceptable daily sucrose equivelent 241
accreditation 81, 83–5
accreditation agencies 90
accuracy 98
acesulfame K 232, 234–6
acidity regulators 250, 252, 256, 258–65
      passim, 266–8, 274–81 passim
acids 250, 252, 256, 258–65 passim, 266,
      274–81 passim
ad hoc analysis 88
additives other than colours and
      sweeteners 20–2, 249–82
   ADI 255, 256, 266–81
   classification 249–54
   JECFA safety evaluation 255–7
   regulation 254–5
   specifications 254–5, 257, 258–65
adulteration 173–4
Adverse Reaction Monitoring System
      (ARMS) 148–50
   complaints received 150, 151

adverse reactions 145–70
   aspartame 151, 152–4
   consumer attitudes 146–7
   controversial additives 150–62
   food dyes 150–2
   future trends 165
   MSG 151, 156–8
   olestra 150, 151, 154–6
   reporting 147–50
   sodium nitrite 151, 158–9
   sulphites 151, 159–62, 162
   see also health effects
agar 122
aglycones 188
agricultural commodities, raw 48–9
algae 198, 202
alitame 232, 236
alkanet 194
allergy 146
   flavourings and 213–15
   see also adverse reactions
allophycocyanins 198
Allura Red (FD&C No. 40) 175, 177,
      178
Amaranth 177
analytical methods 7, 92, 95–103, 111–44
   biosensors 127–9, 131
   CE 123–5, 132
   enzymatic methods 126–7
   flow-injection analysis 126
   future trends 102–3, 131–2

gas chromatography 125
HPLC 112–21
ICT-AES 126
immunoassay 127
infrared techniques 130
ion chromatography 121–3
legislative requirements 95–9
photometric methods 131
rapid enzymatic and test kit methods 130–1
rapid methods 112, 127–31
reference and research methods 111–12, 112–27
standardised methods for contaminants 99–101
sweeteners 114–15, 123, 125, 244
TLC 123
valid 98–9, 108–9
X-ray fluorescence 129–30
see also laboratory quality systems
analytical runs 86–7, 88
anatase 201
ankaflavin 185, 195
annatto 183–4
anoxomer 293, 294
anthocyanins 187–90
anthraquinones 194
anti-caking agents 250, 252, 256, 258–65 *passim*, 268, 274–81 *passim*
anti-foaming agents 250, 252, 256, 258–65 *passim*, 268, 274–81 *passim*
antioxidant synergists 286
antioxidants 21, 165, 250, 252, 283–99
 analytical methods 116–17
 anthocyanins 189
 Australian regulation 295
 Codex Alimentarius 286–8
 EU regulation 288–91, 294
 future trends 296–8
 Japanese regulation 296, 297
 toxicology 284–5
 US regulation 291–4
AOAC *see* IUPAC/AOAC/ISO
approval process, US 50–7
ascorbic acid 292–4
ascorbyl palmitate 292–4, 295
ascorbyl stearate 292–4
aspartame 26, 130, 232, 237
 adverse reactions 151, 152–4
 assigned value ('true' result) 92
astaxanthin 180
asthma 160–1, 162
atomic absorption spectrophotometry 126

attention deficit hyperactivity disorder (ADHD) 150–1
Australia 295

balsam 208
benzoic acid 117, 118
beta-8-carotenal 179, 186, 187
beta-carotene 178–80, 181, 186–7
betacyanins 190
betalains 190–1
betanine 188, 190
betaxanthins 190
BIOQUANT kits 130
biosensors 127–9, 131
biotechnology 27, 28, 215–16
biotransformation 234
bisdemethoxycurcumin 188, 192, 193
bixin 179, 183
bladder tumours 238–9
blends, sweetener 240, 241
Blue Book 222–4
brain cancer 153, 154
bread 70–1
brown polyphenols 200–1
browning 231
budget method 67–8, 69
bulk sweeteners 228, 229, 231, 240, 242, 244–5
 analytical methods 115
 available 232, 233
 functionality 230–1
bulking agents 231, 250, 252, 256, 258–65 *passim*, 268, 274–81 *passim*
butylated hydroxyanisole (BHA) 290, 292–4, 295
butylated hydroxytoluene (BHT) 71–4, 290, 292–4

cancer 153, 154, 159, 238–9
canthaxanthin 179, 186, 187
capillary electrophoresis (CE) 123–5, 132
capsanthin 184, 185
capsorubin 184, 185
caramel 199–200
carbon black 202
carbonated soft drinks 68–71, 72
carmine 193–4
carminic acid 127
carotenoids 114, 155–6, 178–81
 synthetic 186–7
carrageenans 121, 122
carrier solvents 21, 250
carriers 21, 250

categories of additives 25–6, 249–54
  ADI by category 255, 256, 266–73
cations 126
cayenne 184
CEN (European Committee for Standardization) 96–7, 99–101
Center for Science in the Public Interest (CSPI) 154–5
chemical contaminants *see* contaminants
chemical data 51–4
chemical structure
  annatto and saffron 179, 183
  anthocyanins 188–9
  betalains 190
  brown polyphenols 200
  caramel 199
  carbon black 202
  carotenoids 178–9
  chlorophylls 191
  cochineal and carmine 193–4
  FD&C colours 175, 177
  iridoids 196–7
  lutein 182
  lycopene 179, 181
  Monascus 185, 195
  paprika 184, 185
  phycobilins 185, 198
  structure-activity relationships 212
  synthetic carotenoids 179, 186
  titanium dioxide 201
  turmeric 188, 192
children 50
  foods for young children 21, 289, 291
  intake estimates 68–74, 77
chlorophyllide 191
chlorophyllin 191, 192
chlorophylls 114, 191–2
Cholestin 196
classification of additives *see* categories of additives
'coal tar' colourings 173
cochineal 193–4
cocoa 200–1
Codex Alimentarius 286–8
Codex Alimentarius Commission 46
  analytical methods 95
  labour quality assurance 82–3
Codex Committee on Food Additives and Contaminants (CCFAC) 6, 65, 255–6
Codex Committee on Methods of Analysis and Sampling (CCMAS) 82
Codex Standard for Food Additives 6, 65, 242

collaborative trials 96, 98–9
collusion 93
colour adjuvants 256, 258–65 *passim*, 268, 274–81 *passim*
colour retention agents 256, 258–65 *passim*, 268, 274–81 *passim*
colourings 173–206, 250, 252
  adverse reactions 150–2
  analytical methods 113–14, 123, 124
  annatto and saffron 183–4
  anthocyanins 187–90
  betalains 190–1
  brown polyphenols 200–1
  caramel 199–200
  carbon black 202
  carotenoids 114, 155–6, 178–81
  chlorophylls 114, 191–2
  cochineal and carmine 193–4
  EU regulation 16–20
  food, drug and cosmetic colours *see* food, drug and cosmetic (FD&C) colourings
  iridoids 196–8
  lutein 182
  lycopene 179, 181, 181–2
  miscellaneous colourings 202–3
  Monascus 185, 195–6
  outlook 203
  paprika 184, 185
  phycobilins 185, 198–9
  synthetic carotenoids 186–7
  titanium dioxide 126, 201
  turmeric 188, 192–3
  US regulation 47–8, 174–8
combination scores 94
concrete flavours 208
conditionally permitted antioxidants 288, 290
consumers
  attitudes to additives 145, 146–7
  perceptions of risk 76, 162–4
consumption data 64, 66–7
  intake estimates 68–74, 210–11
consumption ratio (CR) 213
contaminants 56–7, 64
  standardised methods of analysis for 99–101
continuous monitoring 87
control materials 86
co-ordinating laboratory/organisation 91
corn endosperm oil 202
cosmetics 173, 174–8
cottonseeds 202

Council of Europe (CE) Committee of
    Experts on Flavouring Substances
    209, 222–4
crocins 179, 183, 196, 198
curcumin 188, 192, 193
cyclamate 232, 237–8
cyclohexylamine 237
cyclophosphamide 192

decision tree 217–19
demethoxycurcumin 188, 192, 193
detailed dietary analysis 211–12
diabetes 235
diketopiperazine (DKP) 237
dilaurylthiodipropionic acid 292–4
direct additives 51, 52, 56–7
distributional analysis 72–4
dose-response characterisation 61, 62,
    63–4
dried algal meal 202

ELISA technique 127
emulsifiers 119–20, 250, 252, 256,
    258–65 *passim*, 268–9, 274–81
    passim
emulsifying salts 250, 252, 256, 258–65
    passim, 269–70, 274–81 *passim*
EN 45000 series 81, 84
Environmental Protection Agency (EPA)
    44–5, 45–6, 57
  pesticide residues 48–9
enzymatic methods 126–7
  rapid methods 130–1
enzyme biosensors 129
enzyme preparations 54–5
epidemiological evidence 164
erythorbic acid 290, 292–4
erythritol 245
essential oils 208
Estimated Daily Intake (EDI) 61, 64
ethoxyquin 116
European Committee for Standardization
    (CEN) 96–7, 99–101
European Food Authority 28
European Union (EU) 5–6, 12–41, 62,
    65, 254–5
  additives other than colours and
    sweeteners 20–2
  ADIs 255, 256, 266–81
  AMFC Directive 80–1
  analytical methods 95–6
  antioxidants 288–91, 294
  categorisation of additives 249–51,
    252–3

colourings 16–20
'comprehensive' directive 15
Directive 89/107/EEC 13–14
extraction solvents 24–5
flavourings 22–4, 26, 221–4
future developments 27–9
implementation of directives 27
key directives 13–27
labelling requirements 25–7
laboratory quality assurance 80–1
list of E numbers of permitted
    additives 32–41
purity criteria 22
SCF 12, 209, 221–2, 236
specifications 258–65
sweeteners 15–16, 17, 26, 242, 243,
    244
White Paper on Food Safety 28–9
exempt colours 175, 176
exposure analysis 61, 62, 64–75
  consumption data 66–7
  estimating intakes 67–75
  estimation for flavourings 210–12
  intake calculations based on
    consumption data 68–74
  probabilistic intake modelling 74–5
  simple intake estimates 67–8
  US approval process 55–7
  usage data 65–6
extraction solvents 24–5
extracts 208

falsification of results 93
FD&C colours *see* food, drug and
    cosmetic (FD&C) colourings
Federation of American Societies for
    Experimental Biology (FASEB)
    157, 160
firming agents 250, 252, 256, 258–65
    passim, 270, 274–81 *passim*
flavonoids 196, 197
flavour enhancers 250, 252–3, 256,
    258–65 *passim*, 270, 274–81
    passim
Flavour and Extract Manufacturers'
    Association (FEMA) Expert Panel
    209, 215–16, 219–21
flavourings 207–27
  allergies and intolerances 213–15
  basic principles of safety evaluation
    209–16
  biotechnology and flavour production
    215–16
  Council of Europe 209, 222–4

flavourings *continued*
  definition and use 207
  EU regulation 22–4, 26, 221–4
  exposure estimation 210–12
  FEMA Expert Panel 209, 215–16, 219–21
  JECFA 209, 212, 216–19
  natural occurrence 212–13
  range and sources 208–9
  regulators 216–24
  SCF 209, 221–2
  structurally related substances 212
  toxicology 209–10
flour treatment agents (FTAs) 250, 253, 256, 258–65 *passim*, 271, 274–81 *passim*
flow-injection analysis (FIA) 126
foam stabilisers 256, 258–65 *passim*, 270, 274–81 *passim*
foaming agents 256, 263, 270, 279
folded oil 208
Food Advisory Committee (UK) 76
Food Advisory Committee (US) 154, 155, 156
Food and Agriculture Organization (FAO) 150, 215
  see also Joint FAO/WHO Expert Committee on Food Additives (JECFA)
Food Balance Sheets (FBSs) 66
Food Chemicals Codex (FCC) 244
  specifications 258–65
food consumption data *see* consumption data
Food and Drug Administration (FDA) 45, 57–8, 161, 215
  approval of olestra 154–6
  ARMS 148–50
  flavours 215, 220–1
  premarket testing 44
  websites and approval process 50–7
food, drug and cosmetic (FD&C) colourings 174–8
  Red No. 2 (Amaranth) 177, 203
  Red No. 40 (Allure Red) 175, 177, 178
  Yellow No. 5 (tartrazine) 151, 152
food irradiation 101
Food Marketing Institute (FMI) 'Trends' survey 146–7
Food Safety and Inspection Service (FSIS) 45
Food Standards Agency (FSA) 5, 6
  research programme on food additives 7–10

surveillance requirements 83, 104–10
fortified functional foods 165
Fourier transform infrared spectroscopy (FTIR) 130
fraction/isolate 208
French paradox 190
functional foods 165

gallates 290, 292–4, 295
gardenia 196–8
gas chromatography (GC) 125
General Standard on Food Additives (GSFA) 6, 65, 242
generally recognized as safe (GRAS) substances 46–7, 284
  approval process 51–3, 54–5
  FEMA 219–21
genetically modified (GM) materials 26–7, 28–9, 215–16
genipin 197, 198
geniposide 197, 198
genotoxicity 235
ginger oleoresin 207, 208
glazing agents 250, 253, 256, 258–65 *passim*, 271, 274–81 *passim*
glutamate 124, 156
glutamic acid 117–18
Guidelines on Internal Quality Control in Analytical Chemistry Laboratories
  basic concepts 85–7
  recommendations 87–9
  scope of guidelines 87

Harmonisation Protocol on Collaborative Studies 99
hazard identification 61, 62–3
health effects
  annatto and saffron 183–4
  anthocyanins 189–90
  carotenoids 180–1
  chlorophylls 192
  lutein 182
  lycopene 181, 182
  Monascus 195–6
  synthetic carotenoids 187
  turmeric 193
  see also adverse reactions
heterocyclic amines (HCAs) 213
high performance liquid chromatography (HPLC) 112–21
human exposure thresholds 63–4, 217
humectants 231, 250, 253, 256, 258–65 *passim*, 271, 274–81 *passim*
hydrocolloids 121

hydroxyaromatic polymers 296
4-hydroxymethyl-2, 6-di-*tert*-butylphenol 292, 294
hyperactivity 150–1

immunoassay 127
in-line analysis 87
indirect additives 51, 53–4, 56
inductively coupled plasma atomic emission spectroscopy (ICP-AES) 126
infants 50
  infant foods 21, 289, 291
infrared analytical methods 130
ingredients 25–6
intake estimates 7, 64–5, 67–75
  based on individual consumption data 68–74, 210–11
  probabilistic modelling 74–5
  simple intake estimates 67–8
  sweeteners 240–1
intense sweeteners 228, 229, 242, 243, 244–5
  analytical methods 114–15, 244
  available 232, 233
  functionality 229–30
  uses 232–3
  see also sweeteners
internal quality control (IQC) 85–9, 108
International Food Biotechnology Council (IFBC) 215
International Harmonised Protocol for Proficiency Testing of (Chemical) Analytical Laboratories 91–3
International Numbering System (INS) 251, 252–3, 286, 287
Internet 164
intolerances 213–15
  see also adverse reactions
ion chromatography (IC) 121–3
iridoids 196–8
Irish National Food Ingredient Database (INFID) 65
iron oxides 202
irradiated food 101
ISO *see* IUPAC/AOAC/ISO
isolate/fraction 208
IUPAC/AOAC/ISO
  Guidelines for IQC 85–9
  Harmonisation Protocol on Collaborative Studies 99
  International Harmonised Protocol for Proficiency Testing 91–3

Japan 296, 297
Joint FAO/WHO Expert Committee on Food Additives (JECFA) 150
  ADIs for antioxidants 284, 285
  classification of additives 251–4
  flavourings 209, 212, 216–19
  safety evaluation 255–7
  specifications 258–65
  sweeteners 236, 244
JSSFA (Japan's Specifications and Standards for Food Additives) 258–65

kermes 194
kinetics 234

labelling
  EU requirements 25–7
  GM foods 216
laboratory quality systems 79–110
  accreditation 81, 83–5
  analytical methods 95–9
  CAC 82–3
  EU legislation 80–1
  FSA surveillance requirements 83, 104–10
  future direction for analytical methods 102–3
  IQC 85–9, 108
  legislative requirements 80–3
  proficiency testing 89–95, 107–8
  standardised analytical methods 99–101
lac 194
lakes 177
LC-MS/DAD systems 131–2
lecithin 119, 120, 130–1, 295
luminometric methods 131
lutein 182
lycopene 179, 181, 181–2

marigold, extracts of 180
measurement uncertainty 102
membrane biosensors 129
metabolism 234
miniaturised instruments 131
modified starch 250, 253
monascin 185, 195
monascorubin 185, 195
monascorubramine 185, 195
Monascus 185, 195–6
monosodium glutamate (MSG) 151, 156–8
  MSG system complex 156, 157–8
multivariate IQC 87

National Advisory Committee on Hyperkinesis and Food Additives (USA) 150–1
National Food Authority (NFA) (Australia) 295
National Institutes of Health (NIH) Consensus Development Panel (USA) 151
National Pesticide Telecommunications Network (USA) 49
natural antioxidants 285, 296
natural colours 113–14, 203
natural occurrence 212–13
neohesperidin dihydrochalcone 232, 238
neotame 245
nitrates 100–1, 122, 158, 159
nitrite 122, 151, 158–9
nitrosamines 158–9
no observed (adverse) effect level (NOAEL or NOEL) 49–50, 63, 209–10
non-thresholded end-points 64
norbixin 183

octopus ink 202
oleoresin 207, 208
olestra 150, 151, 154–6

p-hydroxybenzoic acid esters 117, 118
packaging gases 26, 250, 253, 256, 258–65 *passim*, 271, 274–81 *passim*
palm oil 178–80
paprika 184, 185
pesticide residues 48–9
petitions, guidance for submitting 50–7
phenylketonurics (PKU) 237
pheophorbide 191
pheophytin 191
phospholipids 119, 120
photochemiluminescence 131
phycobilins 185, 198–9
phycocyanins 198
phycoerythrins 198
polydimethylsiloxane 126
polyols 26
polyphenols, brown 200–1
polyphosphates 122
polysaccharides 124–5
potassium bromate 118
potential contractors, information for 83, 104–10
poundage method 211, 212
precision 96, 97, 98

preservatives 253, 258–65 *passim*
ADI 256, 271–3, 274–81 *passim*
analytical methods 117, 118, 125
EU regulation 21
Prevention Index 147
prior sanctioned substances 47
probabilistic intake modelling 74–5
process flavours 213
processed foods 49
processing aids 13
proficiency testing 89–95, 107–8
accreditation agencies 90
importance 90
International Harmonised Protocol 91–3
statistical analysis of results 93–5
propellant gases 251, 253, 256, 258–65 *passim*, 273, 274–81 *passim*
propyl gallate 290, 292–4
purity criteria 22, 244

qualitative synergism 230
quality systems *see* laboratory quality systems
quantitative synergism 230
quantum satis 15

raising agents 251, 253, 256, 260, 264, 273, 276, 280
Rapid Alert System 28
rapid analytical methods 112, 127–31
raw agricultural commodities 48–9
recovery 102–3
red rice 195
Reference Dose 49–50
reference and research analytical methods 111–12, 112–27
regulatory authorities
flavourings 216–24
USA 45–6
see also under individual regulators
repeatability 97
reporting results 93
reproducibility 97
reproductive toxicity 235
resinoid 209
results
analysis of 93–5
falsification of 93
reporting 93
'true'(assigned value) 92
riboflavin 202
risk analysis 61–78
dose-response characterisation 61, 62, 63–4

exposure analysis 61, 62, 64–75
future trends 76–7
hazard identification 61, 62–3
risk communication 62, 76
risk evaluation 62, 75
risk management 62, 75–6
risk perception
  consumers 76, 162–4
  lay groups and scientists 10
RQ flex test kit 130
rubropunctamine 185, 195
rubropunctatin 185, 195
runs, analytical 86–7, 88

saccharin 232, 238–9
saffron 183–4
samples 91–2
  quality control of sampling 87
  sample distribution frequency 92
Scientific Committee on Food (SCF) 12, 209, 221–2, 236
secondary direct additives 51, 56
sequestrants 251, 253, 256, 258–65 *passim*, 273, 274–81 *passim*, 286
shellac 202
single laboratory method validation 102
sodium nitrite 151, 158–9
soft drinks, carbonated 68–71, 72
sorbic acid 117, 118
sorbitan tristearate 120
specifications 254–5, 257, 258–65
spirolina 199
squid ink 202
stability 236
stabilizers 121, 251, 253, 256, 258–65 *passim*, 273, 274–81 *passim*
standard budget method 67–8, 69
standardised products 292, 293–4
Standing Committee on Foodstuffs 13, 14
stannous chloride 290, 292–4
starch 126–7
  modified starch 250, 253
statistical analysis 93–5, 99
Stevia rebaudiana 239
stevioside 232, 239
structure-activity relationships 212
  see also chemical structure
substantial equivalence 215
sucralose 232, 239–40
sucrose 229
sucrose ester group 119–20
sugars 26, 115, 121–2
sulphites 100, 289, 291

adverse reactions 151, 159–62, 162
analytical methods 122–3, 124
sulphur dioxide 289, 291
surface plasmon resonance (SPR) 129
surveillance requirements, FSA 83, 104–10
sweeteners 100, 228–48, 251, 253
  acesulfame K case study 234–6
  alitame 232, 236
  analytical methods 114–15, 123, 125, 244
  aspartame *see* aspartame
  available sweeteners 232–3
  bulk sweeteners *see* bulk sweeteners
  cyclamate 232, 237–8
  definitions 229
  EU regulation 15–16, 17, 26, 242, 243, 244
  functionality 229–31
  intake considerations 240–1
  intense sweeteners *see* intense sweeteners
  neohesperidin dihydrochalcone 232, 238
  outlook 244–5
  purity criteria 244
  regulatory status 242–4
  saccharin 232, 238–9
  safety testing 233–6
  stevioside 232, 239
  sucralose 232, 239–40
  sweetener blends 240, 241
  thaumatin 232, 240
  uses 231–2
synergism 230
synthetic anatase 201
synthetic carotenoids 186–7
synthetic colours 113, 173–8

talc 202
tartrazine 151, 152
TBHQ 292–4, 295
tea 200–1
technological data 51–4
technological need 14
test kit methods 130–1
texture 231
thaumatin 232, 240
THBP 292–4
thickeners 121, 251, 253
thin-layer chromatography (TLC) 123
thiodipropionic acid 292–4
thresholds 63–4, 217
tincture 209

titanium dioxide 126, 201
tocopherol acetate 292–4
tocopherols 292–4, 295
tolerances 48–9
  setting tolerance levels 49–50
toxicology
  annatto and saffron 183
  anthocyanins 189
  antioxidants 284–5
  betalains 191
  brown polyphenols 200–1
  caramel 199
  carbon black 202
  carotenoids 180
  chlorophylls 192
  cochineal and carmine 194
  FD&C colourings 177–8
  iridoids 198
  lutein 182
  lycopene 182
  Monascus 195
  paprika 184
  phycobilins 198
  safety evaluation of flavourings 209–10
  sweeteners 234, 235
  synthetic carotenoids 186–7
  titanium dioxide 201
  turmeric 193
traditional foods 14
turmeric 188, 192–3

United Kingdom
  Food Advisory Committee 76
  FSA see Food Standards Agency
United Kingdom Accreditation Service (UKAS) 81, 84–5
ultramarine blue 202
United States (USA) 42–58
  antioxidants 291–4
  approval process 50–7
  Color Additives Amendment 1960 44, 174–5
  colourings 47–8, 174–8
  colours exempt from certification 175, 176
  Department of Agriculture FSIS 45
  enzyme preparations 54–5
  EPA see Environmental Protection Agency

  estimating exposure 55–7
  FDA see Food and Drug Administration
  federal agencies responsible for enforcement 45–6
  Federal Food, Drug and Cosmetic Act 1938 43–4, 174, 177
  Food Additives Amendment 1958 42–3, 44, 147, 220
  Food Advisory Committee 154, 155, 156
  Food and Drug Act 1906 43, 174
  Food Quality Protection Act 1996 (FQPA) 44–5
  GRAS substances see generally recognized as safe (GRAS) substances
  guidance for submitting petitions and notifications 50–1
  laws and amendments 43–5
  monitoring adverse reactions 147–50
  pesticide residues 48–9
  prior sanctioned substances 47
  recommendations for chemistry data 53–4
  recommendations for submission 51–3
  setting tolerance levels 49–50
  sweeteners 242, 244
usage data 64, 65–6, 71–2

validated methods of analysis 98–9, 108–9
vitamins 101, 118–19

weighed diary method 66–7
wine 188–9, 190
Working Party on Food Additives 10
World Health Organization (WHO) 150, 215
  see also Joint FAO/WHO Expert Committee on Food Additives (JECFA)
World Trade Organization (WTO) 46, 82

X-ray fluorescence (XRF) 129–30
xanthophyll pastes 180

z-scores 93–4
zinc oxide 202